WORK, VACATION AND WELL-BEING

Based on a twenty-year research study, *Work, Vacation and Well-being* delves into the ubiquitous yet often-underestimated issues surrounding vacation and respite. Providing an original outlook on how breaks from work can be beneficial for the well-being of employees, this book also addresses the potential negative impacts of vacation.

Taking into account factors concerning the nature of the break and the person taking it, Etzion delves into the benefits and drawbacks of workplace breaks, from annual leave to maternity leave and sabbaticals. *Work, Vacation and Well-being* looks at breaks from work through various social and cultural lenses, to present a balanced and well-researched perspective on all angles of taking a break.

Perfect for students of Organizational and Health Psychology, *Work, Vacation and Well-being* also widely appeals to those studying Social Policy, Management Studies, Occupational Health and Research Methods.

Dalia Etzion is Associate Professor of Organizational Behavior at Tel Aviv University, Israel, and has held visiting-scholar appointments at various US universities, including MIT, UCLA and UC Berkeley. She is a certified Industrial/Organizational Psychologist and holds membership with SIOP, APA, STAR and the Academy of Management. Her current interests include: job stress and burnout; crossover of strain between life-partners; and the impact of vacation and respite on well-being.

WORK, VACATION AND WELL-BEING

A Psychological Perspective

Dalia Etzion

Routledge
Taylor & Francis Group
LONDON AND NEW YORK

First edition published 2020
by Routledge
2 Park Square, Milton Park, Abingdon, Oxon, OX14 4RN

and by Routledge
52 Vanderbilt Avenue, New York, NY 10017

Routledge is an imprint of the Taylor & Francis Group, an informa business

British Library Cataloguing-in-Publication Data
A catalogue record for this book is available from the British Library

Library of Congress Cataloging-in-Publication Data
A catalog record has been requested for this book

ISBN: 978-1-848-72230-9 (hbk)
ISBN: 978-1-848-72231-6 (pbk)
ISBN: 978-1-315-89002-9 (ebk)

Typeset in Bembo
by Integra Software Services Pvt. Ltd.

CONTENTS

ACKNOWLEDGEMENT

I wish to thank my editor, Ms. Halo Ben Asher of www.editAura.com for her skillful proactive editing of the manuscript and her creative contribution to many aspects of this book.

PREFACE

How it all began

I grew up in Israel toward the middle of the twentieth century in an economically tight family. In 1925, at the age of 21, my father came to the land of Israel from Poland. He came with a Zionist-Socialist pioneering group, ideologically committed to building up a new, proud, productive Jewish nation, where people live by the fruits of their own manual work rather than by the fruits of other people's labor as was the case in the Diaspora. My father taught himself the trade of shoemaking and worked in a shoe factory for many years. At the height of his career, he became self-employed, owning a small shoe factory with two part-time workers. My mother, who was born in Israel, had to leave school early in order to go to work to help support her family. She continued working in a box factory even after I was born, but eventually left to help my father in his shoe business. At the same time, she was an exemplary housewife, doing all the daily chores herself including sewing the clothing of all family members. She managed to run the household on a very tight budget. My sister and I always had decent clothing and healthy food, though luxuries and indulgences were beyond our circumstances.

My parents worked from dawn till dark and never took a day off, not even when sick. I cannot remember them taking even one vacation. Furthermore, they took pride in their Spartan lifestyle and used to mock those "lazy" neighbors who had the habit of leaving home once a year in favor of sightseeing or a recreation resort. Except for medical reasons, going away on vacation was considered – in my family and in many of our friends' families – a waste of money and time. Being idle was the worst crime; being industrious, the best of virtues. In the eyes of the idealist newcomers to the land of Israel – work was sacred. Although most were atheistic, they nevertheless attributed religious value to physical labor. To quote the Zionist philosopher A. D. Gordon (Elon, 1971):

> … So, I shake myself and with all my strength … I throw the old life off. And I start everything from the beginning. And the first thing that opens up my heart to a life I have not known before is labor. Not labor to make a living, not work as a deed of charity, but work for life itself … it is one of the limbs of life, one of its deepest roots.

Little did I know then that ethics of this kind were not exclusive to the time and place into which I happened to be born, but also existed in other parts of the world. It was only in my Sociology studies that I came across Max Weber's (1947) writings and learned how the Protestant ethic became one of the major forces to shape American economic development. 20 years had then passed before I learned how the Protestant ethic was also a major driver in the respite culture. As a scholar of respite, I had come to learn of Cindy Aron (1999), a history professor at the University of Virginia, and her book *Working at Play*. Portraying the history of vacation in America from 1820 to 1940, Aron ascribed the Protestant ethic as a major force shaping the attitude of "discomfort with and suspicion of leisure" in America's contemporary vacation habits. The Protestant ethic, she noted, explained "the relatively limited amount of time that Americans spend vacationing" and their avoidance of "amusement and idleness" during vacation, leading to the perpetuation to "tension between work and play with which American culture struggled in the 19th century and early 20th century," and which we may still detect today not only in the US but globally.

How I came to study vacation

During the course of extensive cross-cultural studies of burnout in Israel and the US (Etzion, Kafry, & Pines, 1982; Etzion & Pines, 1986), my colleagues and I found that contrary to our own prediction, Israelis were systematically less burned out than their American counterparts. We decided to raise this issue in debriefing sessions that we held with the Israeli participants in our studies. In these sessions, many of the participants claimed that annual military reserve service had helped them contain the burning out process. They said that the opportunity to escape temporarily from the pressures of work and family and from the atmosphere of their daily routine augmented their ability to cope with everyday pressures after returning from service. Some participants even said that reserve service was better than a vacation in this regard. What seemed to us a joke at the time, proved to be a substantial theme when it was confirmed in a longitudinal, pre/post-study with a control group (Etzion, Eden, & Lapidot, 1998).

To hone in on the relevant factors that might play an important role in the reserve service respite effect, we conducted preliminary, semi-structured, group interviews with 20 men ranging in age from 25 to 40, who periodically served in the reserves. These interviewees perceived reserve service, a commonplace practice in Israel, to have a number of positive aspects. First, the men perceived it as an opportunity to recharge their batteries. It gave them a sense of moratorium, a perfectly legitimate way to evade their work obligations awaiting them back home. We conceptualized this feature of reserve service as *detachment from work routine*. Furthermore, they described that the men-only experience in many reserve combat units allowed reservists to let off steam and relieve accumulated strain in ways that are acceptable in the military but not in family or work settings. This manifested in somewhat aggressive norms, infantile and frivolous behaviors, not caring how they or their quarters looked, using foul language, and telling off-color jokes. Such behaviors would have been unthinkable back home or at work. The absence of wives and children, bosses, subordinates, and clients in the military context allowed reservists to relax normative self-censoring mechanisms. Additionally, reserve service often has the quality of a class reunion. Interviewees explained they enjoyed meeting old friends and reminiscing about youthful experiences. Moreover, the sporadic return to service imbued the reservists with an idealistic sense of mission and a feeling that they were bearing a portion of the collective burden by contributing to the nation's defense. Finally, for some, reserve service held an aspect of fascination of being privy to classified information, learning intrinsically interesting new tactics or technologies, or spending time outdoors in breathtaking natural surroundings.

Overall, the preliminary interviews with the reservists revealed that their perception of the most significant aspects of reserve service comprised the distinct types of activity and atmosphere, some of which afforded them considerable enjoyment. This change in lifestyle was coupled with a break in routine resulting from the detachment from work life. In short, they described reserve service as a valued getaway. However, as these positive features do not characterize every stint of active service, we chose to focus on two key elements in investigating the impact of reserve service: the degree of detachment from work life and the quality of the experience.

Unlike vacation time, which is frequently optional, military reserve service is usually compulsory. The daily activities involved in reserve service are also different from those applied to vacation time. However, if we think of military reserve service as a respite that affords a substantial reduction in unrelenting job pressure, we can predict similarities between the effects of vacation time and reserve service on burnout. The results of the reserve service study that my colleagues and I carried out started our interest in other types of respites from work and could be designated as the starting point of our Respite Project comprising since some thirty different studies, the results of which are described in this volume.

About this book

A common way for individuals to cope with stress is to take a break, or respite, to remove or at least reduce the stressors for a certain period of time. This allows them to engage in other activities that are relaxing and restorative and enables a return to the stressful situation with renewed strength. Despite the general belief that taking a break helps people cope with stress and replenishes their energy resources, the findings are not unequivocal. For some people, a vacation is itself a source of stress for one reason or another, and they avoid taking the time off due to them.

This book provides an original and unique outlook on the conditions under which a break from work may produce beneficial effects on employees' well-being, highlighting instances in which the beneficial effects cannot be taken for granted, and developing ideas for remedies. The book is based on a lengthy investigation of respite and fadeout effects that resulted in the formation of a comprehensive framework for understanding the effects of different kinds of respites on different kinds of people in different environmental and cultural conditions. Our samples included employees from different industries, such as industrial workers (low-tech and hi-tech), clerical workers, university personnel, students, teachers, hospital nurses, physicians, social workers, and insurance agents, to name a few.

The unique feature of this book is that it is empirically grounded. It is the result of a systematic research program designed to answer questions about respite effects, not only by reporting public opinion polls, as many of the articles on respite often do, but by measuring the moods and attitudes of people before, after, and sometimes also during their vacation. The novelty of this book lies in the extent to which it synthesizes a program of longitudinal empirical research with a theoretical understanding of the nature and determinants of respite and its link with well-being and health. It presents both published and unpublished material and, hopefully, contributes to a new understanding of familiar material by treating it in a stimulating manner.

Our research program was devoted to the greater question of how individuals may integrate their work with other parts of life and whether a retreat from work for specified periods serves individuals and their work output. That many of our findings indicate that taking time off from work is good for both when carefully conditioned might come as a surprise and may well have an effect on how employees and employers think about respite and their relation to them. The purpose of this book is to delineate the conditions under which a break from work may produce beneficial effects on employees' well-being. While it ties to the broad topic of work, health, well-being, and even the overwork culture, it is entirely dedicated to the topic of respite. Conceptually, multiple avenues are explored that are associated with time off – including individual differences, values, and belief, organizational considerations, and potential outcomes associated with respite. Readers will develop an extensive understanding of why individuals may or may not wish to take time off work and how to engage during that time in ways that can promote their well-being. The book also focuses on instances in which the beneficial effects cannot be taken for granted, offering ideas for enhancing the salutogenic respite effect even under such conditions. The respite project aimed to develop a framework for understanding the effects of different kinds of respites on different kinds of people in different environmental and cultural conditions.

Questions answered in the pages of this book are ample:

- Does taking a respite from work affect the level of stress and burnout that people experience upon returning to work?
- Does the time away from work help employees in recovering from stress, revitalizing energy depletion and containing the processes of burnout?
- Does a respite influence other parameters of well-being and various work behaviors (such as absenteeism)?
- What best defines a respite with respect to the objective of alleviating occupational stress and psychological burnout?
- Does it matter how this time off was arranged (planned, collective, or enforced), what was the main theme of the respite, or whether the respite was taken on or off the work site?
- Are all respites equal in their effect?

I began presenting preliminary results of this project in 1995, mostly in academic contexts. The natural readership of this book, then, is clearly scholarly, that is, researchers and students in the areas of industrial-organizational psychology, organizational behavior, and human resource management. Presenting a plethora of both empirical and applied aspects of respites, I also hope to reach practitioners in these areas, as well as those in the fields of occupational health, psychotherapy, social work, and human welfare.

INTRODUCTION

A common way of coping with stress is to get away from the stressors by "taking a break." Such a break, or respite, helps the individual in two ways. First, it removes or at least, reduces the stressors for a certain period of time. Second, it allows the individual to engage in other activities that are relaxing and restorative, thus permitting him or her to return to the stressful situation with renewed strength. Despite the general belief that taking a break helps people cope with stress and replenishes their energy resources, there was little empirical research on the topic until recently. Moreover, the findings on the effect of a respite from work on stress and strain are not unequivocal.

For some people, a vacation itself is a source of stress for one reason or another and they avoid taking the time off due to them. Some worry about the overload that will await them upon their return, others fear that if they take time off, who knows if they will even have a job to which to return. Another type of respite avoidance, according to management lore, is overloading on overtime, sometimes even without extra pay. It seems that giving overworked, ambitious employees the opportunity to rest does not necessarily help them. They themselves must see the respite as legitimate and be convinced that their colleagues perceive it to be so.

Harvard professor Ezra F. Vogel, a sociologist and expert on Japanese society, noted that typical salaried workers in Japan hesitate to realize all the time off-work to which they are entitled by law because "if [they] were to request this full vacation time [they] would be regarded as selfish and disloyal by [their] co-workers and [...] superiors" (see Rybczynski, 1991, pp. 156–157). Similarly, deep-seated characteristics such as Type A personality, workaholism, and inability to relax and enjoy a vacation or leisure (see Vingerhoets, van Huijgevoort, & van Heck, 2002) are other reasons some workers shun vacations.

Since the beginning this research activity in 1995, we have been collecting data in Israel and in other countries, especially the US, using parallel research tools in Hebrew and in English. All the studies within the framework of our respite project, whether published or not were longitudinal, involving two, three, or even more measurements of each respondent over a period of time. The underlying strategy of the reported and planned studies detailed herein was to examine the impact of respite on the well-being of workers. The design comprises at a minimum a pre-respite measurement and a post-respite measurement to which we had added a third measurement taken one month or so after employees' return to work assuming such a period represented their return to the routine of their work or studies. In recent studies we extended measurement to five and six occasions, including up to two measurements during the respite itself.

Most of the empirical work on this subject was conducted at the Faculty of Management at Tel Aviv University, in collaboration with colleagues and students of the Organizational Behavior Program as well as at UC Berkeley's Psychology Department. This project was mostly funded by the HC-IBRI – The Henry Crown Institute of Business Research in Israel. Some financial aid has been also provided by tourist enterprises and foreign universities.

Respite includes various forms of time out from work such as annual vacations, days off, or even stints of military reserve duty. The project focus was on the conditions under which employees recover from job stress, restore depleted energy, and contain the processes of burnout. Thus, we examined the nature of the respite itself (personal/organizational), as well as its length, the satisfaction people derived from it, and the personal and demographic characteristics of these people.

Conceptually the concept of respite is studied in a variety of situations and countries and includes such examples as vacations as well as military service. It is theoretically embedded in a number of current psychological theories. The methods are carefully designed field studies, all longitudinal, and analyzed with careful statistical criteria.

The first chapter sets forth an exposition of the respite concept and its development, from being conceptualized simply as an opportunity to recharge batteries (Maslach, 1976; Pines, Aronson, & Kafry, 1981) to its grounding within prominent, theoretical frameworks, such as The Conservation of Resources (COR) theory (Hobfoll, 1989; 2001; Westman, 1999; Westman, Hobfoll, Chen, Davidson, & Laski, 2005) and the job demands-resources (JD-R) model (Demerouti, Bakker, De Jonge, Janssen, & Schaufeli, 2001). In the second chapter, we consider various aspects of the respite that have been conceptually thought to influence the degree to which a respite might be beneficial to employees. In the third chapter we explore personal characteristics that have been found to enhance or hinder the respite effect, showing that the fit between the respite, its contents, and the individual on respite is crucial in achieving a salutogenic respite effect. In the fourth chapter, we delineate legal, cultural, and organizational conditions affecting both practice and perceptions regarding leisure time and time off work. In the fifth chapter, we discuss key outcomes of the respite on employees, including psychological and physiological well-being, the crossover effect of stress, burnout, and the respite effect, as well as the fadeout process of the respite effect. Finally, in the sixth and closing chapter of the book, we delve deeper into the meaning of recovery and coping, how organizations might promote employees' well-being through the vehicle of respite, and how employees may best capitalize on their respite opportunities to generate the greatest beneficial respite effect possible.

Three appendices appear at the end of the book. The first appendix presents the methodology and results of a meta-analysis conducted upon 23 studies that were included in the respite project (Etzion & Boklis, 2006). This analysis presents an overview of respite and fadeout effects, described throughout this book in reference to individual studies. The second appendix presents an annotated bibliography containing a comprehensive review of the literature on the subject of respite from work. Lastly, in the third appendix are presented the main questionnaire instruments used throughout the 20 years of research conducted during the respite project.

Dalia Etzion

References

Aron, C. S. (1999). *Working at play: A history of vacation at the United States*. New York: Oxford University Press.

Demerouti, E., Bakker, A. B., de Jonge, J., Janssen, P. P. M., & Schaufeli, W. B. (2001). Burnout and engagement at work as a function of demands and control. *Scandinavian Journal of Work and Environment and Health, 27(2)*, 279–286.

Elon, A. (1971). *The Israelis, founders and sons*. New York: Holt, Rinehart and Winston.

Etzion, D. & Boklis, V. (2006). *Meta-analytic study of respite and fade-out effects*. Report by: Tel Aviv University, Henry Crown Institute of Business Research, Faculty of Management, Tel Aviv, Israel.

Etzion, D., Eden, D., & Lapidot, Y. (1998). Relief from job stressors and burnout: Reserve service as respite. *Journal of Applied Psychology, 83(4)*, 577–585.

Etzion, D., Kafry, D., & Pines, A. (1982). Tedium among managers: A cross-cultural American – Israeli comparison. *Journal of Psychology and Judaism, 7(1)*, 30–41.

Etzion, D. & Pines, A. (1986). Sex and culture in burnout and coping among human service professionals. *Journal of Cross – Cultural Psychology, 17(2)*, 191–209.

Hobfoll, S. E. (1989). Conservation of resource – A new attempt at conceptualizing stress. *American Psychologist, 44(3)*, 513–524.

Hobfoll, S. E. (2001). The influence of culture, community, and the nested self in the stress process: Advancing Conservation of Resources Theory. *Applied Psychology: An International Review, 50(3)*, 337–421.

Maslach, C. (1976). Burnout. *Human Behaviour, 5(9)*, 16–22.

Pines, A., Aronson, E., & Kafry, D. (1981). *Burnout: From tedium to personal growth*. New York: Free Press.

Rybczynski, W. (1991). *Waiting for the weekend*. New York: Penguin Books.

Vingerhoets, A. J. J. M., van Huijgevoort, M., & van Heck, G. L. (2002). Leisure sickness: A pilot study on its prevalence, phenomenology, and background. *Psychotherapy and Psychosomatics, 71(6)*, 311–317.

Weber, M. (1947). *The theory of social and economic organization.* New York: Oxford University Press.

Westman, M. (1999). Applying Hobfoll's conservation of resources (COR) theory to respite research. Paper presented at The Academy of Management Symposium, Chicago, IL.

Westman, M., Hobfoll, S. E., Chen, S., Davidson, B. O., & Laski, S. (2005). Organizational stress through the lens of conservation of resources (COR) theory. In P. Perrewé & D. Ganster (Eds.), *Research in occupational stress and well-being* (pp. 167–220). Oxford: JAI Press/Elsevier Science.

1

THEORETICAL BACKGROUND OF RESPITE

> And on the seventh day God ended his work which he had made; and he rested on the seventh day from all his work which he had made.
>
> Genesis 2:2 (King James Version)

The problem

When people complain of pressure or fatigue, the advice most commonly dispensed to them is to "take a break" (Cherniss, 1981; Etzion, 1995; Freudenberger, 1974; Mattes, 1992; Pines & Aronson, 1988). It is often assumed that such breaks allow people to relax, to distract themselves from their day-to-day responsibilities, to engage in more pleasurable activities, and to replenish their energy resources. Despite the general belief that breaks help people to cope with stress, there has been little empirical research on the topic until recent decades (Etzion, 1999).

Since the establishment of the modern work practices of the Industrial Age, the nature and demands of work have continuously changed. Economic crises, wars, globalization, and technological advancement are just a few markers of transformation for the employed population. Since the early days of organized work, employers had been interested in increasing productivity thereby increasing pressures on employees. It was thought that working longer and harder would yield greater productivity. In response to World War I, for example, British workers were required to work longer hours, including on their day off. The surprising outcome, however, was a decrease in overall output, which was alleviated shortly after, when Sunday was reinstated as a rest day and the work day was restored to ten daily hours. It became clear, then, that productivity was "not only a function of how long people worked but also of how long they did not work" (Rybczynski, 1991, p. 56). Despite this experience, the last few decades saw a continuous rise in working hours throughout the world, and especially in developed countries in response to the global economy that is operational 24/7, 365 days of the year. The world of employment is rapidly changing. Globalization, mergers, acquisitions, privatization, downsizing, and technological advancements all influence the workplace. Traditional industries are transforming, and industries that were once considered stable and long-lasting are exposed to economic crises. In an interconnected world, crises occurring in one location affect other localities. Employers additionally opt for working contracts, such as outsourcing and human resource contractors, that reduce organizational obligation toward employees. It is thus no wonder that the industrial world is plagued with waves of employee dismissal resulting from organizational downsizing flood different industries, leading to the voluntary addition of working hours as a sign of organizational commitment (Fox & Chaney, 1998; Leise, Mundt, Dell, Nagy, & Demure, 1997; Moore, Grunberg, & Greenberg, 2004; Sparks, Faragher, & Cooper, 2001; Tennat, 2001). This, according to some researchers, had also led to the development of the phenomenon of presenteeism (Avni, 2007).

At the social level, changes in employment in recent decades have increased employees' need to cope with conflicting work and family demands. On the one hand, employees are required to work longer hours, travel more extensively, accept relocation, and generally be more available to address needs of employers and customers who are often located at a different time zone. On the other hand, social norms regarding family and child rearing involve greater financial and time resources. Additionally, the lengthening lifespan results in a growing population that requires attention due to age and age-related illness. This is known as the work–family or work–life conflict that employees often find challenging to resolve satisfactorily. This is a major social concern addressed by society at large as well as organizations, employers, and employees (Frenkel, 2011). An important facet of these social and organizational conditions is the dilemma faced by many women, who are traditionally responsible for child rearing and therefore refrain from entering the work force. Not only does this entrench gender inequality, but it also robs the workforce of an important pool of talent. An imminent care crisis is unavoidable with parental absence in children's lives due to work engagements, leading to a growing behavioral social problem involving violence (which further burdens nationalities with correctional costs) (Frenkel, 2011).

At the organizational level, the work–family conflict results in high employee turnover that robs the organization of both talent and experience that is specific to the organization. Uncertainty, dissatisfaction, and burnout are among key factors negatively influencing employee productivity. To avoid gender bias, a primary trait of the work–family–life conflict, employees (mostly in the US) have recently begun to introduce policies that facilitate better management of conflicting needs within and outside the organization (Frenkel, 2011).

At the personal level, two coping strategies of employees are to settle for undesirable and underpaid jobs or quit the workforce altogether. Such employees may lose their personal autonomy by becoming dependent on spouses and other family members or the State. When electing to stay an active part of the workforce filling jobs that are adequate to their talent, experience, and education, such employees face physical and psychological stress that often lead to fatigue and compromise health. Diminished job satisfaction and productivity often ensue (Frenkel, 2011). The phenomenon of absenteeism may also be related to the various pressures placed upon employees as a stress-reducing coping mechanism (Westman & Etzion, 2001).

As a result of these problems, job-related stress and burnout as well as the work–family–life conflicts increasingly grow as research areas. It is unsurprising, then, that researchers are also seeking ways to attenuate these phenomena. Personal welfare and the quality of leisure time have gained growing attention in the current millennium as crucial resources for employees' quality of life and success (Levi, 2002). Work, non-work, and their relations with well-being have thus become of central theoretical and empirical interest in the past several decades, continuously developing and branching into related issues. Research on respite, usually conceptualized as any form of absence from the work setting when job stressors are presumably absent or diminished (Westman & Etzion, 2002), may be considered as such a branch, which has focused on the manifold effects of respite from work on well-being. Therefore, most of the research on the effects of a respite that has been carried out has focused on the workplace and has addressed such general issues as its effect on job-related stress and burnout, as well as on job and general life satisfaction (Etzion, 1999).

Initial studies of respite effect focused on job stress (mainly physiological) symptoms and related to time off work as a "control occasion" to be compared with time on the job, as a field "experimental manipulation" of stressful occasions (Eden, 2001). In the past decade, respite research has progressed and developed an extensive perspective, integrating physiological, psychological, and behavioral aspects. The annotated bibliography offered at the end of this volume reflects this wide scope of the field.

Occupational stress

Given the nature of human conditions worldwide, both within and outside workplaces, stress appears to be an inseparable phenomenon of our lives. A certain degree of stress could be seen as beneficial, driving the individual to mobilize energy in order to meet a challenge at hand. The study of occupational stress, however, had gained momentum due to the relationship between stress and organizational productivity, including absenteeism, turnover, job satisfaction, and outputs (Shirom, 1982). An additional reason for this growing interest lies in the extent of time individuals invest in their work and hence the significance of their work in establishing their self-identity and self-worth (Schaufeli & Greenglass, 2001). Finally, research is driven by a disconcerting relationship between stress and chronic illness (Yoffe, 1997).

Stress

Most studies on vacation fall under the heading of respite research and are imbedded in stress theory (Eden, 2001; Etzion et al., 1998; Westman & Eden, 1997). The study of stress nevertheless evolved over the years, taking turns according to the discipline in which it was studied. Stress was first defined by Walter Bradford Cannon (1932), author of the "fight-or-flight" adage. Coming from life sciences, Cannon perceived stress as a physiological reaction of the organism to external, environmental stressor stimuli, leading to breakdown the further away from homeostasis it grew. Still in life sciences, Selye (1956), viewing stress as a biological response to noxious stimuli, noted three phases of response: alarm, resistance, and exhaustion, each characterized by a dwindling in coping resources facing a stressor of longer duration. Following Selye's General Adaptation Syndrome model, the body's initial response to a stressor is a general arousal known as the *alarm* stage. Under continuously straining conditions, the body adapts to the persistent stressor while maintaining elevated levels of physiological arousal known as the *resistance* stage. Finally, under chronically straining conditions, while the body remains chronically overactive and resistance fails, the body then transitions to the *exhaustion* stage, in which it becomes vulnerable to disease and even death.

Researchers then sought to define stress as a stimulus rather than a response, which would allow them to seek preventive measures. Stress acts as a stimulus based on the intensity and duration of an individual's different life events. *Acute* events are typically short-lived, intensive, and significant events that have a clear start and end; despite their short duration, the outcomes of these events may be long term. *Chronic stress* is a continuous state that is characterized by the difficulty to terminate it. Finally, *hassles* are low in intensity but high in frequency (Lazarus & Folkman, 1984; Pratt & Barling, 1988). Chronic stress may be further divided into *stressful event sequences* that are either *continuous* or *intermittent* (Elliot & Eisdorfer, 1982). Findings are inconclusive regarding the significance of one type of stress over the other (Holmes & Rahe, 1967; Kanner, Coyne, & Lazarus, 1981).

Criticism against the generality of these two models led researchers to investigate stress from a more personalized perspective, as a function of personality, perceptions, and contextual settings, allowing for a reciprocal relationship between the person and their environment (Cox & Mackay, 1981; Griffiths, 1981; Hobfoll, 1989). The person-environment (P-E) fit approach views stress as the outcome of a gap between personal traits, such as values and competencies, and environmental conditions, such as demands and supply (Beehr & Newman, 1978; Caplan, 1983; Cummings & Cooper, 1979; Edwards, 1988; Etzion, 1987; 1988; Hobfoll, 1989; Hobfoll & Shirom, 1993; House, 1980;

Lazarus & Folkman, 1984; Loftquist & Dawis, 1969; McGrath, 1976; Schuler, 1980; Shirom, 1982). Where a gap exists, negative psychological, physiological, and behavioral emerge, invoking stress and straining the individual. Specifically, stress emerges when the individual is faced by environmental demands, which they are unsure they can meet, whether because of their shortcomings or because the situational factors do no provide adequate resources for the completion of the task (French, Caplan, & Van Harrison, 1982). According to one approach, for stress to emerge demands must be higher than ability (Shirom, 1989). Thus, it is not only the objective environmental conditions that invoke feelings of stress but also the individual's cognitive, perceptual, and interpersonal characteristics that determine the extent of stress felt (McGrath, 1976). An individual would opt to cope with the stressing stimulus when facing the challenge is perceived to be more rewarding than turning away from it. Stress would thus emerge when the individual's needs is incongruent with the demands and the reward offered by the environment or when the individual perceives incongruence with the resources available to them and the environmental demands. This view of stress confers an advantage upon the person-environment fit model in predicting variance among responses to workplace stressors (Caplan, 1983; Etzion, 1987; Schuler, 1980; Shirom, 1982).

Stress occurs when the individual perceives the environmental demands as a threat over their own psychological, physiological, or social well-being; it has been defined as the "relationship between the person and the environment that is appraised by the person as taxing or exceeding his or her resources and endangering his or her well-being" (Lazarus & Folkman, 1984, p. 19). Individuals thus first appraise whether the situation with which they are faced is threatening to them and then assess the resources available to them to resolve the situation. Stress is influenced by the degree of frustration (past factors of the individual), conflict (current situation), and threat (future factors).

The fit, however is not all subjective. Subjectivity of the person compared to the objectivity of the environmental conditions is based on self-awareness, whereas subjectivity of the environment compared to the objectivity of the person is based on environmental awareness (Caplan, 1983). The intersectional person-environment model stresses the structural components of the person–environment interaction, the environmental factors, and response outcomes. A broader view of this interaction is represented by the transactional person-environment model that highlights the dynamic psychological process involves in cognitive assessment and adaptation. Chronic stress emerges from incongruence between personal traits and the environment that is unrecognized by the individual (Etzion, 1987).

Stress at the workplace

Occupational stress has been defined as the general expression of any negative emotional experience that is the outcome of an imbalance between job demands and the employee's ability to respond (Schaufeli & Enzmann, 1998). Stress is thus a reflection of the manner by which the individual and the organization adapt to varying circumstances, new situations, goal attainment, high quality production (Quick, Quick, Nelson, & Hurrel, 1997). Toward the end of the twentieth century, over 75% of employees had complained about stress in their workplace and 55% of work absences were ascribed to occupational stress. Also on the rise were rates of employees deemed incompetent due to health outcomes emerging from stress (Laski, 2002). In the US during the 1980s, workers' compensation claims for stress-related diseases tripled (Schor, 1992). Occupational stress has been a growing phenomenon due to increasing individualism and interpersonal alienation, increasing mental and emotional overloads, perpetual undermining of professional authority, change in traditional psychological contracts, and a growing demand for employee's investment in work in return for diminished remuneration from their employer (Schaufeli & Enzmann, 1998).

In the work environment, stress emerges in the interaction between the *focal person* (employee), the *role senders* (supervisors) in the *role set* (the organization); job demands and specific straining traits of the job and the role cause stress by leading to overload, job conflict, or role ambiguity (Aviv, 1998; Kahn, Wolfe, Snoek, Quinn, & Rosenthal, 1964; Schuler, 1980). Environmental demands may include job description, physical settings, and social settings (McGrath, 1976). Work conditions will be considered threatening only if the employer is able to punish the employee in the case of failure. Under these circumstances, stress will emerge even when the resources available to the individual are greater than the environmental demands (McGrath, 1976). An application of the person-environment approach that is specifically relevant to the workplace focuses on the congruence between job demands and decision latitude, that is, the extent of authority the individual possess in making decisions and the extent of their decision-making competency (Karasek, 1979). Stress emerges when the demands are high but the decision latitude is limited. Stress may also emerge when negative affect is directly related to doubts concerning one's coping capability (King, Stanley, & Borrows, 1987).

While each occupation presents its unique stressors (Pines & Aronson, 1988), stressors in the work environment can generally be classified into five groups. One group pertains to stressors at the workplace such as impaired physical settings, straining job demands such as lengthy or inconvenient working hours, and physical danger. A subset within this group comprises organizational factors such as organizational and management cultures, budget allocation, increased competitiveness, low decision latitude, poor communication, impaired engagement, reduced sense of belonging, and excessively dominant intra-organizational politics (Aviv, 1998; Cooper, Dewe, & O'driscoll, 2001; Cooper & Marshall, 1976). Another group comprises role-related stressors, including responsibility over others, complexity, diversity, autonomy, control, qualitative and quantitative overload or underload, job ambiguity, or role conflict. In a third group are workplace relationships and in particular social support from superiors, colleagues, and subordinates. In a fourth group are threats to potential career development such as employment insecurity, restricted advancement options, and inability to realize personal goals (Cooper et al., 2001; Cooper & Marshall, 1976). Finally, stress may stem from conflicting and incongruent demands of work and non-work sources such as family issues, financial difficulties, or crises that instill anxiety, worry, resentment and anger which might influence one's state while at work (Cooper & Marshall, 1976; Etzion et al., 1982; Kafry & Pines, 1980). Notwithstanding, not all employees resist organizational stressors, perceiving a necessity to protect their values, beliefs, and self-identity; some employees perceive them as opportunity for growth and success (Lazarus & Folkman, 1984). Stress may be perceived positively when successfully resolved, in which case it is considered as challenging, enriching, and enhancing self-satisfaction (see Aviv, 1998).

The environmental demands or challenges that the employee faces often comprise a negative facet that employees feel must be solved, changed, absolved, or otherwise resolved. However, the absence of positive, or desired, working conditions also leads to stress through feelings of failure, alienation, disappointment, or distress (Caplan, 1983; Etzion, 1984; Hackman & Oldham, 1975; Kahn, 1964; Kanner, Kafry, & Pines, 1978; Lapidot, 1992; Sapir, 1995; Schuler, 1980). In fact, the absence of positive working conditions may bear greater risk due to its covert nature compared to the obvious negative working conditions to which the employee is exposed (Etzion, 1987). Stress will emerge when work-related factors, positive or negative, reciprocally interact with the employee to alter the employee's psychological or physiological state so that the employee is required to diverge from normal functionality (Beehr & Newman, 1978; Etzion, 1984). One's quality of life decreases as a result of the absence of positive psycho-social stimulus or the presence of a negative stimulus; the greater the absence of positive stimuli (e.g., lack of diversity, self-actualization, complexity, significance, influence, material and emotional rewards, or little autonomy, guidance, advice, and performance feedback) and similarly – the greater the presence of negative stimuli (e.g., environmental disruption or threat, physical danger, organizational policies, remuneration, bureaucratic hassles, competitiveness, hurdles to decision-making, relationships) – the greater the stress (Caplan, Cobb, French, Van-Harrison, & Pinneau, 1975; Cooper et al., 2001; Eisenstat & Felner, 1983; Etzion, 1984; Farber, 1991; Hackman & Oldham, 1975; Kanner et al., 1978; Levi, 1972; Schuler, 1980). Based on Herzberg's (1996) two-factor theory (also known as motivation-hygiene theory), the presence of negative factors and the absence of positive factors at one's job operate on two separate dimensions: negative, extrinsic stressors affect dissatisfaction with work while positive, intrinsic stressors affect satisfaction with work. To reduce stress, the work environment should offer both a high level of positive work features and a low level of negative ones (Kanner et al., 1978)

Physiological impact of stress

Stress is a catalyst of negative health outcomes among workers, including heart disease, depression, and aggression. Lengthy exposure to high levels of stress causes physical, psychological, and behavioral overloads, placing employees at risk of physical or psychological injury (Etzion, 2003). Physical strain could be reflected in high blood pressure, high levels of cholesterol, cardiovascular diseases, gastric diseases, loss of muscle functionality, or allergies; psychological strain could be reflected in anxiety and burnout; and behavioral strain could be reflected in absenteeism, adverse decision-making, low productivity, and generally low contribution to the organization (Boyd, 1997; Bruess & Tevis, 1985; Jex & Beehr, 1991; Price & Hooijberg, 1992).

An out-of-balance employee is detrimental to the organization manifested in schedule delays, financial costs, and additional consequences that harm co-workers and the organization at large (Quick et al., 1997). Taken together, stress is a costly phenomenon with respects to its health outcomes; it entails medical expenses, absence from work, and consequences of professional injuries (Abouserie, 1994).

Many studies have reported a relationship between stress and anxiety. Anxiety describes an emotional state on discomfort accompanied by some apprehension of an undefined threat (Taylor & Arnow, 1988). While it may be subjective, developing as a function of personality (Spielberger, 1972; 1976), it cannot be isolated from the environment (Lazarus & Launier, 1978) so that the degree of anxiety experienced is related to the degree of threat perceived in a given situation. Anxiety may be explained as the outcome of the person-environment misfit, when discrepancies are present between the individual's (real or perceived) abilities and the environmental demands (Csikszentmihalyi, 1990).

Employee dismissal is among the greatest stress factors that have been studied in recent decades, and it has been to negatively impacting employees both individually and organizationally. Individuals who had been dismissed or notified about upcoming dismissal reported higher levels of depression, a greater number of impaired health symptoms (headaches, heart issues, ulcers, upset stomach, high blood pressure, and back pains), dietary changes, and increased consumption of alcohol (Grunberg, Moore, & Greenberg, 2001). Even perceived job insecurity affected individuals' physical and psychological well-being, increasing distress, sleep disorders, weight gain, blood pressure, and consumption of medication (Ferrie, Shipley, Marmot, Stansfeld, & Smith, 1998; Johoda, 1982; Kuhnert, Sims, & Lahey, 1989; Leana & Ivancevitch, 1987; McDonough, 2000). Employees exposed to the dismissal of others in their organization, developed negative attitudes toward management, they invested less in production and exhibited reduced productivity, their support of organizational goals was diminished, and they sought more occupational alternatives compared to employees reporting higher job security (Brockner, Grover, Reed, Dewitt, & O'Malley, 1987; King, 2000). Employees' turnover intentions do not directly stem from job insecurity, but are rather mediated by job satisfaction and organizational commitment (Davy, Kinicki, & Scheck, 1991). Such turnover intentions may also be the result of perceived lack of control (Brockner, Grover, Reed, & Dewitt, 1992).

The psychological impacts of job loss are diverse. The financial loss leads to increased levels of anxiety (Warr & Jackson, 1985), a decrease in quality of life (Vinokur, Price, & Caplan, 1996); the loss of daily structure leads to diverse complications (Fryer & McKenna, 1987); diminished quality and quantity of social interactions leads to a process of estrangement from co-workers (Swimburn, 1981); removal from the organization dims the sense of belonging and through removal from one's peer group, it also hurts the individual's self-image and social image and increases feelings that emerge from stigmatization of the unemployed (Briar, 1978; Johoda, 1982; Warr & Jackson, 1987). Unemployment has been found to be related to social pathologies including aggressiveness, increased drinking, rallying, avoidance of civil duties, and even immigration (Warr & Jackson, 1985). The centrality of work as the individual's core values will influence the degree to which job loss will influence psychological and physiological responses (Horovitz, 2002). Negative health outcomes associated with job loss are similar to those of occupational stress, finding an expression during unemployment even if they were absent during employment (Briar, 1978). Job loss, however, is not all negative. A silver lining shows that unemployed who believe their unemployment is merely temporary, approach their unemployment as they would a vacation (O'Brien, 1986).

Another highly researched stress factor that may be tightly related to the threat of job loss as both an instigator and an outcome is long hours (Schor, 1992; Simpson, 1998; Stewart & Swaffield, 1997; Worrall & Cooper, 1997). Long hours are one outcome of organizational downsizing, which leaves surviving employees with greater workloads. They may also be the result of a significant decrease in wages over the years (Bell & Freeman, 1996; Bosch, 1999; Figart & Golden, 1998; Schor, 1992; 1994; Stewart & Swaffield, 1997). A non-causal correlation was found between heart disease mortality and long working hours, so that the risk of heart disease mortality is doubled when the workweek was longer than 40 hours. However, this relationship may be moderated or mediated by numerous other factors such as personality, job type, or unhealthy behaviors such as cigarette smoking or uncontrolled eating (Buell & Breslow, 1960). An addition of 12 weekly working hours for seven months had reduced employees' response time, increased subjective levels of fatigue, and impaired deductive functions (Rosa, Colligan, & Lewis, 1989). Exceptionally long shifts combined with little sleep impaired employees' ability to accomplish simple tasks requiring alertness and focus while also affecting psychological markers of stress-anxiety, depression-moodiness, vigor-activity, fatigue-indifference, and confusion (Leonard, Fanning, Attwood, & Buckley, 1998). Employees working long hours over a period of five years reported higher blood pressure, muscle issues, chronic headaches, and fatigue, which persisted even after a reduction in working hours (Sparks et al., 2001). Among the psychological influences of long working hours are greater levels of anxiety, frustration, and desire for career transformation, as well as reduced job satisfaction (Spector, Dwyer, & Jex, 1988).

Working in shifts, similar to working long hours, may also instill stress in employees (Cooper & Marshall, 1976). The altered circadian cycle of shifts negatively impacts the physiological, psychological, and social rhythm of workers while also creating conflict between the worker's circadian cycle and the circadian cycles in the surrounding environment. Adaptation to shifts is thus challenging, leading to fatigue and frailty, depleting the workers' resources at hours in which they are supposed to be replenished (Aschoff, Hoffman, Pohl, & Weaver, 1975). Jobs involving travel may also increase stress and negative health outcomes (Sharkey, 2000).

Burnout

Burnout is often confused with stress or used interchangeably with it when, in fact, burnout is a chronic outcome of stress under certain conditions (Etzion, 1984; Maslach & Jackson, 1981; Pines et al., 1981; Shirom, 1989; Zeevi-Ohana, 2000). Indeed, various studies have linked between the two phenomena (Cherniss, 1980a; 1980b; Etzion, 1984; Etzion et al., 1998; Etzion et al., 1982; Freudenberger & North, 1985; Hobfoll, 2001; Hobfoll & Shirom, 1993; Kafry & Pines, 1980; Kanner et al., 1978; Magen-Neuhaus, 1993; Maslach, 1982a; 1982b; Shirom, 1989; Westman & Eden, 1997; Westman & Etzion, 2001). Burnout is a "psychological strain caused by continuous stressors which erode individuals' coping resources over the long term, until they reach the point of physical and emotional and mental exhaustion" (Etzion, 2003). The phenomenon first came to scholarly attention in the 1970s, and the term *burnout* was first coined by Freudenberger (1974) to describe psychological withdrawal from occupational relationships, especially in the therapeutic professions.

Burnout, like stress, is the result of incongruence between personality traits, needs, and competences of the individual and the characteristics of the environment in which the individual functions, and the meaning assigned by the individual to the general array of their life (Etzion, 1984). This is a phenomenon experienced at the individual level, involving chronic difficulties, discomfort, negative affect, and other negative outcomes (Shirom, 1989). Burnout affects healthy individuals in a broad range of professions; it is unrelated to one's profession but rather reflects a mismatch between the individual and their occupation, of demands that are greater than competencies and skills, or compensation that is smaller than one's needs. It thus reflects a breach in the psychological contract between the employee and their job. Short-term reactions to job overload prior to burnout include anxiety, stress, and agitation (Ezrachi, 1985). Burnout may be regarded as the unsuccessful outcome of the individual's attempts to cope with stress or the disillusionment from such attempts (Cherniss, 1981). In other cases, due to the gradual nature of burnout, no stress alert is activated prior to burnout rendering the individual defenseless, unresisting, and unaware of the stress that they had been experiencing (Etzion, 1987).

Incongruences leading to burnout are initially handled directly and actively; when they are unsuccessfully addressed, individuals retreat to passive and defensive modes of coping (such as apathy). Eventually individuals lessen goals, reduce responsibilities, diminished idealism, and increased alienation and self-centeredness (Cherniss, 1980a; 1980b; 1981; 1995). These ineffective coping strategies perpetuate and deepen the state of burnout (Schaufeli & Enzmann, 1998). To avoid burnout, congruence is required in several work facets: work load, control, remuneration, social settings, fairness, values (Etzion, 1984; Maslach & Leiter, 1997), diversity, role identity, role significance, role autonomy, and feedback (Hackman & Oldham, 1974). Additionally, increasing job satisfaction, self-confidence, and role uniqueness have been found to reduce or delay burnout (Cherniss, 1995). It is nevertheless possible that burnout is innate to Western societies, which have grown apart from collective values, reducing interpersonal commitment, weakening the ideological basis and motivation of employees, and leading them to arrive at their jobs with high expectations of fulfillment but diminished resources for handling frustrations (Cherniss & Krantz, 1983; Farber, 1983; Sarason, 1977). Additionally, the transition from a largely self-employed workforce early in the nineteenth century to a mostly salaried-based workforce alienated workers' sources of satisfaction as they no longer identify with their work but rather seek to satisfy growing self-interest extrinsic (salary, vacation, work conditions, consumerism) and intrinsic (personal growth, self-expression) needs to the point of possible narcissism (Dubin, Hedley, & Taveggia, 1976; Lasch, 1979). The introduction of technology to the workplace creates additional isolation and feelings of solitude as interpersonal communications grows more restricted (Farber, 2000).

Burnout emerges in situations that present high demands and chronic stress. Thus, although burnout is not profession related, the extent of burnout varies among different vocations so that individuals in helping professions who are exposed daily to human distress are at risk of experiencing greater burnout than individuals whose work is uncharacterized by such exposure (Cherniss & Krantz, 1983; Etzion, 1984). Additionally, helping professionals may find themselves faced with a reality that they cannot alter. Customer-centric service approaches seem to divert psychological resources of service providers in a one-way route, slowly depleting these resources, stemming from the absence of positive feedback and emotional strain (Maslach, 1982a; Pines et al., 1981). Managerial positions occupied by technology experts (such as engineers), who had not received managerial training may also generate a person-job gap that leads to burnout (Etzion, 1987). A deficiency in any of five fundamental elements of the helping professions could also lead to burnout: clear goals consistent with values and priorities, objective milestones that reflect client prosperity and increased control as service provider, objective measurements that reflect milestones, channels for data collection and analysis for short-term indicators, data-driven adjustment strategies (see Aviv, 1998).

According to one model, employees experience burnout when their occupational reality is incongruent with the individual's universal, professional, and personal goals that define perceived significance and meaning of work. Burnout is experienced when employees have exhausted their attempts at finding existential meaning and significance in their job, especially when sacrifices were made to attain the failed goal, way of life, or relationship. As such, individuals exhibiting the greatest levels of engagement and motivation are at greatest risk of occupational burnout (Freudenberger & Richelson, 1980; Pines, 1993). This notion is resonated in a different model positing that individuals experience burnout based on their personality such that emotional persons working in service and helping professions experience burnout due to emotional demands of their job whereas more cognitive individuals are strained by mental demands of their jobs (Garden, 1991). Similarly, dynamic, charismatic, and task-oriented individuals are prone to burnout due to an unrealistic level of expectations and consistent investment in their attainment (Freudenberger, 1974). According to a different model, burnout occurs in four stages: following a stage of excitement and idealism, comes a stage of stagnation in which expectations are lowered to match more realistic goals, then comes a stage of frustration where expectations for intrinsic and extrinsic rewards are unmet and emotional, cognitive, and physical symptoms emerge. Finally, apathy develops, accompanied by detachment and alienation from work and related relationships (Edelwich & Brodesky, 1980).

Regardless of its source, burnout is depicted by gradual loss of energy and general functionality (Etzion, 1987). The accumulation of unidentified or latent person-environment incongruences slowly eat through the individual's emotional resources. Minor stress emerging from such incongruences requires little coping resources, leaving the process of burnout concealed for long periods of time (Etzion, 1987). Because the warning signs of burnout remain obscure for a long period of time, they are often overlooked by supervisors, colleagues, or even the individual themselves (Etzion, 1987). The greater and more intensive the chronic stress experienced by the employee and the smaller perceived control over the situation is, so will burnout be graver and more severe (Cherniss, 1981). Due to its gradual evolution, when exhaustion erupts, it can no long be associated with a triggering event; employees gradually lose interest in their work or colleagues and their passion is diminished (Etzion, 1987).

Burnout is conceptualized as a three-dimensional experience that includes, professional efficacy, cynicism, and emotional exhaustion (Maslach, 1976; Maslach & Leiter, 1997; Maslach, Schaufeli, & Leiter, 2001), which although were surmised to develop linearly, possibly develop independently (Golembiewski & Munzenrider, 1988). Cynicism denotes an attitude of detachment and indifference toward one's job whereas professional efficacy relates to individual and social aspects of professional accomplishments, the individual's expectations of self-efficacy, and their confidence in their abilities. Exhaustion, however, reflects an investment of physical and psychological efforts beyond one's ability. Exhaustion and cynicism both develop at the backdrop of overload and interpersonal conflict, while professional efficacy is reduced following a lack of resources. Job related stressors (role conflict, role ambiguity, and overwork) have been found to predict professional efficacy, cynicism, and emotional exhaustion (Peiro, Gonzalez-Roma, Tordera, & Manas, 2001). Of its three facets, exhaustion has been found as the most important dimension underlying burnout, while job stress has been shown to increase emotional exhaustion (Cooper et al., 2001; Etzion, 2003; Lee & Ashforth, 1996; Maslach et al., 2001; Shirom, 1989), whereas professional efficacy may comprise an element of engagement (Schaufeli, Martinez, Marques Pinto, Salanova, & Bakker, 2002). Under burnout, initiative and responsibility make way for routine, disrespect, indifference, pessimism, and feelings of incompetence (Etzion, 1987; Pines & Aronson, 1988; Pines et al., 1981).

As such, burnout impairs the physical, psychological, and cognitive well-being of both employees and the organization, leading to negative consequences and affect (Etzion, 1984; Maslach, 1982a; 1982b; Maslach & Jackson, 1981; Pines et al., 1981; Shirom, 1989). Physical exhaustion (fatigue) often involves depleted energy, chronic weariness and tedium, headaches, feelings of helplessness, hopelessness, a sense of entrapment; psychological exhaustion (burnout) relates to negative attitudes toward oneself, work, and life in general, pessimism, and feelings of emptiness; and cognitive exhaustion is reflected in confusion, disorganization, lack of focus, and difficulty in decision-making. Physiological health hazards of burnout include a predisposition toward ailments and chronic ailments, indigestion, lack of sleep, and shortness of breath while behavioral hazards of burnout include a tendency for anger, tantrum, suspicion to the point of paranoia, excessive confidence, stubbornness, and rigidness (Freudenberger, 1974; Pines & Aronson, 1988; Pines et al., 1981; Shirom, 1989). Other health outcomes of burnout include depression and chronic fatigue (Cooper et al., 2001). Burnout has additionally been correlated with low self-esteem, forgetfulness, and sensorimotor symptoms such as agitation, inability to calm, involuntary muscle cramps, insomnia, lack of appetite, propensity for accidents (Cordes & Dougherty, 1993; Jackson & Maslach, 1982; Kahill, 1988; Pines et al., 1981; Schaufeli & Buunk, 1996). A disturbing outcome related to burnout is that of the Japanese phenomenon termed *karoshi* (literally: death by overwork), a phenomenon that has taken a high toll among highly overworked personnel (Maruyama & Tokuda, 2017; Rienhold, 1996). American's, too, have been said to be "literally working themselves to death" (Schor, 1992, p. 11). Psychological disabilities emerging from burnout may be considered as a final stage of the burnout process, that is, burnout in itself is not a clinical mental disorder but its extended presence can lead to such an outcome (Paine, 1982). What originates as a professional identity crisis may eventually turn into a personality crisis, leading to greater life crises (Rudow, 1999).

Also, like stress, burnout appears to have aggravated over the years following the alterations that characterize the workplace and particularly in response to the time-squeeze phenomenon, in which younger, urban, and highly competitive employees work 60, 80, or even 100 weekly hours. The work-and-spend cycle of employees is tightening as productivity increases, technology advances, and the potential for greater income exists. The resulting outcomes for the organization are burnout, reduced productivity and performance, increased interpersonal conflict, low morale, low working ethics, ineffective, indifferent, and even hostile customer service, absenteeism, and high turnover (Cherniss, 1981; Etzion, 1984; 1987; Firth & Britton, 1989; Jackson, Schwab, & Schuler, 1986; Schor, 1992). At the organizational and societal levels, burnout can have serious implications, for example, in the case of medical care, expressions of severe burnout might lead to the untimely death of patients (Aviv, 1998). Perceived workload, bureaucratic pressures, organizational climate, organizational structure, and organizational layoffs have been positively related to all burnout facets (Farber, 1983; Greenglass & Burke, 2002; Hobfoll & Shirom, 1993; Pines & Aronson, 1988; Westman & Etzion, 1995). Notwithstanding, research suggests that negative attitudes of employee toward their job may be more strongly related to high demands and lack of control compared to objective organizational factors (Sagie & Krausz, 2003). Organizational promotion based on principles of fairness rather than personal competence catalyzes burnout (Gaines & Jermier, 1983).

Burnout seems to be contagious, so that a burned-out employee may trigger burnout among co-workers. Due to the difficulties in reversing burnout, prevention is the best course of action (Cherniss, 1981; 1982; Etzion, 1987; Lee & Ashforth, 1993). Additionally, exhaustion often spills over to different walks of life, negatively affecting relationships with colleagues, friends, and family (Pines et al., 1981). Similarly, occupational burnout may erupt as a result of non-work strain or a life crisis such as disease, death, daily stress of family life, and so forth. One approach to burnout equates its high prevalence with evidence of normalcy and desirability for the purposes of development and competitiveness, thus eliminating the need to control it (Meyerson, 1994). Moreover, burnout may serve as a first step toward increased self-awareness and enriched understandings and therefore it may serve as a preliminary stage for growth and personal development. Individuals who have learned to overcome burnout then may lead better future lives (Pines et al., 1981).

Work and non-work

In Ancient Greece, work was defined as "non-leisure," filling up what remained after leisure had been exhausted as citizens were free individuals who were not required to work (Anderson, 1964; Miller & Form, 1980). This changed with the advent of the Protestant work ethos (Parker & Smith, 1976). Nowadays, most people regard work as important and valuable component of their life; it defines their lifestyle, social frame, and self-image so that work is often considered a chore while non-work is considered "play" (Kabanoff, 1980; Pines, Kafry, & Etzion, 1979). All workers nevertheless have a life that stretches beyond work, involving multiple roles that must be reconciled, such as parental, spousal, homemaker, and leisure roles (Koekemoer & Mostert, 2010). These roles often support each other as a significant portion of income is meant to provide for family while status, prestige, and remuneration drive the energy and enthusiasm individuals bring home. At the same time, the love and support individuals gain at home facilitate coping with work challenges (Aviv, 1998). Work, home, and leisure are intertwined in the form of a triple helix, where any advancements in one sphere is both influenced by and influencing the other spheres; finding compatibility between these spheres is of significant interest to researchers (see Aviv, 1998). Role bearers must perpetually cope with their different roles in order to meet their carrying demands (Kahn et al., 1964).

A conflict may nevertheless arise between work and non-work spheres when the demands of one sphere are incompatible with the other or when one the demands of one sphere interfere with the demands of the other (Greenglass, Scwarzer, Jakubiec, & Taubert, 1999; Kattenbach, Demerouti, & Nachreiner, 2010; Schieman, Mikkie, & Glavin, 2009). The conflict intensifies in as much as the conflicting roles are meaningful and central to one's self-image, as well as in as much as negative sanctions for incongruence between the roles are stronger (Beutell & Greenhouse, 1985). The conflict also intensifies with increased working hours (Richter, 1992).

Such conflicts between work and non-work impose stress on the individual, potentially contributing to feelings of burnout and negative health outcomes, threatening time, energy, and other personal resources (Haynes & Feinlieb, 1980). Stressful life events, both positive and negative, in either sphere (divorce, death, job loss, marriage, promotion) affect the individual's well-being and require them to personally and socially readjust (Holmes & Rahe, 1967). Such events not only cause stress but they may also contribute to burnout.

The ecological approach to burnout (Carroll & White, 1982) emphasized both personal and environmental factors leading to burnout. Burnout is thus regarded as an ecological dysfunction that is the result of complex interactions within four ecological systems: the micro system, comprising the smallest social unit at the employing organization; the intermediate systems, comprising all working units at the employing organization; the external system, comprising the environment outside of the work place that directly influences the employee such as family or laws; and the macro system, comprising the culture and the world at large. Burnout, according to this approach, cannot be isolated from one's complete life array.

The inter-sphere conflict is typically attributed to three sources. Conflicts based on time are most prevalent as different roles compete on the individual's time and the time dedicated to one role cannot be diverted to the other role. Conflicts based on stress occur when the mental and emotional effort required for the fulfillment of one role taxes the mental burden required to fulfill another role. That is, stress accumulating in one role hinder functionality in the other role. Such stress includes, strain, anger, fatigue, depression, and apathy that can develop both at work due to ambiguous job roles, a changing work environment, or mundane tasks or at home due to deficient family or spouse support.

Behavior-based conflicts reflect a gap between behavioral patterns that are required by the two roles, such as objectivity in the workplace and empathy at home (Beutell & Greenhouse, 1985). Two facets of the work–non-work conflict are pertinent for the purpose of this book: family-related conflict and women's struggle with work–non-work based on gender inequality and parenting demands; and job involvement and work satisfaction as an outcome of the work–non-work interface.

Family-related and gender-based work–non-work conflict

Family–work and work–family research has gained attention in recent years due to the increased participation of women, and particularly mothers, in the workforce as well as the increasing role of men in family life. The conflict between family and work arises due to the contradicting expectations, or the double burden, of the two roles, developing dilemmas among working women regarding prioritization in resource allocation (Arvey & Gross, 1977; Hardesty & Betz, 1980; Marx Ferree, 2006; Pines et al., 1981; Stapp & Pines, 1976). While technological advancements have facilitated the incorporation of women in the workforce, they have not lessened household demands (see Aviv, 1998).

More often than not it is the work sphere that interferes with the family sphere than vice versa (Frone, Russel, & Cooper, 1992). Women are often exposed to stress in the workplace due to gender inequality. Traditionally, women have been placed with the onus of childrearing and home caring, directly leading to the career–family conflict (Etzion & Bailyn, 1994). Working women are often considered to be carrying two jobs: one at their workplace and another at home, leaving little free time for their own personal needs (Hall, 1972; Kafry & Pines, 1980). Challenged by their feminine identity in the conflict between life and work roles, some women embrace the image of a superwoman, which further drives the ongoing, chronic onus of their double burden (Hall, 1972). Maladjustment to the conflicting roles can result in reduced communication and other negative outcomes within the family life and spousal relationship (Bedeian, Burke, & Moffett, 1988).

Indeed, mothers have been found to experience a greater extent of conflict compared to fathers and parents of young children have been found to experience a greater extent of conflict than parents of older children. Parents, especially to young children, are often the most susceptible to time-based role conflicts (Beutell & Greenhouse, 1985) in addition to increased rate of health problem attributed to the double burden (Haynes & Feinlieb, 1980). The conflicting demands of family and life often result in women's absence from work due to family demands, such as a sick child or a child's activity during working hours. Women were absent more than men and mothers were absent more that women who were not mothers. Additional family strains may originate in the role of employees as children of aging parents. The greater the experience of conflict, the greater the resulting conflict. Although evidence between the presence of the conflict and high burnout or low satisfaction has been inconsistent (Burke, 1988; Etzion & Bailyn, 1994; Israeli, 1988), for women this has been suggested as the most significant factor in burnout. Evidence suggests that perhaps men are better able to integrate work and related success with their non-work lives compared to female counterparts. For men, work and non-work success are compatible compared to their female counterparts, for whom work and non-work success seems to conflict (Etzion, 1988). Some work conditions may attenuate the experience of work and life conflict, such as working in shifts, however, crossover may occur between spouses, such as depression developed among men whose wives worked long hours (Galambos & Walters, 1992 and see Aviv, 1998).

Economic stress, an outcome of perceived dwindling income due to significant loss of resources rather than a chronic condition, negatively affects not only the individual but also intra-family conflicts (Kessler, Turner, & House, 1987; Vinokur et al., 1996). Unemployment may be considered a breach in the economic contract of marriage and unmet expectations; the stress accompanying unemployment increases separation and divorce (Kraft, 2001). While recession increases depression among men and women alike, spousal support attenuates the relationship between economic stress and emotional distress (Conger, Lorenz, Elder, & Simon, 1993). Not only one's family is susceptible to the individual's economic stress, but so are farther societal spheres (Conger et al., 1990).

Job involvement and work satisfaction

Job involvement, sometimes termed participation, is mostly a psychological trait, reflecting the extent of influence the individual has over job-related decisions and perceived personal contribution to the job or workplace. Involvement may be considered as congruence between personal and organizational goals, and it requires an extent of information, power, and influence sharing between managers and subordinates. One approach to job involvement views involvement as an outcome of participation and a potential of influence at the job, describing the psychological engagement of the individual in their daily work and the centrality of work in the individual's self-image or self-concept compared to their roles in other spheres of life (Lodahl & Kenjer, 1965). Involvement may also be an issue of self-identity and the identification of the individual with their work (Dubin, 1956; Lawler & Hall, 1970; Lodahl & Kenjer, 1965). Inversely, involvement may be the driver of self-esteem (Vroom, 1962). The job is thus an important factor in job-involved individual who is personally influenced by their job situation (Rabinowitz & Hall, 1977).

Increased involvement at the workplace seems to aggravate the work and life conflict (Burke, Weir, & Dunwors, 1980). For example, while a direct relationship was found between salary and rank on the one hand and job involvement on the other, these variables were inversely related to involvement in one's family. However, involvement in work or in the family had a positive effect over burnout and enjoyment. Women ranking higher and earning more at their workplace experienced a greater work and life conflict, they exhibited high involvement at work and lower involvement in the family (Etzion & Bailyn, 1994). Possibly, high job involvement leads to unaccommodating family patters as high demands of the job interfere with family life (Bailyn, 1977). While job involved significantly contributed to self-concept, self-identity, and life satisfaction, family was perceived a recharging haven (Evans & Bartolome, 1986).

Personal, situational, and environmental, characteristics are equally related to job involvement, which appears to remain stable over time. Age and Protestant work ethics have been found mostly related to job involvement while performance was unrelated to it. At the workplace, the strongest predictors of job involvement were participatory leadership and job stimulation and the strongest outcome were turnout and work satisfaction (Rabinowitz & Hall, 1977).

Satisfaction is inversely related to stress and burnout (Maslach et al., 2001) while dissatisfaction is positively related to them. Dissatisfaction of employees from their job seems to be a prevalent phenomenon (DeFrank & Stroup, 1989). Work satisfaction can be defined as a psychological and emotional response to one's assessment of their work environment and job content (Locke, 1976). Burnout has been tied to deficient work satisfaction, and more so to intrinsic satisfaction compared to extrinsic satisfaction (Schaufeli & Enzmann, 1998). While intrinsic factors tend to explain work satisfaction, extrinsic factors tend to explain work dissatisfaction (Hackman & Oldham, 1975; Herzberg, 1996). The absence of positive characteristics seems to influence both life and work satisfaction to a greater degree than the presence of negative characteristics (Kanner et al., 1978). Life satisfaction can be defined as the general emotional array of the individual toward their life in general. Dissatisfaction leads to the development of negative attitudes toward work, life, and other individuals (Pines, 1981) and has thus been significantly related to burnout, stress, and emotional exhaustion (Etzion, 1988; Friedman & Lotan, 1985) while also being suggested to be their outcome (Lee & Ashforth, 1993; Maslach & Leiter, 1997; O'Driscoll & Beehr, 1994).

Relationships have been established between life satisfaction and work satisfaction (e.g., Schaubroeeck, Ganster, & Fox, 1992). Several effects may be found between the two: a spillover effect where feelings from one sphere spill over and influence the other sphere; a compensation effect, where dissatisfaction in one sphere is compensated by another sphere; a generalization effect, where dissatisfaction from one domain is generalized to other domains; and segregation, where individuals completely divide their work life from their non-work life (Dubin, 1956; Kabanoff & O'Brien, 1980; Spector, 1997). Additionally, work and life spheres may be instrumental to one another such that one sphere contributes to a goal in the other sphere, or the two life spheres are found in a conflict that cannot be easily reconciled (Evans & Bartolome, 1980; Richter, 1992). Life satisfaction and work satisfaction increase with years of working and with age, possibly as a result of a better choices that lead to better person-environment fit for the individual (Cherniss, 1995).

The resolution

Generally speaking, employees cope with the effects of stress and burnout on two dimensions. One dimension represents the directness of their response: directly or indirectly while the other represents the extent of activity. When employees cope directly with the source of their discomfort, they may act actively to change or confront the source of stress and underscore its positive aspects or they may act passively by ignoring, avoiding, or departing the source of stress. When employees cope indirectly with the source of their discomfort, they may act actively by discussing the source of stress, diverting their energy to self-transformation, or otherwise keeping engaged with it. Alternatively, they may act passively by turning to external substances such as alcohol and narcotics, developing disease, or simply collapsing. Burnout is perceived to negatively impact the employees' resources, thereby rendering it difficult to effectively cope with this phenomenon. Employees, however, not only lose their psychological and environmental resources during their burnout experience, but they typically find it difficult to replenish, or recover, their resources during this time.

Conservation of resources

Individuals possess different resources with which they are able to contend with stress. The Theory of Conservation of Resources stipulates that individuals are motivated to obtain, preserve, and maintain valued resources, resources to regain lost resources (Hobfoll, 1989; Hobfoll & Freedy, 1993). Psychological stress emerges in the face of threats to the stability or existence of resources, when individuals or groups are unable to nurture and protect their resources, and when resources are indeed compromised. The threat of dwindling resources jeopardizes self-esteem, time, and money. Burnout emerges in light of unsuccessful attempts to regain resources while a spiraling loss of resources is likely to lead to feelings of despair, helplessness, and depression. Resources may be classified into four groups: material resources known as objects (e.g., possessions), conditions (e.g., marital status), personal characteristics (e.g., personality traits or competencies) that help resist stress, and energy (e.g., time and knowledge) that can facilitate the acquisition of other resources (Hobfoll & Shirom, 1993). Such resources may be valued for their own worth, or they may be valued because they serve to gain or preserve other valued resources (Hobfoll, 2001).

Individuals thus aim to create and preserve resource reservoirs. One goal is to prevent the loss of resources under threatening environmental conditions while another is to use extant resources in order to enhance and multiply the resource reservoir for a time when it might be required. The Theory of Conservation of Resources thus regards burnout as the outcome of the absence of gains and profits. Individuals require an excess of resources to be used to regain lost resources. An investment of resources that fails to achieve gains of other desired resources is experienced as a loss. When the individual lacks the resources required to compensate for lost resources, a cycle of losses may be triggered where one loss leads to additional losses. The loss of resources may be considered graver, then, than the unsuccessful acquisition of resources as individuals' resilience is impaired against the potential loss of future resources. The loss of resources may also be considered more acute than the absence of resources, as the former has been found to lead to depression while the latter did not have such implications (Ennis, Hobfoll, & Schroder, 2000).

When resources are lost, the individual's ability to cope with stress diminishes, increasing their vulnerability to negative outcomes of stress, which leads to further loss of resources. Individuals drawn into the resources loss cycle fail to replenish resources thereby experiencing burnout. Some resistance resources are ascribed to personality traits that are effectively facilitate stress reduction (Hobfoll & Shirom, 2000; Kobasa, 1979), including self-efficacy, perceived control, personal resilience, and achievement motivation (Silverberg, 2007).

Similar to the gradual, lengthy process of resource loss, coping with the loss of resources is also a lengthy process that develops over time. Coping is situational, so that a strategy that is effective under certain conditions may prove ineffective under different conditions. Nevertheless, individuals combine passive and active coping mechanisms to stop the loss of resources while also gaining new resources, increasing growth, and enhancing life satisfaction (Greenglass, 2002; Westman, 2004). Identifying the source of stress is a first step in its containment. However, effective containment is dependent on other competences such as planning, goal setting, organization, and visualization or simulation. An individual who desires change will indeed create that change. Active contention with the source of stress and resource depletion may include the acquisition and preservation of resources; an early action taken when the first buds of a problem are experienced; choosing a contention mechanism that is aligned with the individual's resources within their social setting (see Silverberg, 2007).

The proactive coping style is future-facing rather than past-oriented. Proactive individuals plan ahead, set goals, and assess hurdles and opportunities found within their environment. Their focus is set upon building coping resources and acquiring tools that will facilitate personal growth and achievement of personal goals. In this sense, individual seek to manage objectives rather than risk. Their approach is positive and optimistic so that the stressful event is experienced more as a challenge than a threatening risk. Combining emotional and problem-solving orientations, the proactive attitude helps the individual to cope with events prior to the accumulation of stress. In that, the proactive attitude prevents the buildup of stress as individuals change their environment before the event turns stressful. Individuals who are characterized by this attitude are typically aware of their resources and are able to mobilize their resources to different layers of their lives as needed. Possession of the quality or approach affects individuals in their daily lives, the manner by which they perceive events, and their belief regarding the extent of control they have in changing circumstances. (Greenglass, 2002).

Economic, social, and energetic resources have the power to attenuate the effect of burnout (Hobfoll & Shirom, 1993). When work-related resources are diminished, then, energy, time, or intimacy characteristic of non-work life domains may be enlisted to cope with occupational stress and its outcomes. Additionally, resources found in others may be translated by the individual to their own resources through behavioral, cognitive, or interpersonal competences (Hobfoll, 2001). However, such resources may not be easily accessible as individuals exhibiting burnout symptoms are often unattractive to others (Hobfoll & Freedy, 1993).

Effective proactive coping addresses seven behaviors. Goal-oriented engagement integrates the achievement of autonomic goals while regulating achievement of cognitive and behavioral objectives. Reflexive behavior allows for the simulation, review, and comparison of numerous alternatives while assessing their respective effectiveness. Reflection also allows for brainstorming, analysis of problems and resources, and the evaluation of different modes of action. Strategic planning ensures appropriate translation of needs into a sequence of properly prioritized and planned actions. Proactive prevention involves anticipating potential stressors as well as identifying them and being prepared for them prior to their materialization. Proactively approaching potential stressors also involves seeking support, advice, information, and feedback from other individuals in one's network. The social network can also be used to facilitate regulation of temporary stress by discussing with other conditions and emotions, drawing empathy and company. Finally, stressors may be inhibited so as to postpone the need for coping while formulating a plan of action. Effective proactive coping has been found to directly reduce work-related burnout and depression, especially emotional exhaustion, depersonalization, cynicism, and anger. Moreover, effective proactive coping is associated with beneficial outcomes such as increased professionalism, fairness at the workplace, and greater satisfaction (Greenglass, 2002).

The loss of resources requires a greater investment of coping resources compared to reactive strategies involving only a small investment of resources. This seems to be in line with the presence of negative job characteristics that lead to the chronic loss of resources as opposed to the absence of positive characteristics that bear different psychological impact on the individual's resources (Laski, 2002). Indeed, emotional exhaustion was found to be more related to job demands than to resources and that job demands are identified with resource loss thereby insinuating that burnout is related more to resource loss due to job demands rather than fluctuations in resources (Lee & Ashforth, 1996). Consequently, organizations highly characterized by stressors such as demanding schedules or extreme organizational commitments suffer to a greater extent from employee burnout compared to organizations boasting milder conditions

(Shirom, 2003). In such organizations, individuals seeking to protect and preserve self-esteem, the ultimate goal according to Narcissistic Psychology, by achieving idealistic objectives, consume their resources and reach the state of burnout faster (Schaufeli & Enzmann, 1998).

According to Conservation of Resources Theory, the self is a derivative of primary associations so that the self and its related behavioral alternatives, including one's perceptions, are but a reflection of cultural processes and are indicated through cultural scripts and dictions. Faced with critical stressors, different people might hold similar assessments based on their cultural context or social outcomes. It has thus been suggested that organizational characteristics bear greater weight on the development of burnout compared to other factors (Hobfoll & Shirom, 1993).

Job demands and resources

In the organizational context, two characteristics of the job connect between burnout, stress, and vacation as respite: job demands and resources. Job demands are physical, social, and organizational dimensions of the job that tax physical and psychological (cognitive and emotional) resources. Within the job demands-resources model of burnout, job demands are perceived as stressors that lead to strain when exceeding the individual's adaptive capabilities. Job resources are identical facets of the job that have the power to attenuate job demands and associated physical and psychological prices. That positive job characteristic, that is resources, may serve to buffer the negative effects of job demands has been suggested in the past (e.g., Kahn & Byosiere, 1992; Karasek, 1979); the innovation of the job demands-resources model is in suggesting that the interaction between different types of job demands and job resources allow for the prediction of job strain (Laski, 2013).

Organizational resources may include components such as salary and career, social resources may include social support and team climate, work-related resources may include role clarity and participation in decision making processes, and task-related resources may include performance feedback and skill variety. They are identical to core job dimensions noted in the seminal work of Hackman and Oldham (1976) who emphasized the motivational force of job characteristics that instill meaningfulness in employees. Such resources may also play a role in obtaining job-related objectives as well as facilitate personal growth, learning, and development. Under the job demands-resources model, the impact of job resources is particularly noted in the face of high job demands so that strain is lower for employees who enjoy extensive job resources (Bakker & Demerouti, 2007; Bakker, Hakanen, Demerouti, & Xanthopoulou, 2007; Demerouti, Bakker, Nachreiner, & Schaufeli, 2001).

Job demands are empirically distinguished from job resources and the two are moderately and negatively correlated. High job demands may inhibit the mobility of job resources while the presence of job resources may attenuate the impact of job demands (Bakker et al., 2007). The job demands-resources model suggests that whereas job demands instigate health-impairing strain processes following depletion of energies, job resources instigate motivational processes. Strained employees are continuously required to choose between the preservation of performance objectives (gains) and the mental effort that must be invested in the attainment of such goals (costs). When job demands increase, strain must be regulated to maintain performance levels. Such strain may lead to physiological and psychological costs such as accelerated activity of the sympathetic nervous system, fatigue, and agitation. Continuous, ongoing enlistment of compensatory resources diminishes the employee's energy reservoirs thereby promoting burnout and negative health outcomes (Schaufeli & Bakker, 2004; Schaufeli, Bakker, & Van Rhenen, 2009). The motivational influence of job resources, on the other hand, leads to high work engagement, low cynicism, and outstanding performance. This is true whether job resources are intrinsic or extrinsic.

Intrinsic resources address fundamental human needs such as autonomy, relatedness, and competence. A learning-driving resource may increase work-related competencies while extrinsic resources are found in collegiate support, supervisor feedback, or performance feedback. In the balance between job demands and job resources, the former predicts burnout whereas the latter predicts work engagement. Job demands are negatively related to burnout but positively related to job engagement while aggressive job demands are positively related to burnout and showed no influence on job engagement (Kim, Shin, & Swanger, 2009; Schaufeli, Salanova, Gonzalez-Roma, & Bakker, 2002).

In organizational work environments in which resources are abundant employees feel encouraged to investment efforts and capabilities in the job, leading to task completion and goal attainment (Schaufeli & Bakker, 2004; Xanthopoulou et al., 2007a; Xanthopoulou, Bakker, Demerouti, & Schaufeli, 2007b). Resources such as autonomy and social support have been found to enhance commitment and extra-job performance (Bakker, Demerouti, & Verbeke, 2004) whereas insufficiently awarded resources lead to long-term withdrawal and diminished motivation and commitment (Bakker, Demerouti, De Boer, & Schaufeli, 2003). Autonomy moderates the influence of job demands of burnout as employees have the freedom to determine the manner by which they respond to job demands. Feedback offers employees vital information that crucial for task completion and maintenance of health. Collegial and supervisory support provide employees with instrumental and emotional help that allows them to meet their goals. Collegial support may also be directly enlisted to the completion of tasks and reduction of workload (Bakker, Demerouti, & Euwema, 2005; Van der Doef & Maes, 1999). Job resources contribute most to the preservation of work engagement in the face of high job demands. Resources that are particularly linked to work engagement are supervisory support, innovation, esteem, and adequate organizational climate (Hakanen, Bakker, & Demerouti, 2005; Xanthopoulou et al., 2007a; 2007b). Work engagement further enhances resources, for example, by strengthening competence-related beliefs (Bakker et al., 2010; Llorens, Schaufeli, Bakker, & Salanova, 2007).

Work engagement is not time dependent on the availability of resources so that the availability of resources at one point in time may be related to work engagement at a different point in time (Xanthopoulou, Bakker, Demerouti, & Schaufeli, 2009). From this perspective, burnout may be considered as the erosion of work engagement on a single continuum leading from one state to the other. Alternatively, burnout and engagement are perhaps employee prototypes representing dimensions of pleasure and activation (Maslach & Leiter, 1997; Maslach et al., 2001).

When employees experience work-related stress through loss of resources, threatened resources, or perceived inadequate returns on investment in resources, they attempt to contain the influence of that stress by conservation of energy, which demands resources allocation that may further diminish energies and contribute to burnout (Melamed, Shirom, Toker, Berliner, & Shapira, 2006). Nevertheless, job demands and resources may be subject to employees' perception of their work environment (Zapf, Dormann, & Frese, 1996). Employees perceiving their job demands as high due to their own feelings of burnout, may often complain about their workload thereby creating a negative work environment (Bakker & Schaufeli, 2000). At the same time, highly engaged employees might perceive affluence in their job resources thereby distributing their recourses more effectively (Xanthopoulou et al., 2007a; 2007b). Engaged employees have been shown to possess psychological capital in the form of energy, self-efficacy, optimism, goal-orientation, resilience, feelings of positive accomplishment, organizational commitment, and lower turnover intentions; they influence events in their lives and create positive feedback in terms of appreciation, recognition, and success (Bakker, Albrecht, & Leiter, 2011; Demerouti, Bakker, De Jonge, et al., 2001; Gorgievsky, Bakker, & Schaufeli, 2010; Hakanen et al., 2005; Schaufeli & Bakker, 2004; Xanthopoulou et al., 2007a; 2007b; 2009). This is in line with the Conservation of Resources Theory by which individuals for whom the access to strong resource reservoirs is limited experience the cycle of resources loss, whereas those who have access to such resource reservoirs will make an investment toward future resource gains thereby experiencing the cycle of resource gain (Hobfoll & Shirom, 2000). Tenants of positive psychology similarly suggest that positive experiences directly or indirectly broaden resource reservoirs thereby enhancing personal well-being (e.g., Fredrickson, 2001; Salanova, Llorens, & Schaufeli, 2011).

Social support

The term *social support* denotes a multifaceted supportive social environment that helps the individual to effectively handle challenges, stress, and eventual burnout (Cohen, 1988; Cohen & Wills, 1985; Hobfoll, 1989; Kahn et al., 1964; Wallston, Alagna, De-Villis, & De-Villis, 1983; Wethington & Kessler, 1986). This is an important resource defining the extent to which the individual's needs for affection, esteem, relatedness, and security are significantly fulfilled by others (Caplan et al., 1975). Social support is translated into information that confers upon the individual the belief that they are liked and that they are partner to a mutually reciprocal binding relationship (Cobb, 1976). The extent of social support experienced by the individual is thus subjective (House, 1981). Not only does social support mediate certain hardships but it also adds complimentary resources to one's life (Kanner et al., 1978). Different perspectives may be found in the literature regarding the essence of social support and whether it relates to quality, quantity, and other structural and functional aspects of support (Wills & Fegan, 2001).

Social support buffers the relationship between stress and burnout yet individuals suffering from diminished resources reservoirs seek less after social ties (Etzion, 1984; House, 1981; Seers, Mcgee, Serey, & Graen, 1983). Social support is extended both cognitively and practically through emotional support, relevant information, and a sense of belonging that is fostered in individuals through both formal and informal relationships with other individuals or groups, including spouses and other family members, supervisors, colleagues, subordinates, and friends, as well as community agents and authorities (House, 1981).

The term social support relates to a broad range of interpersonal behaviors exercised by members of the individual's network, that may assist the individual in successfully handling the different circumstances and events in one's life (Sarason, Sarason, & Pierce, 1990). Such behaviors may fulfil the need of relatedness by providing emotional support through affection, love, and sympathy extended through time spent together in leisure, entertainment, or rest. They may directly address the problem or indirectly supply opportunities to enlist alternative resource by proving material support, services, and advice. They may also help the individual in understanding the origins of stress by providing information regarding the environment which may help calibrate the extent and intensity of the problem. Finally, such behaviors may enhance feelings of control and purposefulness by strengthening the individual's self-esteem. All of these outcomes may attenuate the impact of the strenuous event as well as its consequences including anxiety (Burleson, 2003; Cohen & Wills, 1985; House, 1981; Taylor & Arnow, 1988; Thoits, 1982; Ullah, Banks, & Warr, 1985). The attenuating effect of social support, however, is not unequivocal and be dependent on congruence between the type of problem experienced by the individual, the source of support, and the type of support offered, such as supervisory, collegial, and familial support (Etzion, 1984; Payne, 1980).

The need for social support may well be biological, innate, and essential for the promotion of health and psychological, physical, and emotional well-being of humans (Bowlby, 1969; Ganster & Bart, 1988; Ganster, Fusilier, & Mayes, 1986). The absence of social relations leads to negative psychological outcomes such as anxiety and depression. Social support operates in two ways to influence well-being delineated in the primary effect model and the buffer model. The primary effect model perceives social support as positively influential and helpful regardless of the extent of strain to which the individual is subjected. Here, an extensive social network confers upon the individual a sense of positivity, stability, and recognition in one's self-value and self-image. Support in this sense may attenuate both stress and strain. The buffer model perceives the value of social support to be revealed particularly when the individual is subject to a high extent of stress. Social support in this case defends the individual against potential health impairment and dysfunctionality following strenuous events. The buffer effect is expressed when the individual assesses that they possess few resources than required to cope with a threatening or demanding situation. Here, social support serves as a buffer between stress and strain and may attenuate the perceived threat of the stimulus or enhance the individual's belief of being in possession of adequate vehicles with which they might contend with the stimulus hence reassessing the situation as less strenuous (Cohen & Wills, 1985; Hobfoll, 1989). Social support may involve six distinct functions, including sharing problems (listening), appreciation and affirmation of competence (technical support), getting constructive criticism (technical challenge), recharging one's batteries (emotional support), getting a different point of view from a person one can trust (emotional challenge), and alleviating feelings of mental isolation (sharing social reality) (Pines, 1981; 1983).

At the workplace, social, emotional, and professional support includes trust, esteem, and autonomy extended by supervisors and co-workers, as well as gratitude, recognition, feedback, sharing, and a tolerant attitude of the organization at large toward challenges emerging in the work–non-work conflict. The absence of such public support leads to burnout (Cherniss, 1995; Schaufeli & Enzmann, 1998). Organizational support as well as one's hierarchical level within the organization, and their extent of job autonomy and independency may also influence business travelers' reaction to stress (DeFrank & Stroup, 1989). At the personal level, social support at the workplace involves empathy toward others, listening, and guidance provided by co-workers and supervisors (Mirowsky & Ross, 1989). Social support at the workplace has the power to preserve resilience and professional efficiency in the face of downsizing. Empowering of employees combined with support extended by supervisors and co-workers has been inversely related to reported stress levels (Joiner & Bartram, 2004).

Men and women seem to difference in their needs to receive support, enlist it, or give it. Women emphasize issues of intimacy, empathy, and emotions compared to men. The also devote themselves to a greater degree to their families and friends compared to men. Therefore, they seem to be offering greater social support but receiving less of it compared to men (Greenglass, 1982; Schwarzer & Gutierrez-Dona, 2005). Understandably, then, men appear to prefer work-related sources of support such as supervisory and collegial whereas women seem to prefer familial sources of support (Etzion, 1984).

Spousal support has the power to attenuate stress related to the work–non-work conflict (Wallace, 2005). It also has the power to attenuate stress and burnout among business travelers (DeFrank, Konopaske, & Ivancevich, 2000; Yoffe, 1997). Spousal support is particularly important as stress and burnout seem to crossover between spouses whereas the importance individually assigned by spouses to perceived control also seems to crossover between them (Etzion, 1995).

The impact of social support may vary based on the type of community within which it is extended, for example, support within urban settings tends to buffer anxiety whereas support within the rural settings of the Israeli kibbutz tends to buffer depression. This may be attributed to the size of the supporting family: whereas in most community settings social support is extended by the nucleus family only, in some community settings social support is extended by a more elaborate family constellation including grandparents, uncles, aunts, cousins, and so forth (Ichilov & Bar, 1980). An additional factor in the successful contribution of social support to the individual may be in the extent to which the request or expectation for such support is legitimate and ingrained within the community. In communities that tend toward collectivism, such as the kibbutz, the community is better prepared to facilitate working parents and members are more easily accessible and reciprocally involved in each other's lives thereby attenuating potential work–non-work strain. Possibly, in such communities where equality is greater among gender roles in the family, women in particular may feel free of the conflicting roles experienced by women in other communities (Koffman & Elizzur, 1983; see also Levi, 2002).

Respite

An effective way to stop the spiraling loss of resource cycle is to prevent its onset while facilitating a process of resource gain. While the prevention of resources loss, job demands, and chronic work-related strain, is not always possible, there is nearly always room for intervention. In fact, such interventions were studied in the early twentieth century as part of the renowned Hawthorne productivity studies conducted by Elton Mayo and Fritz Roethlisberger in the 1920s (Roethlisberger & Dickson, 1939). Respite is a way for individuals to break from their investment of resources in their work environment thereby breaking the resource loss cycle. Respites characterized by rest, leisure activities, and time spent with dear ones, may help replenish emotional and physical resources (Hobfoll, 1989; 2001; Westman & Eden, 1997).

Respites may range from a quick break during the work day to the weekend, to a vacation, and to a full year sabbatical – all of which have been found to attenuate stress and burnout due to the absence of daily work-related stressors. Reserve duty and business travel, too, have been found to have such an effect (Etzion et al., 1998; Sapir, 1995; Vinocour-Hai, 2001; Westman, Etzion, & Yoffe, 1998). However, such an effect is achieved only when the respite experience is a positive one. Individuals who are able to detach from work during their respite are more capable of enjoying the present moment, thereby enhancing their respite experience. The greater the detachment from work and the work environment during the respite, the greater its effect on stress and burnout (Etzion et al., 1998; Westman & Etzion, 2002). When respite is forced upon employees, the respite effect may be less than positive. For example, following a computer shutdown that forced respite on employees, blood pressure increased upon return to work, likely due to employees' anticipation of their workload (Caplan & Jones, 1975; Eden, 1990). A forced respite that is considered a positive experience detaching employees from work may nevertheless attenuate employees' levels of stress and burnout (Lapidot, 1992). This influence may also be attributed to the legitimacy of employees' absence from work.

Breaking off from work has a positive effect on burnout reduction over time (Cherniss, 1995; Farber, 1991). Women who have taken breaks throughout their careers were less burnt out as they grew older compared to men. These women also had a greater sense of success and self-confidence (Etzion, 1987). Employees taking more time off have been shown to have a more beneficial balance between work, leisure, and family (Cherniss, 1995). Women, as regrettable as it may be, are likely to feel less relaxed on vacation than men because of the heavy burden of home-related chores that is their lot when the whole family is on vacation (Etzion, 2006).

An important aspect of respites is a pause in stress transfer between colleagues, managers, and subordinates. The theory of stress transfer is based on role theory, which emphasizes the reciprocal nature between the focal individual and their role instigators. In other words, the strain and stress felt by one individual influence others in various circumstantial environments. Stress may be transferred directly between individuals through empathy and identification. It may be transferred vicariously through the mutual impact of environmental strain on individuals sharing that environment. Or, it may be transferred indirectly as a result of reciprocal influences that transpire through interpersonal interactions. In other words, stress and strain may be contagious, especially when passing a certain threshold or with especially stressful vocations (Westman, 2001; Westman & Etzion, 1999; Westman & Vinokur, 1998).

Leisure

The consistency of empirical studies in indicating the ameliorative effect of various forms of respite has raised the need for further comprehension of the nature of respite pursuits that enable psychological and physiological recovery. Hence, daily activities on and off work and their contribution to daily recovery and periodical alleviation of stress have also become important issues in the context of respite effect. Thus, leisure research has been integrated into recovery and respite research, highlighting kinds of activities during employees' time off and their effects on well-being.

Leisure comprises passing experiences throughout life rather than a permanent lifestyle (Gunter & Gunter, 1980). It is characterized by its quality as expressed by the activities carried out during this time, which lead to self-actualization and satisfaction. This is a time that is associated with pleasant expectations and experiences and a feeling of freedom with minimal interruptions by undesired activities or other obligations. Based on two dimensions, work-relatedness and discretion, one might thus define leisure as an activity that is taken by personal choice during a time that is free of any role obligations; it is an activity that is related to cultural values and typical norms and instills pleasant feelings both prospectively (anticipation) and retrospectively (reminiscence); it might be characterized by different degrees of commitment and intensity, offering opportunity for joy, personal growth, and service to others (Kelly, 1972; Parker, 1971; Schor, 1992). Ironically, most individuals aspire to increase their leisure time, yet few ever achieve doing so (Bosserman, 1989; Parker, 1971). This may be unsurprising as free time in the second half of the twentieth century has decreased significantly. In the US, for example, the average worker sustained a cut of 140 hours annually in their free time (Schor, 1992).

Often leisure time is associated with major events, such as vacationing, that are uncharacteristic of one's daily life, are often expensive, and have meaning in their scarcity. From a different perspective, leisure may be considered as daily actions individuals take in their daily life, such as informal interactions with family members or other individuals in one's closest environments, work included. That is, informal interactions at the workplace may be considered as leisure time (Kelly & Godbey, 1992).

Leisure is considered a means to enrich one's non-material aspects of life. It is an opportunity for individuals to materialize their values, express creativity and authenticity, nurture relationships within the community, and invest in one's body and mind. Leisure has been found to positively influence not only satisfaction but also well-being and health as a result of realizing and honing one's skills, expressing oneself, self-actualizing, freedom of choice, autonomy, and decision-making authority. Since these traits may emerge also in the family or work spheres, leisure may perhaps be inseparable from these spheres and tied to them multidimensionally (Kelly & Godbey, 1992; Kelly & Kelly, 1994; Melamed, Meir, & Samson, 1995). In this respect, leisure follows the work–non-work model mentioned earlier, in which individuals' attitudes, beliefs, and behaviors in their various life spheres may spill over, compensate, or be entirely separate (Dubin, 1956; Parker, 1971; Wilensky, 1960). Perhaps it is the nature of one's work that mediates this relationship (Rousseau, 1978).

Leisure may serve as an outlet for friction between different life domains. Individuals seek for alignment between their environment and their personal traits, including personality, demographics, health condition, culture, social relationships, and so forth. The greater the alignment between the two, the greater the satisfaction one feels. Individuals nevertheless choose and remain in spheres that are not in alignment with their personal traits, causing psychological strain that requires resolution especially when that sphere is one's central life interest. Indeed, correlation was found between professional inclinations and leisure activities. Where alignment exists between one's central life interest and one's personal traits, individuals gain greater satisfaction in their central life interest compared to those whose central life interest is in other spheres. Similarly, in cases of misalignment, individuals whose central life interest is in work experience greater dissatisfaction in that sphere whereas individuals whose central life interest is in other sphere experience greater dissatisfaction when leisure activities are misaligned with their personal traits (Dubin, 1956; 1979; Holland, 1973; Meir & Melamed, 1986; Melamed & Meir, 1981).

Leisure may be considered as time, activity, or mental state. As time, leisure is the time left after individuals meet their obligations at work, home, in their family, and in satisfying fundamental life needs (such as sleep). As activity, leisure comprises activities that are not financially remunerated. As a mental state, leisure comprises an experience of freedom, enjoyment, self-expression, and self-enrichment, where the experience is the only goal. Comprehensive definitions of leisure, then, consider all three elements and the dynamics between them (Parker, 1971; Schor, 1992).

In determining the degree of leisure associated with any activity, three assumptions are made. First, that allocated time and activities of individuals are derived from social and psychological powers, that is, the social roles of the individual have the power to restrict or enhance one freedom while psychologically individuals develop positive or negative affect toward certain activities with respect to the extent of freedom ingrained within them. Second, that a certain degree of leisure is found in any situation where individuals enjoy a freedom of choice as well as the activity itself. Third, that activities are perceived as appealing or repellant based on personal experience and tendencies leading to the feelings individuals experience with respect to different activities and the meaning that is assigned to them. The highest degree of leisure, *pure leisure*, would thus entail a strong sense of control over the choice of activity, a great extent of pleasure, and detachment – both physical and emotional – from daily life. *Institutional leisure* refers to a lifestyle in which individual choose to spend their free time doing things that they are obligated to do by their different commitment in life so that activities characterizing duties are satisfying, enjoying, and self-actualizing. In contrast, *anomic leisure* appears in a lifestyle that carries different obligations that spill over to one's free time; it is characterized by little control or joy and is accepted as a compulsory onus. *Alienated leisure* occurs when the individual is unable to instill meaningfulness into their leisure resulting in lack of goals and satisfaction, confusion, anxiety, and anger, all of which are innately in contrast with the notion of leisure. Family, social, or spiritual obligations are thus only partially considered as leisure, whereas activities aimed at self-expression involving freedom of choice as well as expectations of pleasure and personal gains are exclusively considered as full leisure (Gunter & Gunter, 1980).

Leisure has become an important area of social research following the notion that the solution of some of the graver issues of employment may be found in leisure (Shamir, 1988). *Serious leisure* integrates components from both the work and leisure environments suggesting that individuals pursue activities as amateurs, hobbyists, or volunteers. Serious leisure demands a considerable investment on the part of the individual, involving aspects of achievement, success, and failures. Investment in serious leisure builds on specific competences, knowledge, or training, and it involves a set of values, norms, and performance standards. It instills pride in the individual and serves as the basis for self-identity. Serious leisure rewards individuals through self-expression, self-actualization, enrichment, rejuvenation, a sense of achievement, enhanced self-image, social relationships, a sense of belonging, and material rewards. Prices paid in lieu of serious leisure may include stress, disappointment, and conflict between life spheres. Serious leisure helps confer legitimacy to inactive rest. By these standards, non-serious leisure, which aims to serve enjoyment only, is far less important in one's life (Stebbins, 1992). Tangent to serious leisure is *structured leisure*, a well-scheduled activity in which competencies are constantly honed increasing in challenge and complexity. Structured leisure requires active attention and clear performance feedback in contrast to spontaneous leisure activities that take place under no rules or guidance (Mahoney & Stattin, 2000). In such a hierarchy of leisure, *free time* is considered to relate to time but not to content and albeit being a necessary condition for the leisure it is insufficient in its own merit (deGrazia, 1962; Parker, 1982). Free time is at risk of turning into alienated time, especially in Western societies where free time is encouraged for pricey consumer activities. Under such conditions leisure has become an acquired rather than innate skill. Moreover, individuals in such societies seem to have lost their ability to stay present in their leisure to the point of avoidance behavior.

Possibly the onus of work demands taxes individual's reservoirs so that they are unable to assume active leisure turning instead to activities that have very low energetic demands such as watching TV, resting, or reading (Ben-Moshe, 2006; Schor, 1992). Leisure is thus offered as a resolution to the tedium, redundant onus of work. Indeed, leisure has been identified as one of four buffers, along with self-confidence, physical activity, and a sense of purpose, that may attenuate the negative influence of stress on subjective well-being (Wheeler & Frank, 1988). Leisure may serve as a resource for coping with stress and life crises. Leisure that is associated with perceived freedom moderates the negative impact of life strains over the individual's overall health (Coleman, 1993). Appropriate leisure facilitates better interpersonal communication and contributes to the quality of life; it increases perceived serenity, restores energy levels, and offers opportunities to practice skills not used in other spheres, leading to greater self-esteem, creativity, self-expression, and fitness (Leitner & Leitner, 1994). When individuals are able to align their leisure activities with their personal traits, not only do satisfaction and self-esteem increase but burnout, somatic pains, and anxiety decrease (Melamed et al., 1995). In the face of increasing pressures, an active style of leisure enhances physical and psychological health resilience. This effect is not obtained by a passive leisure style, such as watching TV or going to the movies or the mall. The active leisure style in which individuals are being challenged physically or psychologically builds on personal skill that enhance one's sense of control and influence over one's personal destiny. (Coleman & Iso-Ahola, 1993; Iso-Ahola, 1997; Stanton-Rich & Iso-Ahola, 1998). Leisure activities that involve the senses, motion, aesthetics, freedom, and thought, still psychological experience characterized by feelings of cleansing, refinement, excitement, interest, attraction, and fascination.

Whereas career advancement and monetary goals seemed to be primary objectives in the twentieth century, leisure seems to be the status icon of employees in the early twenty-first century. Employees in Western societies seem to be working to secure more enjoyment in their leisure (Mattes, 1992). Leisure has been found to contribute to both physical and psychological well-being. Different leisure activities have been associated with greater perceived physical, mental, and social health, for example, physical leisure activities not only enhance health outcomes but also reduces depression and increase perceived well-being (Caldwell, Smith, & Weissinger, 1992; Coleman, 1993; Iso-Ahola, 1984; McCann & Holmes, 1984; Neulinger, 1982).

Notwithstanding, not all who participate in non-work activities actually experience a feeling of leisure during that activity. Thus, perhaps in addition to the characteristics mentioned thus far, the assessment of non-work as leisure may be subjective. The perceived degree of freedom rather than dictation involved in the choice of action and feelings of enjoyment rather than compulsion are likely to determine how individuals experience their non-work time. One way to assess such perceptions is through the extent of satisfaction experienced through the non-work activity and state (Beard & Ragheb, 1980; Caldwell et al., 1992; Ingham, 1986; Kelly, 1972). Central characteristics that have been shown to produce the greatest psychological benefits and satisfaction in leisure are thus perceived freedom, intrinsic motivation, and leisure experience. Perhaps the most important feature is freedom of choice in choosing activities that are not determined by any specific life roles the individual fulfills (Iso-Ahola, 1980; Kelly & Kelly, 1994; Tinsley & Tinsley, 1986). Drivers for engagement in leisure activities may include an escape from one's routine, an escape from responsibility, creativity, relaxation, social contact, social power, self-actualization, challenge, and so forth. Intrinsic motivation for leisure activities, then, leads to leisure satisfaction as it confers a sense of competence and a reprieve from daily issues.

Leisure satisfaction emerges is several spheres. Psychologically, it confers a sense of freedom, enjoyment, involvement, and challenge. Educationally, it confers intellectual challenge and knowledge gains. Socially, it allows one to build rewarding relationships with others. Mentally, it facilitates relaxation with the relief from strain and stress. Physiologically, it offers an opportunity to enhance fitness, health, weight control, and well-being while enhancing immunity. Finally, aesthetically, it serves as a response to the pleasing design and beauty of the environment (Beard & Ragheb, 1980; Glaser, Kiecilt-Glaser, Speicher, & Holloday, 1985). The direct result of leisure characteristics is the leisure experience which may be enhanced with engagement, reduced focus, a sense of boundarylessness, intense feelings, increased sensory reactions, and reduced awareness to the passage of time and to the self, also known as *flow* (Crandall, 1980; Iso-Ahola, 1980; Mannell, 1980; Neulinger, 1974; Tinsley & Tinsley, 1986). Flow is experienced when activities and challenges are in balance with the individual's competencies. When challenges are greater than competences, individuals experience apprehension whereas when competencies are greater than the challenges, individuals experience boredom. The better the balance between the two, the greater the satisfaction experienced in the activity, whether within the boundaries of work or outside of it (Csikszentmihalyi, 1982; Csikszentmihalyi & LeFevre, 1989).

Shaviv (1998) found that taking part in the decision to participate in an organized vacation did not contribute to variance in post-vacation stress or burnout compared to the influence of the levels of these experiences prior to the vacation, which was crucial and great. It nevertheless served to moderate the influence of the vacation over burnout so that those who did not participate in the decision experienced greater post-vacation burnout compared with those who did participate and for whom post-vacation burnout remained unchanged compared to the pre-vacation baseline. Shaviv's study is unique in that it brought together employees from different organizations going together on the same organized vacation, rather than employees from a single organization, each going on their own respective vacation.

Vacation

Taking a vacation as part of the concept of free time has become the focus of increased attention and debate in recent years to the point where vacations are the most frequently studied type of respite. Vacation is defined as "time away from the job" (Etzion, 2003) or as a break in the continuity of work, where individuals are not actively engaged in their work (Westman & Etzion, 2001). This is a respite from work for the purpose of enjoying one's leisure, and it allows one to engage in activities that require a lengthy block of free time (Eden, 2001; Zeevi-Ohana, 2000). As a type of respite, vacation is assumed to share its characteristics and aid in breaking the resource loss cycle when resources are indeed replenished through it (Hobfoll, 1989; 2001; Westman & Eden, 1997). Indeed, vacation has been associated with reduced anxiety and depression (Caplan & Jones, 1975), feelings of refreshment and comfort (Mattes, 1992), job satisfaction, work engagement, organizational commitment, turnover intentions, and general life satisfaction (Lounsbury & Hoopes, 1986). Time away from work also reduces immediate exposure to work-related stress and strain thereby lowering friction and minimizing gaps between personal and organizational needs and desires (Etzion, 1987).

Vacation satisfaction is tied with work and life satisfaction (Lounsbury & Hoopes, 1986; Westman & Eden, 1997). When measuring its effect on employees, it is assumed that work traits are unchanged during one's vacation, thereby allowing researchers to assess the positive aura of well-being being carried from one's vacation to their workplace (Campbell, Converse, & Rodgers, 1976, p. 99). It is difficult to determine whether the vacation effect stems from the spillover of vacation satisfaction to work and life satisfaction or whether an increase in the latter represents a compensatory influence of desired experience that are typically absent from work (Shaviv, 1998). Vacation fairness has been positively related to performance assessment and negatively related to absenteeism (Utechet & Aldag, 1989). Vacations have been related to increased work efficiency, as well as increased post-vacation perceptions of interest in one's job (Lounsbury & Hoopes, 1986).

The vacation experience is not equal for all. Some individuals may experience their vacation is a stressor in its own right, predicting physical ailments (Holmes & Rahe, 1967). Some employees are apprehensive about taking time off to which they are entitled due to different fears and perceived work insecurities, such as hurting their careers or not meeting work objectives, despite their craving for a break. The Protestant work ethic may also serve as a driver for employees to forgo their due vacation time (Etzion, 1999; Richards, 1999; Schlossberg, 1991). Employees tend to realize their vacation time as a function of higher income and education. Organizational shut downs facilitate employees' decision to leave on vacation void of fears concerning social pressures and possible negative perceptions of colleagues regarding a lengthy vacation (Schlossberg, 1991). Organizational shut downs allow employees and the business to prepare ahead of time to employees' absence from work. Often such periods are short, revolving national holidays, yet some shut downs are considerably longer, such as in educational institutions (Eden, 1990; Magen-Neuhaus, 1993; Ofek, 1997). In that respect, it has been claimed that three months are the ideal time for stress reduction (Eden, 2001). The organizational shut down fosters a sense of legitimacy thereby liberating the employee from fear of social or organizational pressures (Magen-Neuhaus, 1993).

Vacation seems to have increased in desirability as employees professed partiality to vacation over other employment benefits. This is attributed to their desire to rest, enrich themselves intellectually, spend time with the families, experience adventure or seclusion, break away from their routine (escapism), forget about their obligations, find an outlet for strain and release from tension, and improve their mood (Lounsbury & Hoopes, 1986; Mattes, 1992; Richards, 1999; Rubinstein, 1980; Schiff, 1992). Employees report that vacations improve their personal lives, job performance, and marriage while increasing their sportive and friendliness tendencies. They also believe vacations are central in preventing burnout. Since increased productivity at the workplace is one outcome of vacations, they have the potential to contribute to the economy at large (Schlossberg, 1991).

Vacations contribute to the well-being of employees not only by stopping the spiral of resource loss but also by igniting the opposing cycle of resource gain (Hobfoll, 1989; 2001). Some employees require longer vacations for such an effect while others may be satisfied with shorter vacations (Calhoun, 1988; Doerr, Mitchell, Klastorin, & Brown, 1996; Frankenhaeuser, Lundberg, Fredrikson, Melin, Tuomisto, Myrstern, et al., 1989; Kanfer, Ackerman, Murtha, Dugdale, & Nelson, 1994). A shorter vacation, however, runs the risk of not allowing for sufficient work detachment. While this may facilitate an easier reacclimatization at work, it reduces the full potential effect of the vacation on work-related stress and burnout (Sapir, 1995). Among its other benefits, vacation reduces post-vacation absenteeism even after six weeks upon employees' return to their workplace (Westman & Etzion, 2001). A series of well-planned short (weekend) vacations may perhaps be more effective than a single longer vacation (Quick & Quick, 1984; Sapir, 1995; Schaufeli, Maslach, & Mark, 1993; Sonnentag, 2003). One study showed that the optimal length of vacation with respect to reducing stress and burnout is ten days (Etzion, 2003; Etzion & Sapir, 1996).

With the relationship between non-work and work satisfaction, vacations are used by organizations not only to enhance working conditions but also to achieve certain organizational goals. In the 1970s, organizations had begun to recognize the importance of employees' physiological and psychological well-being. They thus began purposefully supporting recreational programs for employees to facilitate physical and intellectual growth while also increasing organizational commitment and loyalty (Mannell & Reid, 1999). Such programs also include professional conventions in recreational resorts meant to alleviate exhaustion by overwork (Etzion, 1999). Vacations seem to reduce absenteeism reflected in a decrease of post-vacation absences due to health and other reasons (Westman & Etzion, 2001).

This approach may nevertheless backfire, as vacations have also been found to decrease job involvement as a central life interest while increasing turnover intention following the vacation (Lounsbury and Hoopes, 1986). Similarly, the more satisfied vacationers were on a long vacation, the more burnt out they were upon their return to work (Sapir, 1995). Only a small fraction of vacationers returning to their jobs report an increase in their productivity or greater perceived interest in their jobs (Lounsbury & Hoopes, 1986). However, the rosy view of vacations faded within a week or so after the vacation ended (Mitchell, Thompson, Peterson, & Cronk, 1997). The connection between vacation satisfaction and job satisfaction and reciprocal influence between work and non-work domains, then, are not self-evident. Detachment from work while on vacation may be a moderating factor (Etzion et al., 1998). In contrast, situational factors left behind at the workplace, such as a temporary replacement, may negate the effect of vacation in reducing stress (Sapir, 1995).

Notwithstanding, vacations, especially when involving travel, may be costly. Vacationers, then, are required to be skilled consumers backed by financial and time resources. Tourist vacationers, then, attempt to 'squeeze in' as many activities and experiences as possible, to acquire new skills, to seek maximal stimulation, and to 'consume' cultural events (Richards, 1999). The optimal level of stimulation one requires during their vacation depends on personality, psychological condition, personal experience, and culture (Berlyne, 1960). A lengthy organizational shut down, however, is not likely to be used in full to vacation. Rather, individuals faced with such extensive leisure time are likely to be spent at home. In fact, the most popular leisure activities are not what individuals fantasize about (e.g., sailing, hiking, or going to the opera) but rather TV, alcohol, tobacco, sex, and gambling, whereas the activities ranked most important involve one's social and familial spheres (Kelly, 1983).

Interestingly, individuals more satisfied with the vacation were found to have felt a significantly greater relief of burnout during the vacation. The same individuals were also more burned out on three out of four occasions of burnout measurement while not on vacation. It may be, then, that the vacation is most helpful for those who need it most (Westman & Eden, 1997). It may also be that an individual's response intensity to their vacation is similar to their response intensity in other spheres of life.

How do we find what works?

Examining the impact of vacation and other types of respite over work-related stress and burnout, the body of research described in this book shows that respite in the form of vacation indeed reduces work-related stress and burnout. This book reports mainly the findings of studies conducted over the past three decades by myself, my team of students, and my partnering colleagues.

Study variables and measurements

Respite research, our own included, has been conducted through longitudinal, repeated measures designs, enabling different characteristics to be intermittently present and absent, with and without control groups. Recognizing the need to study respite and leisure activities within it as a pre-post effect respite moderator, studies are based on two to five measurements taken before, during, and after the respite. The length of studied respites ranged from a day off work, a weekend, voluntary vacations, and involuntary organizational shut downs or business trips of varying duration (from single days to the school summer break), as well lengthy life events such as military reserve. Some studies have additionally addressed the cross-cultural differences aspect of the respite effects.

Studied samples were diverse, encompassing a broad range of vocations, employment status, professional experience, and organizational hierarchy. Among the samples were found low-tech and hi-tech employees, medical staff, teachers, soldiers, and flight attendants to name a few. Research has even expanded to respite effect on crossover of stress and strains.

To deeply understand the respite effect, our work relied primarily on a proprietary survey that we had developed, the validity of which has been established collectively among our samples. We additionally measured moderating effects based on a variety of instruments addressing issues such as vacation satisfaction, quality of experience, content and activities, personal characteristics and traits, and so forth. Below are described some of the variables and instruments we had used (occasionally with adaptations or translated by our own team) over the years listed alphabetically. A representative sample of our instruments is detailed in Appendix 3.

Anxiety was measured based on The State-Trait Anxiety Inventory (STAI) (Spielberger, Gorsuch, & Lushene, 1970).

Burnout was measured based on two alternative instruments. The Maslach Burnout Inventory – General Survey (Maslach, Jackson, & Leiter, 1996) is a questionnaire in which respondents are asked to assess the frequency of different feelings associated with work regarding three dimensions: exhaustion (physical and emotional tiredness, burnout), cynicism (lack of interest in and enthusiasm for work, doubts), and professional efficacy (lack of confidence in one's performance at work). The Burnout Measure (BM) (Kafry & Pines, 1980; Pines et al., 1981) measures how often respondents experience physical exhaustion (weak, tired), emotional exhaustion (depressed, entrapped), and mental exhaustion (worthless, rejected). Similarities and differences of the two instruments were discussed by Schaufeli, Enzmann, & Girault (1993).

Business trips were studied based on Klein (2002), aimed at assessing different traits of this type of respite, including notification time, frequency, degree of control, and overload. A proprietary questionnaire was additionally developed in order to assess the purpose of business trips and the degree of proactiveness (negotiations) *versus* passiveness (visiting an exhibition) required by the traveler in the activity for which they were traveling (Koniak, 2003).

Coping was measured based on The Proactive Coping Inventory (PCI) (Greenglass, Schwarzer, Jakubiec, Fiksenbaum, & Taubert, 1999) aimed at assessing coping strategies (proactive, reflective, strategic planning, preventive, instrumental support seeking, emotional support seeking, avoidance) based on available resources, responsibility, and vision. The instrument reflects one's belief that they may improve themselves as well as the environment in which they act.

Economic stress was measured based on proprietary instruments (Kessler, Turner, & House, 1988; Vinokur & Caplan, 1987; Vinokur & Price, 1996).

Energy depletion was measured based on The Energy Depletion Index developed by Garden (1985) and extended by us in which respondents are asked to rate the frequency at which they feel so depleted of energies that they are unable at the end of their work day to perform different activities such as listening to music, ongoing house chores, or hobbies.

Enjoyment was measured based on Etzion and Bailyn (1994), including items such as "I am enthusiastic to start the new day."

General satisfaction was based on Kuninn's (1995) attitude measure aimed to assess life satisfaction, self-satisfaction, and work satisfaction.

Job involvement was measured based on Lounsbury and Hoopes (1986).

Job satisfaction was measured based on Schnake (1983) addressing issues of extrinsic (job security) and intrinsic (diversity) satisfaction factors.

Mood was measured based on the instrument developed by Watson, Clark, & Tellegen (1988), which addresses negative (distress, hostility, guilt) and positive (interest, excitement, enthusiasm) affect.

Non-work involvement was extracted from Etzion and Bailyn (1994).

Organizational commitment was measured based on The Organizational Commitment Scales (Allen & Meyer, 1990), assessing affective, normative and continuance types of organizational commitment types.

Perceived control was assessed based on items extracted from the Perceived Stress Scale (Cohen, Kamark, & Mermelstein, 1983). An alternate instrument was developed proprietarily addressing both objective and subjective control (Ben-Moshe, 2006).

Perceived health was measured based on a single item in which respondents were required to rate their health from excellent to unsound (Koniak, 2003).

Perceived job stressors were measured based on the job characteristics inventory by Etzion et al. (1982), in which respondents are asked to assess the degree at which various statements might characterize their work. The instrument addresses stress resulting from the presence of extrinsic, negative work characteristics, as well as the absence of intrinsic, positive work characteristics. Negative work characteristics are aspects of work that are generally experienced as stressful and are hypothesized to be directly correlated with burnout, including competitiveness, bureaucratic pressures, administrative hassles, workload frequency, unmet schedule-related stress, conflicting demands, breaking regulations. Positive work characteristics relate to work aspects that, when present, may help alleviate stress, so that their absence is considered stressful, including: variety, autonomy, self-actualization, significance, potential to influence policy-making, performance recognition and appreciation, responsibility, support from different sources, and the clarity of role definition and evaluation criteria.

Personal information was studied based on a large variety of items according to the needs of each study.

Personality was measured using two alternative approaches. One was The Big Five Inventory personality questionnaire (McCrae & John, 1992), which assesses five personality dimensions (extraversion, neuroticism, agreeableness, conscientiousness, and openness to experience). Others related to Type A and Type B personalities (e.g., Matteson, Ivancevich, & Gamble, 1987).

Professional need to keep up-to-date was measured as a stressor based on a proprietary instrument developed by us (Sraya-Wallin, 1999). It addresses eleven emotions employees may feel due to the ongoing need to keep professionally current, including both positive emotions (e.g., serenity) and negative emotions (e.g., stress, worry).

Respite involvement was measured based on a proprietary questionnaire (see Moshkowitz, 2003).

Shift work was studied based on a propriety instrument (Hirschfeld, 1991), which addressed different traits of the shift work (preferred shift, fixed working arrangement, importance to family), the degree of consideration in shift planning and additional advantages and disadvantages to the shift work arrangement (flexible leisure time, time to spend with loved ones, diversity, influence on sleep and food arrangements, etc.).

Social support was measured based on two alternative sources. The support component in the Job Stress Questionnaire (JSQ) (Caplan et al., 1975) was adapted to include not only organizational support (by supervisor and co-workers) but also emotional support (by family and friends) (Koniak, 2003). Similarly, the Social Support (House, 1981) questionnaire was adapted to include support by spouse, extended family members, friends, and colleagues.

Social undermining was measured based on the Interpersonal Undermining Index (Abbey, Abramis, & Caplan, 1985) assessing undermining between spouses addressing both the source and the target of such behaviors (acting unpleasantly or angrily, making one another's life difficult, make others feel unwanted, criticizing them, as well as accepting such behaviors by others).

Student stressors were measured based on a proprietary instrument (Zvi, 2004). Students were asked to rate the degree to which different factors serve as stressors, including the need to integrate work and school demands, financial difficulties, studying and preparing for exams.

The necessity of leisure for detachment and relaxation was measured based on a proprietary instrument (Oron, 1979), aimed at assessing the relationship between leisure and work activities (spillover, compensation, or segregation).

Vacation was studied based on proprietary questionnaires (Etzion & Sapir, 1996; Lapidot, 1992) aimed at assessing experiences and feeling of the vacationing employee aggregated on three dimensions: vacation satisfaction, extent of detachment from work, and type of vacation (errands, family visits, hobbies, rest, active leisure, other). In some occasions, vacation satisfaction was additionally measured based on Lounsbury and Hoopes (1986).

Vacation activities referred to both planned and actual activities, respectively asked about before and after the respite. Activities described in questionnaires included active recreation such as working out, visiting friends and family, working on hobbies, traveling for sightseeing, entertainment (e.g., theatre, concerts, museums), and other activities, such as running errands and attending to personal matters (e.g., shopping, health issues, house maintenance). These were typically measured on a scale of time invested in each listed activity. The degree to which activities may be considered serious leisure was also measured in some studies.

Vacation legitimacy was measured based on three proprietary items aimed to assess perceived legitimacy of the maternity leave by the organization, direct supervisor, and other employees (Ben Yoseff, 2004).

Vacation satisfaction was measured based on two alternatives. The questionnaire by Lounsbury and Hoopes (1986) aimed at assessing vacationers' satisfaction with different aspects of their vacation, including their activities, rest, social interaction, and more objective factors such as the weather. Excerpts of the Leisure Satisfaction Scale (LSS) (Beard & Ragheb, 1980) were also used to assess vacation satisfaction.

Work addiction was based on two alternative instruments (Machlowits, 1980 and Spence & Robbins, 1992, respectively), which classified respondents in job addicts (work addicts, work enthusiasts, enthusiastic addicts) and non-addicts (unengaged workers, relaxed workers, disenchanted workers).

Workload and duties were assessed using a proprietary instrument developed based on Eden, Shirom, Kellerman, Aronson, and French (1977) and Shirom, Eden, and Kellermann (1981) addressing work demands, difficulties, quantity, lack of time, and speed of work (e.g., I am required to work too fast, I am required to complete complex and complicated jobs that require my full faculties to overcome them).

Work–non-work relationship was measured using a number of alternative instruments. The first two were based on Etzion and Bailyn (1994) and Etzion (1988), the third was developed by Frone et al. (1992), and the fourth was developed by Hirschfeld (1991) focusing on scope of employment contract, shift work, and the work–non-work conflict.

Study hypothesis

In respite research, workers' stress and burnout levels are compared in relation to the respite viz-a-viz the work context. It is assumed that employees are exposed to different stressors while at work and are not exposed to such stressors while resting, and specifically – while vacationing. Based on the vast number of variables, hypotheses were plentiful. Below is a short summary of the most central hypotheses guiding our work.

- Stress and burnout will be positively related at all measurements.
- Both stress (regardless of source) and burnout will be lower immediately after the vacation than just prior to it but will then rise again, once the routine has settled in.
- A vacation will moderate the relationship between stress and burnout; the relationship between stress and burnout will be weaker immediately after the vacation than just prior to it but will then strengthen again, once the routine has settled in.
- The greater the stress and burnout, the greater will be the need to detach and relax; the greater the work detachment during vacation, the lower will be the stress and burnout.
- The greater the vacation satisfaction, the stronger will be the inverse relationship between work detachment during vacation and stress and burnout; similarly, the greater the work detachment during vacation, the stronger will be the inverse relationship between vacation satisfaction and stress and burnout.
- Individual's involvement in timing the vacation will moderate the relationship between stress and burnout; this relationship will be weaker after a personalized time choice than after an organization shutdown.
- Vacation satisfaction will moderate the influence of the vacation over stress and burnout; satisfied vacationers will experience greater alleviation in stress

and burnout compared to unsatisfied vacationers and smaller rise in both measures once the routine has settled in and the relationship between stress and burnout will weaken against vacation satisfaction; it will also weaken against a proactive (versus passive) vacation.

- Vacation legitimacy will moderate the influence of the vacation over stress and burnout; vacationers whose vacation is perceived as legitimate will experience greater alleviation in stress and burnout compared to unsatisfied vacationers and smaller rise in both measures once the routine has settled in.
- General life satisfaction will be inversely related to stress and burnout; it will be greater immediately after the vacation than just prior to it but will then diminish again, once the routine has settled in.
- Self-efficacy will moderate the influence of stress on burnout; individuals high in self-efficacy will experience less stress from strenuous events and will exhibit less burnout.
- The work–non-work conflict will be positively related to burnout and negatively related to proactive coping.
- Work and non-work involvement will moderate the relationships between vacation, stress, work and non-work conflict, and burnout.
- The interaction between work involvement and work detachment will moderate the influence of respite over burnout; high involvement at either work or non-work combined with low detachment from work during the respite will lead to lower detachment; the same will be true for low involvement at either work or non-work and high detachment from work during the respite.
- The interaction between non-work involvement and work detachment will moderate the influence of the respite over burnout.
- Individuals whose job traits include responsibility and decision-making will enjoy greater decrease in stress and burnout following the vacation.
- Serious leisure will weaken the relationship between control over working arrangements and burnout.
- Variables inversely related to burnout (and stress) will include: the extent of control (actual or perceived) over working arrangements; perceived freedom; serious leisure; leisure involvement; leisure and vacation satisfaction; social support; job satisfaction; moderate work addiction; organizational commitment; extent of religiousness; proactive coping.
- Moderate work addiction or Type B personality will be inversely related to unused vacation days and positively related to work enjoyment, whereas extreme work addiction or Type A personality will be positively related to unused vacation days and burnout and inversely related to work enjoyment.
- Personality traits will moderate the degree of stress and burnout experienced by different individuals.
- Organizational commitment will be positively related to unused vacation days and work enjoyment.

- The relationship between job satisfaction and degree of burnout or work enjoyment will be stronger for intrinsic job satisfaction than extrinsic job satisfaction.
- Anxiety will crossover between spouses when one is unemployed and more strongly so the lengthier the unemployment period.
- Economic stress and spouses' anxiety will be positively related and more strongly so the lengthier the unemployment period.
- Social support will be inversely related to anxiety among spouses and will inversely moderate the strength of the relationship between economic stress and anxiety.
- Social support will moderate the strength of the positive relationship between frequency of business travel and perceived stress and health.
- Spousal support will moderate the positive relationship between stress and burnout; individuals living with a spouse or significant other will experience less stress and will hence exhibit less burnout both before and after a vacation.
- Perceived stress will be positively related to frequent business travel with shorter notice and greater workload and inversely related to low control; perceived health will have the opposite relationship with these variables; perceived stress and health, then, will also be inversely related.
- The greater the proactiveness required of the business traveler on their trip, the greater will be the stress.
- Social support will moderate the relationship between business travel frequency and purpose and perceived health and stress; the greater the social support, the greater will be perceived health and the weaker will be perceived stress.

This book

The current volume addresses questions concerning the ways by which employees in organizations recover from occupational stress and psychological burnout. It aims to shed light on those factors or circumstances that help employees make the best of respites available. This book also offers employers greater understanding on the opportunities they may offer their employees that will help alleviate occupational stress and psychological burnout.

Questions answered in the pages of this book are ample:

- Does taking a respite from work affect the level of stress and burnout that people experience upon returning to work?
- Does the time away from work help employees in recovering from stress, revitalizing energy depletion and containing the processes of burnout?
- Does it matter how this time off was arranged (planned, collective, or forced), what is the main theme of the respite (e.g., annual vacation,

holiday leave, maternity leave, military reserve duty, business trip, training), or whether the respite is taken on or off the working site?
- Are all respites equal in their effect?
- What best defines a respite with respect to the objective of alleviating occupational stress and psychological burnout?

This book is an exploration of different characteristics that influence the effectiveness of a respite. The first characteristics explored are those of the respite itself, such as duration, location, choice of activities, level of detachment from work, and degree of change from one's routine. The second set of characteristics are personal and situational, involving demographic background, personality styles and preferences, values and beliefs, employment characteristics, and family relations. Next, are explored organizational and environmental characteristics that influence the effectivity of respites on employees. Finally, respite outcomes are discussed.

References

Abbey, A., Abramis, D. J., & Caplan, R. D. (1985). Effects of different sources of social support and social conflict on emotional well-being. *Basic and Applied Social Psychology, 6 (2)*, 111–129.

Abouserie, R. (1994). Sources and levels of stress in relation to locus of control and self-esteem in university students. *Educational Psychology, 14(3)*, 323–330.

Allen, N. J. & Meyer, J. P. (1990). The measurement and antecedents of affective, continuance and normative commitment to the organization. *Journal of Occupational Psychology, 63(1)*, 1–18.

Anderson, N. (1964). *Dimensions of work*. New York: McKey.

Arvey, R. D. & Gross, R. H. (1977). Satisfaction levels and correlates of satisfaction in the homemaker job. *Journal of Vocational Behavior, 10(1)*, 13–24.

Aschoff, J., Hoffman, K., Pohl, H., & Weaver, R. A. (1975). Re-entrainment of circadian rhythms after phase-shifts of the Zeitgeber. *Chronobiologia, 2(1)*, 23–78.

Aviv, N. (1998). The meaning of vacation among women working shifts: The effect of vacation on job stress, burnout and work-home conflict. M.Sc., Tel Aviv University, Israel. [Hebrew].

Avni, T. (2007). The impact of organizational vacation on the amount of regular working hours, overtime, sick days and absenteeism among the organization's employees. M.Sc., Tel Aviv University, Israel. [Hebrew].

Bailyn, L. (1977). Involvement and accommodation in technical careers: An inquiry into the relation to work at mid-career. In J. Van Maanen (Ed.), *Organizational careers: Some new perspective* (pp. 109–133). London: Wiley.

Bakker, A. B., Albrecht, S. L., & Leiter, M. P. (2011). Key questions regarding work engagement. *European Journal of Work and Organizational Psychology, 20(1)*, 4–28.

Bakker, A. B., Boyd, C. M., Dollard, M., Gillespie, N., Winefield, A. H., & Stough, C. (2010). The role of personality in the job demands-resources model. *Career Development International, 15(7)*, 622–636.

Bakker, A. B. & Demerouti, E. (2007). The job demands-resources model: State of the art. *Journal of Managerial Psychology, 22(3)*, 309–328.

Bakker, A. B., Demerouti, E., De Boer, E., & Schaufeli, W. B. (2003). Job demands and job resources as predictors of absence duration and frequency. *Journal of Vocational Behavior, 62(2)*, 341–356.

Bakker, A. B., Demerouti, E., & Euwema, M. C. (2005). Job resources buffer the impact of job demands on burnout. *Journal of Occupational Health Psychology, 10(2)*, 170–180.

Bakker, A. B., Demerouti, E., & Verbeke, W. (2004). Using the Job Demands-Resources model to predict burnout and performance. *Human Resource Management, 43(1)*, 83–104.

Bakker, A. B. & Schaufeli, W. B. (2000). Burnout contagion processes among teachers. *Journal of Applied Social Psychology, 30(11)*, 2289–2308.

Bakker, A. B., Hakanen, J. J., Demerouti, E., & Xanthopoulou, D. (2007). Job resources boost work engagement, particularly when job demands are high. *Journal of Educational Psychology, 99(2)*, 274–284.

Beard, J. G. & Ragheb, M. G. (1980). Measuring leisure satisfaction. *Journal of Leisure Research, 12(1)*, 20–33.

Bedeian, A. G., Burke, B. G., & Moffett, R. G. (1988). Outcomes of work-family conflict among married male and female professionals. *Journal of Management, 14(3)*, 475–491.

Beehr, T. A. & Newman, J. (1978). Job stress employee health, and organizational effectiveness: A facet analysis model and literature review. *Personnel Psychology, 31(4)*, 655–669.

Bell, L. & Freeman, R. (1996). *Earnings inequality and hours of work in Germany and the U.S.* (Vol. working paper 4808). Cambridge: National Bureau of Economic Research.

Ben Yoseff, N. (2004). The effect of maternity leave on burnout among women serving in the I.D.F. M.Sc., Tel Aviv University, Israel. [Hebrew].

Ben-Moshe, L. (2006). Work features, "serious leisure" and burnout among flight attendants. M.Sc., Tel Aviv University, Israel. [Hebrew].

Berlyne, D. E. (1960). *Conflict, arousal and curiosity*. New York: McGraw-Hill.

Beutell, N. J. & Greenhouse, J. H. (1985). Sources of conflict between work and family roles. *Academy of Management Review, 10(1)*, 76–88.

Bosch, G. (1999). Working time: Tendencies and emerging issues. *International Labour Review, 138(2)*, 131–150.

Bosserman, P. (1989). The USA: Modern times and the new solidarity. In A. Olszewski & K. Roberts (Eds.), *Leisure and life-style: A comparative analysis of free time* (pp. 162–177). Beverly Hills, CA: Sage Publications.

Bowlby, J. (1969). *Attachment and loss*. New York: Basic Books.

Boyd, A. (1997). Employee traps – corruption in the workplace. *Management Review, 86(8)*, 9–10.

Briar, K. H. (1978). *The effect of long-term unemployment on workers and their families*. San Francisco, CA: R & E Research Associates.

Brockner, J., Grover, S., Reed, T. F., & Dewitt, R. L. (1992). Layoffs, job insecurity, and survivors' work effort: Evidence of an inverted-u relationship. *Academy of Management Journal, 35(2)*, 413–425.

Brockner, J., Grover, S., Reed, T. F., Dewitt, R. L., & O'Malley, M. (1987). Survivors' reactions to layoffs: We get by with a little help from our friends. *Administrative Science Quarterly, 32(4)*, 526–541.

Bruess, C. E. & Tevis, B. (1985). *Decision for health*. Belmont, CA: Wadsworth.

Buell, P. & Breslow, L. (1960). Mortality from coronary heart disease in California men who work long hours. *Journal of Chronic Diseases, 11(6)*, 615–626.

Burke, R. J. (1988). Some antecedents of work-family conflict. *Journal of Social Behavior and Personality, 3(4)*, 287–302.

Burke, R. J., Weir, T., & Dunwors, R. E. (1980). Work demands on administrators and spouse well-being. *Human Relations, 33*, 253–278.
Burleson, B. R. (2003). The experience and effects of emotional support: What the study of cultural and gender differences can tell us about close relationships, emotions, and interpersonal communications. *Personal Relationships, 10(1)*, 1–23.
Caldwell, L. L., Smith, E. A., & Weissinger, E. (1992). The relationship of leisure activities and perceived health of college students. *Society and Leisure, 15(2)*, 545–556.
Calhoun, H. (1988). Relaxation and you. *Supervision, 50*, 12–13.
Campbell, A., Converse, P., & Rodgers, W. (1976). *The quality of American life*. New York: Russell Sage.
Cannon, W. B. (1932). *The wisdom of the body*. New York: Norton.
Caplan, R. D. (1983). Person-environment fit: Past, present, and future. In C. L. Cooper (Ed.), *Stress research* (pp. 35–78). New York: Wiley.
Caplan, R. D., Cobb, S., French, J. R. P., Van-Harrison, R., & Pinneau, S. (1975). Job demands and worker health: Main effects and occupational differences. Institute for Social Research, The University of Michigan. Retrieved from: https://babel.hathitrust.org/cgi/pt?id=mdp.39015071884410;view=1up;seq=3.
Caplan, R. D. & Jones, K. W. (1975). Effects of work load, role ambiguity, and Type A personality on anxiety, depression and heart rate. *Journal of Applied Psychology, 60(6)*, 713–719.
Carroll, J. F. X. & White, W. L. (1982). Theory building: Integration individual and environmental factors within ecological frame work. In W. S. Paine (Ed.), *Job stress and burnout: Research, theory, and intervention perspectives* (pp. 41–60). London: Sage Publications.
Cherniss, C. (1980a). *Professional burnout in human service organizations*. New York: Praeger.
Cherniss, C. (1980b). *Staff burnout: Job stress in the human services*. Beverly Hills, CA: Sage Publications.
Cherniss, C. (1981). Preventing burnout: From theory to practice. In J. W. Jones (Ed.), *The burnout syndrome* (pp. 172–176). London: London House Management Press.
Cherniss, C. (1995). *Beyond burnout: Helping teachers, nurses, therapists and lawyers recover from stress and disillusionment*. New York: Routledge.
Cherniss, C. & Krantz, D. L. (1983). The ideological community as an antidote to burnout in the human services. In B. A. Farber (Ed.), *Stress and burnout in the human service professions* (pp. 198–212). New York: Pergamon Press.
Cobb, S. (1976). Social support as moderator of life stress. *Psychosomatic Medicine, 38(5)*, 300–314.
Cohen, S. (1988). Psychosocial models of the role of social support in the etiology of physical disease. *Health Psychology*, 7*(3)*, 269–297.
Cohen, S., Kamarck, T., & Mermelstein, R. (1983). A global measure of perceived stress. *Journal of Health and Social Behavior, 24(4)*, 385–396.
Cohen, S. & Wills, T. A. (1985). Stress, social support and the buffering hypothesis. *Psychological Bulletin, 98(2)*, 310–357.
Coleman, D. (1993). Leisure based social support, leisure dispositions and health. *Journal of Leisure Research, 25(4)*, 350–361.
Coleman, D. J. & Iso-Ahola, S. E. (1993). Leisure and health: The role of social support and self-determination. *Journal of Leisure Research, 25(2)*, 111–128.
Conger, R. D., Elder, G. H., Lorenz, F. O., Conger, K. J., Simons, R. L., Whitbeck, L. B., et al. (1990). Linking economic hardship to marital quality and instability. *Journal of Marriage and the Family, 52(3)*, 643–656.

Conger, R. D., Lorenz, F. O., Elder, G. H., & Simon, R. L. (1993). Husband and wife differences in response to undesirable life events. *Journal of Health and Social Behavior, 34 (1)*, 71–88.

Cooper, C. L., Dewe, P. J., & O'driscoll, M. P. (2001). *Organizational stress: A review and critique of theory, research and applications.* Thousand Oaks, CA: Sage Publications.

Cooper, C. L. & Marshall, J. (1976). Occupational sources of stress. *Journal of Occupational Psychology, 49(1)*, 11–28.

Cordes, C. L. & Dougherty, T. W. (1993). A review and an integration of research on job burnout. *Academy of Management Review, 18(4)*, 621–656.

Cox, T. & Mackay, C. (1981). A transactional approach to occupational stress. In E. N. Coorbett & J. Richardson (Ed.), *Stress, work design and productivity* (pp. 91–113). New York: Wiley.

Crandall, R. (1980). Motivations for leisure. *Journal of Leisure Research, 12(1)*, 45–54.

Csikszentmihalyi, M. (1982). Toward a psychology of optimal experience. In L. Wheeler (Ed.), *Review of Personality and Social Psychology.* Beverly Hills, CA: Sage Publications.

Csikszentmihalyi, M. (1990). *Flow, the psychology of optimal experience.* New York: Harper Perennial.

Csikszentmihalyi, M. & LeFevre, J. (1989). Optimal experience in work and leisure. *Journal of Personality and Social Psychology, 56(5)*, 815–822.

Cummings, T. G. & Cooper, C. L. (1979). A cybernetic framework for studying occupational stress. *Human Relations, 32(5)*, 395–418.

Davy, J. A., Kinicki, A. J., & Scheck, C. L. (1991). Developing and testing a model of survivor responses to layoffs. *Journal of Vocational Behavior, 38(3)*, 302–317.

DeFrank, R. S., Konopaske, R., & Ivancevich, J. M. (2000). Executive travel stress: Perils of the road warrior. *Academy of Management Executive, 14(2)*, 58–71.

DeFrank, R. S. & Stroup, C. A. (1989). Teacher stress and health: Examination of a model. *Journal of Psychosomatic research, 33(1)*, 99–109.

de Grazia, S. (1962). *Of Time, work and leisure.* New York: The Twentieth Century Fund.

Demerouti, E., Bakker, A. B., Nachreiner, F., & Schaufeli, W. B. (2001). The job demands resources model of burnout. *Journal of Applied Psychology, 86(3)*, 499–512.

Doerr, K. H., Mitchell, T. R., Klastorin, T. D., & Brown, K. A. (1996). Impact of material flow policies and goals on job outcomes. *Journal of Applied Psychology, 81(2)*, 142–152.

Dubin, R. (1956). Industrial Workers' World: A Study of the "Central Life Interests" of Industrial Workers. *Social Problem, 3(3)*, 131–142.

Dubin, R. (1979). Central life interest: Self integrity in a complex world. *Pacific Sociological Review, 22(4)*, 405–426.

Dubin, R., Hedley, R. A., & Taveggia, T. (1976). Attachment to work. In R. Dubin (Ed.), *Handbook of Work, Organization and Society* (pp. 281–341). Chicago, IL: Rand McNally.

Edelwich, J. & Brodesky, A. (1980). *Burnout: Stages of disillusionment in helping professionals.* New York: Human Science Press.

Eden, D. (1990). Acute and chronic job stress, strain and vacation relief. *Organizational Behavior and Human Decision Processes, 45(2)*, 175–193.

Eden, D. (2001). Vacations and other respites: Studying stress on and off the job. In C. L. Cooper & I. T. Robertson (Eds.), *International review of industrial and organizational psychology* (pp. 121–146). Chichester, UK: Wiley.

Eden, D., Shirom, A., Kellerman, J. J., Aronson, Y., & French, J. R. P. J. (1977). Stress, anxiety and coronary risk in a supportive society. In C. D. Spielberger & I. G. Sarason (Eds.), *Stress and anxiety* (Vol. 4) (pp. 251–268). New York: Wiley.

Edwards, J. R. (1988). The determinants and consequences of coping with stress. In C. L. Cooper & R. Payne (Eds.), *Causes, coping and consequences of stress at work.* (Vol. 8) (pp. 233–263). Chichester, UK: Wiley.

Eisenstat, R. A. & Felner, R. D. (1983). Organizational mediators of the quality of care: Job stressors and motivators in human service settings. In B. A. Farber (Ed.), *Stress and burnout in the human service professions* (pp. 142–154). New York: Pergamon Press.

Elliot, G. R. & Eisdorfer, C. (1982). *Stress and human health: An analysis and implications of research. A study by the Institute of Medicine, National Academy of Sciences.* New York: Springer Publishing.

Ennis, N. E., Hobfoll, S. E., & Schroder, K. E. E. (2000). Money doesn't talk, it swears: Howeconomic stress and resistance resources impact inner-city women's depressive mood. *American Journal of Community Psychology, 28(2)*, 149–173.

Etzion, D. (1984). The moderating effect of social support on the relationship of stress and burnout. *Journal of Applied Psychology, 69(4)*, 615–622.

Etzion, D. (1987). Burnout: The hidden agenda of human distress (Working Paper No. 930/87). Tel Aviv, Israel: Henry Crown Institute of Business Research, Faculty of Management, Tel Aviv University.

Etzion, D. (1988). The experience of burnout and work/non-work success in man and female engineers: A match-pairs comparison. *Human Resource Management, 27(2)*, 163–179.

Etzion, D. (1995). Is a break from work an effective means for combating stress and preventing burnout in organizations? *Innovations in Management* – A Publication of the Faculty of Management, Tel Aviv University. [Hebrew].

Etzion, D. (1999). Work and vacation: Who is afraid to take a break? Paper presented at the 59th Annual Meeting of the Academy of Management, Chicago, IL.

Etzion, D. (2003). Annual vacation: Duration and relief from job stress and burnout. *Anxiety, Stress, and Coping, 16(2)*, 213–226.

Etzion, D. (2006). Respite as a means of recovery from occupational stress and burnout. In P. Buchwald (Ed.), *Stress and anxiety – Application to health, community, work place and education* (pp. 186–195). Cambridge: Cambridge Scholar Press.

Etzion, D. & Bailyn, L. (1994). Patterns of adjustment to the career/family conflict of technologically trained women in the U.S. and Israel. *Journal of Applied Social Psychology, 24(17)*, 1520–1549.

Etzion, D., Eden, D., & Lapidot, Y. (1998). Relief from job stressors and burnout: Reserve service as respite. *Journal of Applied Psychology, 83(4)*, 577–585.

Etzion, D., Kafry, D., & Pines, A. (1982). Tedium among managers: A cross-cultural American – Israeli comparison. *Journal of Psychology and Judaism, 7(1)*, 30–41.

Etzion, D. & Sapir, O. (1996). The effect of an annual vacation in relieving stress and burnout at work (a controlled field study at an industrial enterprise in Israel). Report by: Tel Aviv University, Henry Crown Institute of Business Research, Faculty of Management, Tel Aviv, Israel.

Evans, P. & Bartolome, F. (1980). *Must success cost so much?* New York: Basic Books.

Evans, P. & Bartolome, F. (1986). The dynamics of work-family relationships in managerial lives. *International Review of Applied Psychology, 35(3)*, 371–395.

Ezrachi, Y. (1985). Burnout in managerial and command positions: A construct validation of a concept. Ph.D., Tel Aviv University, Israel.

Farber, B. A. (1983). Current trend, future directions. In B. A. Farber (Ed.), *Stress and burnout in the human service professions* (pp. 242–248). New York: Pergamon Press.

Farber, B. A. (1991). *Crisis in education. Stress and burnout in the American teacher.* San Francisco, CA: Jossey-Bass.

Farber, B. A. (2000). Introduction: Understanding and treating burnout in a changing culture. *Journal of Clinical Psychology, 56(5)*, 589–594.

Ferrie, J. E., Shipley, M. J., Marmot, M. G., Stansfeld, S. A., & Smith, G. D. (1998). The health effects of major organizational change and job insecurity. *Social Science Medicine, 46(2)*, 243–254.

Figart, D. M. & Golden, L. (1998). The social economics of work time: Introduction. *Review of Social Economy, 56(4)*, 411–424.

Firth, H. & Britton, P. (1989). Burnout, absence and turnover amongst British nursing staff. *Journal of Occupational Psychology, 62(1)*, 55–59.

Fox, G. L. & Chaney, D. (1998). Sources of economic distress: Individual and family outcome. *Journal of Family Issues, 19(6)*, 725–749.

Frankenhaeuser, M., Lundberg, U., Fredrikson, M., Melin, B., Tuomisto, M., Myrsten, A.-L., et al. (1989). Stress on and off the job as related to sex and occupational status in white-collar workers. *Journal of Organizational Behavior, 10(4)*, 321–346.

Fredrickson, B. L. (2001). The role of positive emotions in positive psychology: The broaden-and-build theory of positive emotions. *American Psychologist, 56(3)*, 218–226.

French, J. R. P., Caplan, R. D., & Van Harrison, R. (1982). *The mechanisms of job stress and strain*. Chichester, UK: Wiley.

Frenkel, M. (2011). Policy on work-family balance: Findings from global organizations, countries, and employers. Literature review. Report by: Ministry of Industry, Trade, and Labor, Israel.

Freudenberger, H. J. (1974). Staff burnout. *Journal of Social Issues, 30(1)*, 159–165.

Freudenberger, H. J. & North, G. (1985). *Women's burnout*. Garden City, New York: Anchor Press/Doubleday.

Freudenberger, H. J. & Richelson, G. (1980). *Burnout: The high cost of high achievement*. Garden City, New York: Doubleday.

Friedman, I. A. & Lotan, A. (1985). *The psychological burnout of the teacher in Israel in elementary school*. Jerusalem, Israel: Henrietta Szold Institute.

Frone, M. R., Russell, M., & Cooper, M. L. (1992). Prevalence of work-family conflict: Are work and family boundaries asymmetrically permeable? *Journal of Organizational Behavior, 13*(7), 723–729.

Fryer, D. C. & McKenna, S. (1987). The laying off of hands unemployment and the experience of time. In S. Fineman (Ed.), *Unemployment personal and social consequences* (pp. 47–73). London: Tavistock Publications.

Gaines, J. & Jermier, J. (1983). Emotional exhaustion in a high stress organization. *Academy of Management Journal, 26(4)*, 567–586.

Galambos, N. L. & Walters, B. J. (1992). Work hours, schedule inflexibility and stress in dual- earner spouses. *Canadian Journal of Behavioral Science, 24(3)*, 290–302.

Ganster, D. C. & Bart, V. (1988). The impact of social support on mental and physical health. *British Journal of Medical Psychology, 61(1)*, 17–36.

Ganster, D. C., Fusilier, R., & Mayes, B. T. (1986). Role of social support in the experience of stress at work. *Journal of Applied Psychology, 71(1)*, 102–111.

Garden, A. M. (1985). The effect of Jungian type on burnout. *Journal of Psychological Type, 10*, 3–10.

Garden, A. M. (1991). The purpose of burnout: A Jungian interpretation. *Journal of Social Behavior and Personality, 6*(7), 73–93.

Glaser, R., Kiecolt-Glaser, J. K., Speicher, C. E., & Holiday, J. E. (1985). Stress, loneliness, and changes in herpesvirus latency. *Journal of Behavioral Medicine, 8(2)*, 25–49.

Golembiewski, R. T. & Munzenrider, R. (1988). *Phases of burnout: Developments in concepts and applications*. New York: Praeger.

Gorgievsky, M. J., Bakker, A. B., & Schaufeli, W. B. (2010). Work engagement and workaholism: Comparing the self-employed and salaried employees. *Journal of Positive Psychology, 5(1)*, 83–96.

Greenglass, E. R. & Burke, J. R. (2002). Hospital restructuring and burnout. *Journal of Health & Human Services Administration, 25(1)*, 89–114.

Greenglass, E. R. (1982). *A world of differences: Gender roles in perspective*. Toronto, Ontario, Canada: Wiley.

Greenglass, E. R. (2002). Proactive coping. In E. Frydenberg (Ed.), *Beyond coping: Meeting goals, vision and challenges* (pp. 37–62). London: Edward Elgar.

Greenglass, E. R., Schwarzer, R., Jakubiec, D., Fiksenbaum, L., & Taubert, S. (1999). The Proactive Coping Inventory (PCI): A multidimensional research instrument. Paper presented at the 20th International Conference of the Stress and Anxiety Research Society (STAR), Cracow, Poland.

Griffiths, D. (1981). Job evaluation, technical expertise, and dual ladders in research and development. *Personnel Review, 10(4)*, 14–17.

Grunberg, L., Moore, S. Y., & Greenberg, E. (2001). Differences in psychological and physical health among layoff survivors: The effect of layoff contact. *Journal of Occupational Health Psychology, 6(1)*, 15–25.

Gunter, B. G. & Gunter, N. C. (1980). Leisure styles: A conceptual framework for modern leisure. *The Sociological Quarterly, 21(3)*, 361–374.

Hackman, J. R. & Oldham, R. G. (1974). *The job diagnostic survey: An instrument for the diagnosis of jobs and the evaluation of job redesign project.* Report by: Yale University, New Haven, CT.

Hackman, J. R. & Oldham, G. R. (1975). Development of the job diagnostic survey. *Journal of Applied Psychology, 60(2)*, 159–170.

Hackman, J. R. & Oldham, G. R. (1976). Motivation through the design of work: Test of a theory. *Organizational Behavior and Human Performance, 16(2)*, 250–279.

Hakanen, J. J., Bakker, A. B., & Demerouti, E. (2005). How dentists cope with their job demands and stay engaged: The moderating role of job resources. *European Journal of Oral Sciences, 113(6)*, 479–487.

Hall, D. T. (1972). A model of coping with role conflicts: The role behavior of college educated women. *Administrative Science Quarterly, 17(4)*, 471–486.

Hardesty, S. A. & Betz, N. E. (1980). The relationship of career salience, attitudes toward women, and demographic and family characteristics to marital adjustment in dual career couples. *Journal of Vocational Behavior, 17(2)*, 242–250.

Haynes, S. G. & Feinlieb, M. (1980). Women, work and coronary heart disease: Prospective findings from the Framingham Heart Study. *American Journal of Public Health, 70(2)*, 133–141.

Herzberg, F. (1996). *Work and the nature of man*. Cleveland, OH: World Publishing.

Hirschfeld, I. (1991). The relationship between job volume, shift work, and home roles and sleep adjustment and work and life satisfaction. M.Sc., Tel Aviv University, Israel.

Hobfoll, S. E. (1989). Conservation of resource – A new attempt at conceptualizing stress. *American Psychologist, 44(3)*, 513–524.

Hobfoll, S. E. (2001). The influence of culture, community, and the nested self in the stress process: Advancing Conservation of Resources Theory. *Applied Psychology: An International Review, 50(3)*, 337–421.

Hobfoll, S. E. & Freedy, J. (1993). Conservation of resources: A general stress theory applied to burnout. In W. B. Schaufeli, C. Maslach, & T. Marek (Eds.), *Professional burnout: Recent developments in theory and research* (pp. 115–129). Washington, DC: Taylor & Francis.

Hobfoll, S. E. & Shirom, A. (1993). Stress and burnout in the workplace: Conservation of resources. In R. T. Golembiewski (Ed.), *Handbook of organizational behavior* (pp. 41–60). New York: Marcel Dekker.

Hobfoll, S. E. & Shirom, A. (2000). Conservation of resources theory: Applications to stress and management in the workplace. In R. T. Golembiewski (Ed.), *Handbook of organization behavior* (pp. 57–81). New York: Marcel Dekker.

Holland, J. L. (1973). *Making vocational choice: A theory of careers* (2nd ed.). Englewood Cliffs, NJ: Prentice Hall.

Holmes, R. M. & Rahe, R. H. (1967). The social readjustment rating scale. *Journal of Psychosomatic Research, 11(2)*, 213–218.

Horovitz, S. (2002). The relationship of economic stress and social support and crossover anxiety between couples in times of unemployment. M.Sc., Tel Aviv University, Israel. [Hebrew].

House, J. S. (1980). Occupational stress and the mental and physical health of factory workers. Research report series, Institute for Social Research, Survey Research Center, The University of Michigan.

House, J. S. (1981). *Work stress and social support.* Reading, M.A: Addison Wesley.

Ichilov, O. & Bar, S. (1980). Extended family ties and the allocation of social rewards in veteran kibbutzim in Israel. *Journal of marriage and the family, 42(2)*, 421–426.

Ingham, R. (1986). Psychological contributions to the study of leisure-part one. *Leisure Studies, 5(3)*, 255–279.

Iso-Ahola, S. E. (1980). *The social psychology of leisure and recreation.* Dubuque, LA: William C. Brown.

Iso-Ahola, S. E. (1984). Social psychological foundations of leisure and resultant implications for leisure counseling. In E. T. Dowd (Ed.), *Leisure counseling, concepts and applications.* Springfield, IL: Charles C Thomas.

Iso-Ahola, S. E. (1997). A psychological analysis of leisure and health. In J. T. Haworth (Ed.), *Work, leisure and well being* (pp. 131–144). London: Routledge.

Israeli, D. (1988). Burning-out in medicine: A comparison of husbands and wives in dual-career couples. *Journal of Social Behavior and Personality, 3(4)*, 329–346.

Jackson, S. E. & Maslach, C. (1982). After effects of job related stress: Families as victims. *Journal of Occupational Behavior, 3(1)*, 63–77.

Jackson, S. E., Schwab, R. L., & Schuler, R. S. (1986). Toward an understanding of the burnout phenomenon. *Journal of Applied Psychology, 71(4)*, 630–640.

Jex, S. M. & Beehr, T. H. (1991). Emerging theoretical and methodological issues in the study of work-related stress. *Research in Personnel Human Resources Management, 9*, 311–365.

Johoda, M. (1982). *Employment and unemployment.* Cambridge: Cambridge University Press.

Joiner, A. T. & Bartram, T. (2004). How empowerment and social support affect Australian nurses' work stressors. *Australian Health Review, 28(1)*, 56–64.

Kabanoff, B. (1980). Work and nonwork: A review of models, methods and findings. *Psychological Bulletin, 88(1)*, 60–77.

Kabanoff, B. & O'Brien, G. E. (1980). Work and Leisure: A task attributes analysis. *Journal of Applied Psychology, 65(5)*, 596–609.

Kafry, D. & Pines, A. (1980). The experience of tedium in life and work. *Human Relations, 33(7)*, 477–503.

Kahill, S. (1988). Symptoms of professional burnout: A review of the empirical evidence. *Canadian Psychology, 29(3)*, 284–297.

Kahn, R. L. & Byosiere, P. (1992). Stress in organizations. In M. D. T. Dunnette & L. M. Hough (Eds.), *Handbook of industrial and organizational psychology* (pp. 571–651). Palo Alto, CA: Consulting Psychology Press.

Kahn, R. L., Wolfe, D. M., Snoek, J. E., Quinn, R. P., & Rosenthal, R. A. (1964). *Organizational stress: Studies in role conflict and ambiguity* (Vol. 2). New York: Wiley.

Kanfer, R., Ackerman, P. L., Murtha, T. C., Dugdale, B., & Nelson, L. (1994). Goal setting, condition of practice, and task performance: A resource allocation perspective. *Journal of Applied Psychology, 79(6)*, 826–835.

Kanner, A. D., Coyne, C. S., & Lazarus, R. S. (1981). Comparison of two modes of stress measurement: Daily hassles and uplifts, versus major life events. *Journal of Behavioral Medicine, 4(1)*, 1–39.

Kanner, A. D., Kafry, D., & Pines, A. (1978). Conspicuous in its absence: The lack of positive condition as a source of stress. *Journal of Human Stress, 4(4)*, 33–39.

Karasek, R. (1979). Job demands, job decision latitude, and mental strain: Implication for job redesign. *Administrative Science Quarterly, 24(2)*, 285–309.

Kattenbach, R., Demerouti, E., & Nachreiner, F. (2010). Flexible working times: Effects on employee's exhaustion, work-nonwork conflict and job performance. *Career Development International, 15(3)*, 279–295.

Kelly, J. R. (1972). Work and leisure: A simplified paradigm. *Journal of Leisure Research, 4 (1)*, 50–62.

Kelly, J. R. (1983). *Leisure identities and interactions*. London: George Allen & Unwin.

Kelly, J. R. & Godbey, G. (1992). *The sociology of leisure*. State College, PA: Venture Publishing.

Kelly, J. R. & Kelly, J. R. (1994). Multiple dimensions of meaning in the domains of work, family and leisure. *Journal of Leisure Research, 26(3)*, 250–274.

Kessler, R. C., Turner, J. B., & House, J. (1987). Intervening processes in the relationship between unemployment and health. *Psychological Medicine*, 17(*4*), 949–961.

Kessler, R. C., Turner, J. B., & House, J. S. (1988). The effects of unemployment on health in a community survey: Main, modifying, and mediating effects. *Journal of Social Issues, 44*(*4*), 69–86.

Kim, H. J., Shin, K. H., & Swanger, N. (2009). Burnout and engagement: A comparative analysis using the Big Five personality dimensions. *International Journal of Hospitality Management, 28(1)*, 96–104.

King, J. E. (2000). White-collar reactions to job insecurity and the role of the psychological contract: Implications for human resource management. *Human Resource Management, 39(1)*, 79–91.

King, M., Stanley, G., & Borrows, G. (1987). *Stress: Theory and practice*. Sydney, Australia: Grune & Stratton.

Klein, A. (2002). Influence of business trips abroad: Notice time and frequency, over stress and burnout perceptions. M.Sc., Tel Aviv University, Israel.

Kobasa, S. C. (1979). Stressful life events, personality and health: An inquiry into hardiness. *Journal of Personality & Social Psychology, 37*(*1*), 1–11.

Koekemoer, E. & Mostert, K. (2010). Work-nonwork interference: Preliminary results on the psychometric properties of a new instrument. *SA Journal of Industrial Psychology, 36 (1)*, 1–13.

Koffman, M. & Elizzur, E. (1983). Bereavement of kibbutz and non-kibbutz children following the death of the father. *Journal of Child Psychology and Psychiatry, 24(3)*, 435–442.

Koniak, M. (2003). Overseas business trips: The impact of business trip purpose, frequency of business trip and warning time before business trip on stress and health. M.Sc., Tel Aviv University, Israel. [Hebrew].

Kraft, K. (2001). Unemployment and the separation of married couples. *Kyklos, 54(1)*, 67–88.

Kuhnert, K. W., Sims, R. R., & Lahey, M. A. (1989). The relationship between job security and employee health. *Group and Organization Studies, 14(4)*, 399–410.

Kunin, T. (1995). The construction of a new type of attitude measures. *Personnel Psychology, 8(1)*, 65–77.

Lapidot, Y. (1992). The effects of a military-reserve service on work stress and burnout. M. Sc., Tel Aviv University, Israel. [Hebrew].

Lasch, C. (1979). *The culture of narcissism: American life in an age of diminishing returns.* New York: Norton.

Laski, S. (2002). The effect of summer break activities and personality variables on stress and burnout among high school teachers in Israel. M.Sc., Tel Aviv University, Israel. [Hebrew].

Laski, S. (2013). Recovery during vacation and its fadeout – A longitudinal study. Ph.D., Tel Aviv University, Israel.

Lawler, E. E. & Hall, D. T. (1970). Relationship of job characteristics to job involvement, satisfaction & intrinsic motivation. *Journal of Applied Psychology, 54(4)*, 305–312.

Lazarus, R. S. & Folkman, S. (1984). *Stress appraisal and coping*. New York: Springer.

Lazarus, R. S. & Launier, R. (1978). Stress related transactions between person and environment. In L. A. Pervin & M. Lewis (Ed.), *Perspective in International Psychology* (pp. 287–329). New York: Plenum.

Leana, C. R. & Ivancevitch, J. (1987). Involuntary job loss: Institutional interventions and research agenda. *Academy of Management Review, 12(2)*, 301–312.

Lee, R. T. & Ashforth, B. E. (1993). A further examination of managerial burnout: Toward an integrated model. *Journal of Organizational Behavior, 14(1)*, 13–20.

Lee, R. T. & Ashforth, B. E. (1996). A meta-analytic examination of the correlates of the three dimensions of burnout. *Journal of Applied Psychology, 81(2)*, 123–133.

Leise, B., Mundt, K. A., Dell, L. D., Nagy, L., & Demure, B. (1997). Medical insurance claims associated with interpersonal business travel. *Occupational and Environmental Medicine, 54(7)*, 499–503.

Leitner, M. J. & Leitner, S. F. (1994). *How to improve your life through leisure.* Salt Lake City, UT: Northwest Publishing.

Leonard, C., Fanning, N., Attwood, J., & Buckley, M. (1998). The effects of fatigue, sleep deprivation and onerous working hours on the physical and mental well- being of pre-registration house officers. *Irish Journal of Medical Science, 167(1)*, 22–25.

Levi, L. (1972). *Stress and distress in response to psychosocial stimuli.* Oxford: Pergamon Press.

Levi, M. (2002). The impact of summer break and personal variables on job stress and burnout among teachers in a Kibbutz regional school. M.Sc., Tel Aviv University, Israel. [Hebrew].

Llorens, S., Schaufeli, W., Bakker, A., & Salanova, M. (2007). Does a positive gain spiral of resources, efficacy beliefs and engagement exist? *Computers in Human Behavior, 23(1)*, 825–841.

Locke, E. A. (1976). The nature and causes of job satisfaction. In M. D. Dunnette (Ed.), *Handbook of industrial and organizational psychology* (pp. 1297–1349). Chicago, IL: Rand McNally.

Lodahl, T. M. & Kenjer, M. (1965). The definition & measurement of job involvement. *Journal of Applied Psychology, 49(1)*, 24–33.

Loftquist, L. H. & Dawis, R. V. (1969). *Adjustment of work*. New York: Appleton-Century-Crofts.

Lounsbury, J. W. & Hoopes, L. L. (1986). A vacation from work: Changes in work and non-work outcomes. *Journal of Applied Psychology, 71(3)*, 392–401.

Machlowits, M. (1980). *Workaholics, living with them, working with them*. Massachusetts: Adison Wesley Publishing.

Magen-Neuhaus, T. (1993). The effect of vacation and feelings of stress at work on emotional burnout. M.Sc., Tel Aviv University, Israel. [Hebrew].

Mahoney, J. L. & Stattin, H. (2000). Leisure activities and adolescent antisocial behavior: The role of structure and social context. *Journal of Adolescence, 23(2)*, 113–127.

Mannell, R. C. (1980). Social psychological techniques and strategies for studying leisure experience. In S. E. Iso-Ahola (Ed.), *Social psychological perspectives on leisure and recreation* (pp. 62–88). Springfield, IL: Charles C Thomas.

Mannell, R. C. & Reid, D. G. (1999). Work and leisure. In L. E. L. Jackson & T. L. Burton (Eds.), *Leisure studies prospects for the 21st century*. State College, PA: Venture Publishing.

Maruyama, I. & Tokuda, Y. (2017). Changing work styles among Japanese healthcare professionals. *Journal of General and Family Medicine, 18(4)*. Retrieved from: http://onlinelibrary.wiley.com/doi/10.1002/jgf2.117/full.

Marx Ferree, M. (2006). Globalization and feminism: Opportunities and obstacles for activism in the global arena. In M. Marx Ferree & A. M. Tripp (Eds.), *Global feminism. Transnational women's activism, organizing, and human rights* (pp. 3–23). New York: NYU Press.

Maslach, C., Jackson, S. E., & Leiter, M. P. (1996). *MBI-GS manual* (3rd ed.). Palo Alto, CA: Consulting Psychologist Press.

Maslach, C. & Leiter, M. P. (1997). *The truth about burnout*. San Francisco, CA: Jossey-Bass.

Maslach, C., Schaufeli, W. B., & Leiter, M. P. (2001). Job burnout. *Annual Review of Psychology, 52*(February), 397–422.

Maslach, C. (1982a). *Burnout: The cost of caring*. Englewood Cliffs, NJ: Prentice Hall.

Maslach, C. (1982b). Job stress and burnout. In W. S. Paine (Ed.), *Understanding burnout: Definitional Issues in analyzing a complex phenomenon* (pp. 29–40). Beverly Hills, CA: Sage Publications.

Maslach, C. (1976). Burnout. *Human Behaviour, 5(9)*, 16–22.

Maslach, C., Jackson, S. E., & Leiter, M. P. (1996). *MBI-GS manual* (3rd ed.). Palo Alto, CA: Consulting Psychologist Press.

Maslach, C., Schaufeli, W. B., & Leiter, M. P. (2001). Job burnout. *Annual Review of Psychology, 52*(February), 397–422.

Mattes, K. (1992). In pursuit of leisure: Employee want more time off. *HR Focus, 69(5)*, 1, 7.

Matteson, M. T., Ivancevich, J. M., & Gamble, G. O. (1987). A test of the cognitive social learning model of Type A behavior. *Journal of Human Stress, 13(1)*, 23–31.

McCann, L. & Holmes, D. (1984). Influence of aerobic exercise on depression. *Journal of Personality and Social Psychology, 46(5)*, 1142–1147.

McCrae, R. R. & John, O. P. (1992). An introduction to the five-factor model and its applications. *Journal of Personality, 60(2)*, 175–215.

McDonough, P. (2000). Job insecurity and health. *International Journal of Health Services, 30 (3)*, 453–476.

McGrath, J. E. (1976). Stress and behavior organizations. In M. D. Dunnette (Ed.), *Handbook of industrial and organizational psychology* (pp. 1351–1397). Chicago, IL: Chicago Rand McNally.

Meir, E. I. & Melamed, S. (1986). The accumulation of person-environment congruencies and well-being. *Journal of Occupational Behavior, 7(4)*, 315–323.
Melamed, S. & Meir, E. I. (1981). The relationship between interest-job incongruity and selection of avocational activity. *Journal of Vocational Behavior, 18(3)*, 310–325.
Melamed, S., Meir, E. I., & Samson, A. (1995). The benefits of personality-leisure congruence: Evidence and implications. *Journal of Leisure Research, 27(1)*, 25–40.
Melamed, S., Shirom, A., Toker, S., Berliner, S., & Shapira, I. (2006). Burnout and risk of cardiovascular disease: Evidence, possible causal paths, and promising research directions. *Psychological Bulletin, 132(3)*, 327–353.
Meyerson, D. E. (1994). Interpretation of stress in institutions: The cultural production of ambiguity and burnout. *Administrative Science Quarterly, 39(4)*, 628–653.
Miller, D. C. & Form, W. H. (1980). *Industrial sociology.* New York: Harper and Row.
Mirowsky, J. & Ross, C. E. (1989). Explaining the social patterns of depression: Control and problem solving – or support and talking? *Journal of Health and Social Behavior, 30(2)*, 206–219.
Mitchell, T. R., Thompson, L., Peterson, E., & Cronk, R. (1997). Temporal adjustments in the evaluation of events: The "rosy view". *Journal of Experimental Social Psychology, 33 (4)*, 421–448.
Moore, S., Grunberg, L., & Greenberg, E. (2004). Repeated downsizing contact: The effects of similar and dissimilar layoff experiences on work and well- being outcomes. *Journal of Occupational Health Psychology, 9(3)*, 247–257.
Moshkowitz, S. (2003). The relationship between work-overload, leisure activities and burnout among high-tech employees. M.Sc., Tel Aviv University, Israel. [Hebrew].
Neulinger, J. (1974). *The psychology of leisure: Research approaches to the study of leisure.* Springfield, IL: Charles C Thomas.
Neulinger, J. (1982). Leisure lack and the quality of life: The broadening scope of the leisure professional. *Leisure Studies, 1(1)*, 53–63.
O'Brien, G. E. (1986). *Psychology of work and unemployment.* Chichester, UK: Wiley.
O'Driscoll, M. & Beehr, T. (1994). Supervisor behaviors, role stressors and uncertainty as predictors of personal outcomes for subordinates. *Journal of Organizational Behavior, 15 (2)*, 141–155.
Ofek, A. (1997). The effect of organizational vacation and personal vacation on the connection between stress and burnout. M.Sc., Tel Aviv University, Israel. [Hebrew].
Oron, S. (1979). The relationship between vocational fit and satisfaction from vocational choice and leisure patterns and significance. M.A., Tel Aviv University, Israel.
Paine, W. S. (1982). Job stress and burnout. In W. S. Paine (Ed.), *Research, theory and intervention perspectives.* Beverly Hills, CA: Sage Publications.
Parker, S. R. (1971). *The future of work and leisure.* New York: Praeger Publishers.
Parker, S. R. (1982). *Work and retirement.* London: George Allen & Unwi.
Parker, S. R. & Smith, M. A. (1976). Work and leisure. In R. Dubin (Ed.), *Handbook of work, organization and society* (pp. 37–62). Chicago, IL: Rand McNally.
Payne, R. (1980). Organizational stress and social support. In C. L. Cooper & R. Payne (Eds.), *Current concerns in occupational stress* (pp. 269–298). New York: Wiley.
Peiro, J. M., Gonzalez-Roma, V., Tordera, N., & Manas, M. A. (2001). Does role stress predict burnout over time among health care professionals? *Psychology and Health, 16(5)*, 511–525.
Pines, A. (1983). On burnout and the buffering effects of social support. In B. A. Farber (Ed.), *Stress and burnout in the human service professions* (pp. 155–173). Elmsford, NY: Pergamon Press.

Pines, A. (1993). Burnout: An existential perspective. In W. B. Schaufeli, C. Maslach, & T. Marek (Eds.), *Professional burnout: Recent development in theory and research* (pp. 33–51). Washington, DC: Taylor & Francis.

Pines, A. & Aronson, E. (1988). *Career burnout: Causes and cures*. New York: Free press.

Pines, A., Aronson, E., & Kafry, D. (1981). *Burnout: From tedium to personal growth*. New York: Free Press.

Pines, A., Kafry, D., & Etzion, D. (1979). Tedium in managers: A cross-cultural Israeli-American perspective. Paper presented at the Western Psychological Association Convention, San Diego, CA.

Pratt, L. I. & Barling, J. D. (1988). Differing between daily events, acute and chronic stressors: A framework and its implications. In J. J. Hurrel, L. R. Murphy, S. L. Sauter, & C. L. Cooper (Eds.), *Occupational stress: Issues and development in research* (pp. 41–53). London: Taylor & Francis.

Price, R. H. & Hooijberg, R. (1992). Organizational exit pressures and role stress: Impact on mental health. *Journal of Organizational Behavior, 13(7)*, 641–651.

Quick, J. C. & Quick, J. P. (1984). *Organizational stress and preventive management*. New York: McGraw-Hill.

Quick, J. C., Quick, J. P., Nelson, D. L., & Hurrel, J. J. (1997). *Preventive stress management in organizations*. Washington, DC: American Psychological Association.

Rabinowitz, S. & Hall, D. T. (1977). Organizational research on job involvement. *Psychology Bulletin, 84(2)*, 265–288.

Richards, G. (1999). Vacations and the quality-of-life patterns and structures. *Journal of Business Research, 44(3)*, 189–198.

Richter, J. (1992). Balancing work and family in Israel. In S. Zedeck (Ed.), *Work, families and organizations* (pp. 362–394). San Francisco, CA: Jossey-Bass.

Rienhold, B. B. (1996). *Toxic work*. New York: Plum/Penguin.

Roethlisberger, F. J. & Dickson, W. J. (1939). *Management and the worker*. Cambridge, MA: Harvard University Press.

Rosa, R. R., Colligan, M. J., & Lewis, P. (1989). Extended work days: Effects of 8 hour and 12 hour rotating schedules on performance, subjective, alertness, sleep patterns and psychological variables. *Work and Stress, 3(1)*, 21–32.

Rousseau, D. M. (1978). Relationship of work to nonwork. *Journal of Applied Psychology, 63(4)*, 513–517.

Rubinstein, C. (1980). Vacations, expectations, satisfactions, frustrations, fantasies. *Psychology Today, 14(5)*, 62–66, 61–76.

Rudow, B. (1999). Stress and burnout in the teaching profession: European studies, issues, and research perspectives. In R. Vandenberghe & A. M. Huberman (Eds.), *Understanding and preventing teacher burnout: A sourcebook of international research and practice* (pp. 38–58). Cambridge: Cambridge University Press.

Rybczynski, W. (1991). *Waiting for the weekend*. New York: Penguin Books.

Sagie, A. & Krausz, M. (2003). What aspects of job have most effect on nurses? *Human Resource Management Journal, 13(1)*, 46–62.

Salanova, M., Llorens, S., & Schaufeli, W. (2011). "Yes, I can, I feel good and I just do it!" On gain cycles and spirals of efficacy beliefs, affect and engagement. *Applied Psychology - An International Review, 60(2)*, 255–285.

Sapir, O. (1995). The effect of vacation on stress and burnout at work. M.Sc., Tel Aviv University, Israel. [Hebrew].

Sarason, S. B. (1977). *Work, aging, and social change: Professionals and the one life – One career imperative*. New York: The Free Press.

Sarason, B. R., Sarason, I. G., & Pierce, G. R. (1990). *Social support*. New York: Wiley.

Schaubroeeck, J., Ganster, D., & Fox, M. (1992). Dispositional affect and work related stress. *Journal of Applied Psychology, 77(3)*, 322–335.

Schaufeli, W. B. & Bakker, A. B. (2004). Job demands, job resources and their relationship with burnout and engagement: A multi-sample study. *Journal of Organizational Behavior, 25(5)*, 293–315.

Schaufeli, W. B., Bakker, A. B., & van Rhenen, W. (2009). How changes in job demands and resources predict burnout, work engagement, and sickness absenteeism. *Journal of Organizational Behavior and Human Decision Processes, 30*(*7*), 893–917.

Schaufeli, W. B. & Buunk, B. P. (1996). Professional burnout. In M. J. Schabracq, J. A. Winnubst, & C. L. Cooper (Eds.), *Handbook of work and health psychology*. Chichester, UK: Wiley.

Schaufeli, W. & Enzmann, D. (1998). *The burnout companion to study and practice*. London: Taylor & Francis.

Schaufeli, W. B., Enzmann, D., & Girault, N. (1993). Measurement of burnout: A review. In W. B. Schaufeli, C. Maslach, & T. Marek (Eds.), *Professional burnout: Recent developments in theory and research* (pp. 199–215). Washington, DC: Taylor & Francis.

Schaufeli, W. B. & Greenglass, E. R. (2001). Introduction to special issues on burnout and health. *Psychology and Health, 16(5)*, 501–510.

Schaufeli, W. B., Martinez, I. M., Marques Pinto, A., Salanova, M., & Bakker, A. B. (2002). Burnout and engagement in University students – A cross-national study. *Journal of Cross-Cultural Psychology, 33(5)*, 464–481.

Schaufeli, W. B., Maslach, C., & Mark, T. (Eds.). (1993). *Professional burnout: Recent developments in theory and research*. Washington, DC: Taylor & Francis.

Schaufeli, W. B., Salanova, M., Gonzalez-Roma, V., & Bakker, A. B. (2002). The measurement of engagement and burnout: A confirmative analytic approach. *Journal of Happiness Studies, 3*, 71–92.

Schieman, S., Mikkie, A. M., & Glavin, P. (2009). When work interferes with life: Work-nonwork interference and the influence of work-related demands and resources. *American Sociological Review, 74*(*6*),923, 966–988.

Schiff, M. (1992). Vacation. In A. Katz, H. Haas, M. Gurevitz, S. Witz, H. Adoni, M. Schiff, & D. Goldberg (Eds.), *Israel's leisure culture: The changing patterns of cultural activities 1970/1990*. Jerusalem, Israel: Guttman Institute for Applied Social Research. [Hebrew].

Schlossberg, H. (1991). Hyatt Survey: If you take a vacation, do it the right way. *Marketing News, 25*, 21–22.

Schnake, M. E. (1983). An empirical assessment of the effects of affective response in the measurements of the organizational climate. *Personal Psychology, 36*(*4*), 791–807.

Schor, J. B. (1992). *The overworked American: The unexpected decline of leisure*. New York: Basic Books.

Schor, J. B. (1994). Decline of leisure time in America. *Vital Speeches of the Day, 60(24)*, 748–752.

Schuler, R. J. (1980). Definition and conceptualization of stress in organizations. *Organizational Behavior and Human Performance, 25*, 184–215.

Schwarzer, R. & Gutierrez-Dona, B. (2005). More spousal support for men than for women: A comparison of sources and types of support. *Sex Roles, 52(78)*, 523–532.

Seers, A., Mcgee, B. W., Serey, T. T., & Graen, G. B. (1983). The interaction of job stress and social support: A strong reference investigation. *Academy of Management Journal, 26 (2)*, 273–284.

Selye, H. (1956). *The stress of life*. New York: McGraw-Hill.

Shamir, B. (1988). Commitment and leisure. *Sociological Perspective, 31(4)*, 238–258.

Sharkey, J. (2000, May 10). The World Bank gauges the toll travel takes on employees and looks for ways to soften the effect, *New York Times*, p. C00008.

Shaviv, T. (1998). The impact of an organized tour vacation on work stress and burnout among managerial and non-managerial vacationers. M.Sc., Tel Aviv University, Israel. [Hebrew].

Shirom, A. (1982). What is organizational stress? A facet analytic conceptualization. *Journal of Occupational Behavior, 3*, 21–37.

Shirom, A. (1989). Burnout in work organizations. In L. Cooper & I. Robertson (Eds.), *International review of industrial and organizational psychology* (Vol.*2*, pp. 25–48). New York: Wiley.

Shirom, A. (2003). Job-related burnout. In J. C. Quick & L. E. Tetrick (Eds.), *Handbook of occupational health psychology*. Washington, DC: American Psychological Association.

Shirom, A., Eden, D., & Kellermann, J. T. (1981). Effects of population changes on psychological strain in kibbutz communities. *American Journal of Community Psychology, 9(2)*, 27–43.

Silverberg, S. (2007). Home/work conflict and burnout among working mothers after maternity leave – Situational and personal characteristics as moderators. M.Sc., Tel Aviv University, Israel. [Hebrew].

Simpson, R. (1998). Presenteeism, power and organizational change: Long hours as a career barrier and the impact on the working lives of women managers. *British Journal of Management, 9*(s*1*), 37–50.

Sonnentag, S. (2003). A new look at the interface between non-work and work recovery, work engagement, and proactive behavior recovery, work engagement, and proactive behavior. *JAP, 88(3)*, 518–528.

Sparks, K., Faragher, B., & Cooper, C. L. (2001). Well-being and occupational health in the 21st century workplace. *Journal of Occupational and Organizational Psychology, 74(4)*, 489–509.

Spector, P. E. (1997). *Job satisfaction: Application, assessment, causes and consequences*. Thousand Oaks, CA: Sage Publications.

Spector, P. E., Dwyer, D. J., & Jex, S. M. (1988). Relations of job stressors to affective, health, and performance outcomes: A comparison of multiple data sources. *Journal of Applied Psychology, 73(1)*, 11–19.

Spence, J. T. & Robbins, A. S. (1992). Workaholism: Definition, Measurement, and Preliminary Results. *Journal of Personality Assessment, 58(1)*, 160–178.

Spielberger, C. D. (1972). *Conceptual and methodological issues in anxiety research. Anxiety: Current trends and research* (Vol.*2*, pp. 483–493). New York: Academic Press.

Spielberger, C. D. (1976). The nature and measurement of anxiety. In C. D. Spielberger & R. Diaz-Guerred (Eds.), *Cross-cultural anxiety*. Washington, DC: Hemisphere.

Spielberger, C. D., Gorsuch, R. L., & Lushene, R. E. (1970). *Manual for the state-trait anxiety inventory*. Palo Alto, CA: Consulting Psychologists Press.

Sraya-Wallin, O. (1999). The effect of vacation characteristics and the fit between job and career-anchors on stress and burnout among employees in a hi-tech environment. M. Sc., Tel Aviv University, Israel. [Hebrew].

Stanton-Rich, H. M. & Iso-Ahola, S. E. (1998). Burnout and leisure. *Journal of Applied Social Psychology, 28(21)*, 1931–1950.

Stapp, J. & Pines, A. (1976). Who likes competent women? *Human Behavior, 5(11)*, 49–50.

Stebbins, R. A. (1992). *Amateurs, professionals, and serious leisure.* Montreal and Kingston, Canada: McGill-Queen's University Press.
Stewart, M. B. & Swaffield, J. K. (1997). Constraints on the desired hours of work of British men. *The Economic Journal, 107*(March), 520–535.
Swimburne, P. (1981). The psychological impact of unemployment on managers and professional staff. *Journal of Occupational Psychology, 54(1)*, 178–192.
Taylor, B. C. & Arnow, B. A. (1988). *The nature and treatment of anxiety disorder.* New York: The Free Press.
Tennant, C. (2001). Work-related stress and depressive disorders. *Journal of Psychosomatic Research, 51(5)*, 697–704.
Thoits, P. A. (1982). Conceptual, methodological and theoretical problems in studying social support as a buffer against life stress. *Journal of Health and Social Behavior, 23(2)*, 145–159.
Tinsley, H. E. A. & Tinsley, D. J. (1986). A theory of the attributes, benefits and causes of leisure experience. *Leisure Sciences, 8(1)*, 1–45.
Ullah, P., Banks, M., & Warr, P. (1985). Social support, social pressures and psychological distress during unemployment. *Psychological Medicine, 15(2)*, 283–295.
Utechet, K. M. & Aldag, R. J. (1989). Vacation discrepancy: Correlates of individual differences and outcomes. *Psychological Reports, 65(3)*, 867–882.
van der Doef, M. & Maes, S. (1999). The job demand-control (-support) model and psychological well-being: A review of 20 years of empirical research. *Work & Stress, 13(2)*, 87–114.
Vincour-Hai, V. (2001). Stress and burnout before and after vacation and their crossover between spouses. M.Sc., Tel Aviv University, Israel. [Hebrew].
Vinokur, A. D. & Caplan, R. D. (1987). Attitudes and social support: Determinants of job-seeking behavior and well-being among the unemployed. *Journal of Applied Social Psychology, 17(12)*, 1007–1024.
Vinokur, A. D., Price, R. H., & Caplan, R. D. (1996). Hard times and hurtful partners: How financial strain affects depression and relationship satisfaction of unemployed persons and their spouses. *Journal of Personality and Social Psychology, 71(1)*, 166–179.
Vroom, V. H. (1962). Ego-involvement, job satisfaction, and job performance. *Personnel Psychology Bulletin, 15(2)*, 159–177.
Wallace, J. (2005). Job stress, depression and work-to-family conflict. *Relations Industrielles/ International Relations, 60(3)*, 510–539.
Wallston, B. S., Alagna, S. W., De-Villis, B. M., & De-Villis, R. F. (1983). Social support and physical health. *Health Psychology, 2(1)*, 367–391.
Warr, P. & Jackson, P. (1985). Factors influencing the psychological impact of prolonged unemployment and re-employment. *Psychology Medicine, 15(4)*, 795–807.
Warr, P. & Jackson, P. (1987). Adapting to the unemployed role: A longitudinal investigation. *Social Science Medicine, 25(11)*, 1219–1224.
Watson, D., Clark, L. E., & Tellegen, A. (1988). Development and validation of brief measure of positive and negative affect: The PANAS scales. *Journal of Personality and Social Psychology, 54(6)*, 1063–1070.
Westman, M. (2001). Stress and strain crossover. *Human Relations, 54(6)*, 717–751.
Westman, M. (2004). Strategies for coping with business trips: A qualitative exploratory study. *International Journal of Stress Management, 11(2)*, 167–176.
Westman, M. & Etzion, D. (1995). Crossover of stress, strain and resources from on spouse to another. *Journal of Organizational Behavior, 16(2)*, 169–181.
Westman, M. & Eden, D. (1997). Effects of a respite from work on burnout: Vacation relief and fade-out. *Journal of Applied Psychology, 82(4)*, 516–527.

Westman, M. & Etzion, D. (1999). The crossover of strain from school principals to teachers and vice versa. *Journal of Occupation and Health Psychology, 4(3)*, 269–278.

Westman, M. & Etzion, D. (2001). The impact of vacation and job stress on burnout and absenteeism. (Special issue on burnout). *Psychology & Health, 16(5)*, 595–606.

Westman, M. & Etzion, D. (2002). The impact of short overseas business trips on job stress and burnout. *Applied Psychology: An International Review, 51(4)*, 582–592.

Westman, M., Etzion, D., & Yoffe, Y. (1998). The effect of overseas business trips on job stress and burnout (Working Paper). Tel Aviv University, Henry Crown Institute of Business Research, Faculty of Management, Tel Aviv, Israel.

Westman, M. & Vinokur, A. (1998). Unravelling the relationship of distress levels within couples: Common stressors, empathic reaction, or crossover via social interaction? *Human Relations, 51(2)*, 137–156.

Wethington, E. & Kessler, R. C. (1986). Perceived support, received support, and adjustment to stressful life events. *Journal of Health and Social Behavior, 27(1)*, 78–89.

Wheeler, R. J. & Frank, M. A. (1988). Identification of stress buffer. *Behavioral Medicine, 14 (2)*, 78–89.

Wilensky, H. (1960). Work careers, and social integration. *International Social Sciences Journal, 12(4)*, 543–560.

Wills, T. A. & Fegan, M. F. (2001). Social network and social support. In A. Baum, T. A. Revenson, & J. E. Singer (Eds.), *Handbook of health psychology* (pp. 209–234). Mahwah, NJ: Lawrence Erlbaum.

Worrall, L. & Cooper, C. L. (1997). *IM-UMIST, Quality of Working Life Survey*. London: Institute of Management.

Xanthopoulou, D., Bakker, A., Dollard, M. F., Demerouti, E., Schaufeli, W., & Taris, T. W. (2007a). When do job demands particularly predict burnout? *Journal of Managerial Psychology, 22(8)*, 766–786.

Xanthopoulou, D., Bakker, A. B., Demerouti, E., & Schaufeli, W. (2007b). The role of personal resources in the Job Demands-Resources model. *International Journal of Stress Management, 14(2)*, 121–141.

Xanthopoulou, D., Bakker, A. B., Demerouti, E., & Schaufeli, W. B. (2009). Work engagement and financial returns: A diary study on the role of job and personal resources. *Journal of Organizational and Occupational Psychology, 82(1)*, 183–200.

Yoffe, Y. (1997). The effects of overseas business trips on job stress and burnout. M.Sc., Tel Aviv University, Israel. [Hebrew].

Zapf, D., Dormann, C., & Frese, M. (1996). Longitudinal studies in organizational stress research: A review of the literature with reference to methodological issues. *Journal of Occupational Health Psychology, 1(2)*, 145–169.

Zeevi-Ohana, A. (2000). The impact of summer break activities and personality variables on job stress and burnout among high school teachers. M.A., The Hebrew University, Jerusalem, Israel.

Zvi, E. (2004). Stress and burnout among students before and after the Passover break – Vacation activities and personality variables as moderators. M.Sc., Tel Aviv University, Israel.

2

CHARACTERISTICS OF THE BREAK

Respite is defined by the Merriam-Webster dictionary as "an interval of rest or relief" (www.merriam-webster.com/dictionary/respite). In literature, respite has been called vacation (vacation relief, respite vacation, annual vacation, holiday vacation, leisure vacation), holiday, leave (maternal leave, parental leave, sickness leave, leave of absence), break (summer/spring/winter break, semester break, lunch, or coffee break), time out, time off, relief, let-up, and getaway. Different forms of respites serve as different ways to disrupt or interrupt the domination of work routines, aimed mostly at relaxation, restoration, recovery, renewal, revitalization, recuperation, and revival. Respite is considered any form of absence from the work setting, when everyday pressures of the job are absent or diminished including days off work (e.g., Frankenhaeuser et al., 1989; Sluiter, Frings-Dresen, Van der Beek, & Meijman, 2001), weekends (e.g., Chen, Westman, Etzion, & Dach-Zahavi, 2009; Fritz & Sonnentag, 2005), holidays (e.g., Eden, 1990; Etzion & Ofek, 1998; Etzion & Zvi, 2004; Westman & Etzion, 2001), and vacations (e.g., Caplan & Jones, 1975; de Bloom et al., 2009; Eden, 1990; Etzion, 2003; 2005; Fritz & Sonnentag, 2006; Lounsbury & Hoopes, 1986; Strauss-Blasche, Ekmekcioglu, & Marktl, 2000; 2003; Strauss-Blasche, Muhry, Lehofer, Moser, & Marktl, 2004; Westman & Eden, 1997). Rubinstein (1980) suggested that respites can be profiled based on goals and needs ascribed to them. Our own research shows that respites can additionally be profiled based on duration, location, the respite experiences, and level of detachment. Finally, some types of respites comprise a case of their own.

Duration

Respites range in length from very brief to very long. Conceptually, a longer break from work would allow employees sufficient recovery from their workload and set stronger boundaries between work and non-work domains. That is, we might expect that a longer respite (e.g., vacation) would present an opportunity for a more powerful respite thereby moderating the effect of vacation on stress and burnout. Notwithstanding, it is likely that when a vacation is too long, individuals might replace leisure activities with household chores thereby impairing the replenishing of personal resources in preparation of going back to work (Etzion, 2003; Etzion & Boklis, 2006; Nawijn, de Bloom, & Geurts, 2013). It is also possible that employees find it easier, and hence less stressful, to return to work after short respites, whereas longer respites foster greater feelings of detachment form work (Levi, 2002). Tightly related to the length of the respite is its frequency. The overall effects of a break from work (such as a vacation) on employee well-being may be insignificant if employees are only allowed one or two vacation periods a year. It must be noted, however, that duration may be related to personal characteristics (discussed in the next chapter), as some employees are able to psychologically detach from work as soon as they set their foot out the door while others may require a lengthier process to psychologically detach from work (see Sapir, 1995).

Annual leave is probably the longest respite most employees would have during the year (Deem, 1996). In many countries, the eligibility or length of annual leave is determined by length of tenure (Etzion, 2003). The length of vacation was studied with respect to its influence on job satisfaction, job involvement, organizational commitment, turnover intention and life satisfaction, happiness (de Bloom et al., 2010), positive and negative work affect (Horsman & Kelloway, 2011), mood (Nawijn, Marchand, Veenhoven, & Vingerhoets, 2010), as well as happiness and well-being (de Bloom et al., 2010). Vacations of two to 17 days long investigated in these studies generally showed no influence over the studied variables (Nawijn et al., 2013). In contrast, Horsman and Kelloway (2011) found that long weekends were more beneficial than regular two-day weekends in reducing job-related stress because the shorter weekend is insufficient for individuals to effectively engage recovery processes.

In our own studies, the length of vacation was measured by days in a single questionnaire item. Summarizing our entire body of research (Etzion & Boklis, 2006), vacations are most beneficial in terms of reducing burnout when they are limited in time to six to 18 days, whereas weekends or highly extended vacations (over two months long) are not as beneficial (e.g., Laski, 2002). The length of vacation did not influence the level of stress, whereas over 50% of variability in post-vacation stress could be attributed to pre-vacation stress levels (e.g., Shaviv, 1998). Notwithstanding, within this ideal timeframe, respondents reported lower levels of stress after returning from a longer short vacation (over ten days) compared to those who returned from a shorter (seven to ten day) vacation (Etzion, 2003). Additionally, the extent to which individuals are satisfied with their vacation appeared to be a function of the length of the vacation (Etzion, 2003). Finally, employees returning from longer vacations reported lower levels of not only general stress but also extrinsic stress compared to those returning from shorter vacations (Sapir, 1995).

Location

The effect of respite location was assessed in only few studies. Summarizing 29 available studies at the time that assessed the effect of travel on perceived psychological well-being, Chen and Petrick (2013) found that nearly all confirmed the health and wellness benefits of travel across different geographical locations. More specifically, Sonnentag (2004), whose diary study of 47 flight attendants was not included in this summary, found that staying in a hotel did not impair well-being at bedtime.

While Strauss-Blasche and colleagues (Strauss-Blasche et al., 2004) found no differences among vacationers in moderate vs. low altitudes with respect to their post-vacation health gains, his subsequent study (Strauss-Blasche, Reithofer, Schobersberger, Ekmekcioglu, & Marktl, 2005) showed that recuperation could be facilitated by warmer, sunnier vacation locations.

The respite experience

The importance of the respite experience is based on Parker's (1971) model, which defined true leisure time as comprising activities freely chosen by the individual (Etzion & Boklis, 2006). The contribution of a positively perceived respite experience to one's health and well-being is based on the *Broaden and Build Theory* (Fredrickson, 2001), the *Self-determination Theory* (Ryan & Deci, 2000), and the fulfillment of the needs for relatedness (de Bloom, Geurts, & Kompier, 2013). From the perspective of the Theory of Conservation of Resources (Hobfoll, 1989; Hobfoll & Freedy, 1993), some respites have the potential to replenish resources while others tend to deplete them (Westman, 1999). Non-leisure maintenance activities such as errands and house chores could live up to the potential or respite leisure as they were found to be positively related to extrinsic job satisfaction (Sapir, 1995).

Qualitative leisure theories are concerned with the quality of the respite experience more than actual activities carried out during the respite. It appears that the quality may also be more important than the quantity. While an insignificant relationship suggests that the more time employees invest in leisure, the lesser the burnout they experience, no such effect was found in any of the burnout dimensions when tested against the number of hours invested in leisure (Moshkowitz, 2003). Perceived freedom of choice and autonomic decision-making are considered of highest order, immediately followed by intrinsic motivation for that choice and intrinsic satisfaction with the choices that have been made. It is important that these elements product a psychological gain for the individual experiencing them (Caldwell & Smith, 1988; Iso-Ahola, 1984; Kelly, 1982; Mannfredo, Driver, & Tarrant, 1996; Neulinger, 1974; Tinsley & Tinsley, 1986). Perhaps these factors explain why in one of our studies, perceived freedom was not found to be related to burnout (Raviv, 2002). Sometimes, even one's intention to take a respite contributes to vacation satisfaction although such plans may increase pre-vacation stress (Zvi, 2004).

The manner by which individuals perceive their environment and respond to it derives from the extent to which they feel they may influence that environment, or the amount of control individuals perceive to have over their circumstances. Control may be divided into three dimensions: behavioral control, referring to one's ability to directly influence the environment so as to generate desirable results or avoid undesired consequences; decision-making control where individuals have the choice of different actions, tasks, or outcomes; and cognitive control where individuals control their perception and interpretation of circumstances (Averill, 1973). The need for control is fundamentally shared by human beings, serving as an essential component in one's motivation (Ganster & Fusilier, 1989), and hence – perceived satisfaction. In this sense, leisure is the goal while respite is the means to achieve it (Neulinger, 1974). With respect to respite, the question of control over timing and duration has already been discussed. There is additionally the question of control over the choice of activities during the respite, which are discussed below, as well as the question of how the experience is interpreted based on personal characteristics (e.g., Csikszentmihalyi, 2000), which is further discussed in Chapter 3. Vacationers perceiving greater leisure control experienced lower stress resulting from the presence of negative job characteristics. In contrast, no relationship was found between leisure control and the absence of positive job characteristics (Raviv, 2002).

Studies show that different recovery experiences offer incremental validity in predicting strain and well-being outcomes among employees involving psychological detachment from work demands, relaxation (low tension typically associated with low-effort activities), non-work challenging activities, control over activities, physical activity, social affiliation, fun and humor, self-reward, and planning. Presumably, these activities contribute to health and well-being by freeing demands on functional systems that are typically used during work (Day & Stevens, 2011; Etzion et al., 1998; Etzion, Eden, & Lapidot, 1998; Sonnentag & Fritz, 2007).

Leisure activities that offer an opportunity for self-expression, enrichment, social engagement, and a body-mind integration, fulfil intrinsic needs (Caldwell & Smith, 1988). The preference for leisure activities has shifted over the years. For example, between the 1970s and the 1990s in Israel, collective activities such as involvement in canonical culture outlets (the theater, cinema, or museums) has suffered a decline while individualistic, hedonistic activities were on the rise (see Laski, 2002). One such example is the rise in TV consumption, reported to have averaged in the 1980s in the US at a quarter of individuals' free time (Kelly, 1983). Notwithstanding, these trends, even with respect to TV consumption alone, may be changing again with the advent of social media and digitized communications (Conlin, Billings, & Averset, 2016; Dhoest & Simons, 2016; Webb, Kohlbacher, & Prieler, 2017). Our own studies show family time as a favored leisure activity (Moshkowitz, 2003; Zeevi-Ohana, 2000), which is in line with other research showing that social activities and time spent with family and friends may facilitate recovery, as well as psychological and physical health (Byrne, 1993; Kelly, 1983; Sonnentag, 2001; Sonnentag & Niessen, 2008). These activities may foster social support instilling in individuals the feeling of being valued, appreciated or loved, and of being embraced by a system or relationship(s) upon which they may rely when in need (Pines, 1983). Notwithstanding, in some cases, planning and executing leisure time involving family and personal errands corresponds to increased stress due to the absence of positive job characteristics (Laski, 2002).

Individuals vary in the amount of relief they experience as a result of respite (Etzion et al., 1998; Westman & Eden, 1997; Westman, Etzion, & Yoffe, 1998). Enjoyment of the respite may possibly be augmented by interventions aimed at instilling respiting individuals with self-fulfilling prophecies concerning a positive experience (Eden, 2001). Studying three quite different respite effects, Mitchell et al. (1997) found, however, that expectations are typically more positive than the event itself. They also found that memories of the experience are more positive than the actual respite event. In this rosy view of respite, retrospective perceptions about the vacation were closer to perceptions during the respite compared to prospective perceptions prior to the respite. Possibly, then interventions aimed at prolonging the memory of the event would similarly prolong the ameliorative effect of the respite on work-related burnout and stress (Eden, 2001). The study of Mitchell et al. (1997) emphasizes the importance of the emergent practice (e.g., Westman, 2004) of measuring stress during the vacation in addition to the pre-test and post-test measurements in order to fully understand the effect of the respite experience.

Respite activities may be characterized as passive or active. Active leisure includes workout, hobbies, entertainment and outdoor activities, whereas passive leisure includes reading, reflection, internet surfing, TV viewing and social time with family and friends. Active leisure tends to better characterize young, unmarried employees, whereas more restful activities, family time included, equally characterize young, unmarried employees and older, married employees (Moshkowitz, 2003). While active leisure is thought to influence stress and burnout based mostly on freedom of choice, passive leisure activities are thought to have that influence based on the conservation of resources. Thus, some studies show that leisure involving physical activities enhances both physical and psychological well-being (McCann & Holmes, 1984), others found passive activities to have a greater ameliorative effective on stress (Kabanoff & O'Brien, 1986).

Our studies show that stress and sometimes also burnout are higher among passive vacationers compared to active ones. This seems to be related to active vacationers' reports of lesser intrinsic and general stress as well the fact that leisure characteristics do not seem to influence self-efficacy or cynicism – two of the three measured dimensions of burnout (Moshkowitz, 2003; Raviv, 2002; Sapir, 1995; Sraya-Wallin, 1999). Among teachers, more than half of their summer vacation time was spent on various respite and recreational activities, whereas approximately a third of their vacation time was spent on work-related activities, and 15% of their vacation time was spent on personal matters and errands. Those who spent more time attending to personal matters and errands during the vacation revealed lower perception of their professional efficacy after the summer break. Perceived success, at work and especially in non-work spheres, was rated significantly higher after the summer vacation, whereas the importance of work insignificantly declined after the break. Post-vacation success at work was highly and negatively correlated with work-life conflict (Zeevi-Ohana, 2000).

A daily respite in the form of after-work sports activities was found to enhance personal resources in the evening (mediated by positive affect) and increase work engagement on the next day (Naegel & Sonnentag, 2011). The ameliorating effect of workout was also noted by Strauss-Blasche et al. (2005) and Raviv (2002), although this was among the least preferred leisure activities (Moshkowitz, 2003). The greater the stress experienced from the absence of positive work characteristics and the lower the stress from the presence of negative work characteristics, the greater the engagement in workout activities. In fact, workout explains 5% of the variability found in stress that originates in the absence of positive work characteristics. Workout is also inversely related to burnout, and specifically to its self-efficacy dimension, explaining 9% of the variability in reported burnout (Raviv, 2002). Notwithstanding, leisure activities such as workout may be culturally influenced. Kirkcaldy and Cooper (1993), for example, found a negative relationship between regular workout in non-work hours among British executives but a positive relationship between these variables among German executives.

A weekend involving Sabbath keeping may also make a positive contribution to the respite experience: first, through active life segmentation; second, by prescribing positive meaning to this life segmentation; and third, as an integrated time of daily rest, reflection, and relationship development (Diddams, Surdyk, & Daniels, 2004). Interestingly, research shows that the more time spent on leisure activities *before* the vacation, the lower the burnout level after it (Ron, 2003).

Numerous studies show that respites filled with work-related activities impairs well-being, whereas leisure activities such as sports or social activities enhances well-being while decreasing burnout, stress, and exhaustion (e.g., Moshkowitz, 2003; Raviv, 2002; Sonnentag, 2004). Studying the respite effect of an organized tour on job stress and burnout among 72 members of the Israeli Engineers Association, Shaviv (1998) found no respite effect on stress and burnout yet specific vacation characteristics served as moderators in this relationship. Thus, she noted lower post-vacation stress levels for vacationers who experienced higher vacation satisfaction.

Research conducted among Israeli employees aged 25–50 listed the following as preferred activities: time spent with different family members and different manners, errands for which there was no time, trekking in nature, at home with a good book, spa alone or with company, and working out (Kelim Shluvim Research Ltd., 2010). A study among 228 Israeli undergraduate students showed a discrepancy between planned activities and activities carried out during vacation, where students planned to study more but ended spending time with family and friends. Students spending their vacation studying reported more stress and burnout and less satisfaction from their respite experience, whereas students who spent time with friends, spent time in nature, or engaged in entertainment reported less stress and burnout and more satisfaction from their respite experience. In fact, the mere intention to engage in these activities during their vacation produced a similar effect (Etzion & Zvi, 2004; Zvi, 2004). In the case of the 47 female flight attendants, however, Sonnentag (2004) reported that time spent on social activities increased depression at the end of the day.

While leisure was found to be a significant buffer against stress (Moen, Dempster-McClain, & Williams, 1989; Rook, 1987; 1990), the ways by which this mechanism operates are not always clear. Intensively carrying out pleasant leisure activities seems to have an ameliorating effect on mood and stress even at the face of negative life events (Reich & Zautra, 1981), whereas positive life events, including leisure, plays a central role in conferring resistance to pathological influence of negative life events and maintaining psychological well-being (Cohen & Hoberman, 1983). Leisure was found to be one of four useful means of protection against stress and general well-being. The study of positive life events such as leisure, then, is helpful in better understanding our ways of coping with pressure (Folkman & Moskowitz, 2000; Folkman, Moskowitz, Ozer, & Park, 1997). Studies show a positive relationship between life and job satisfaction in different life domains and vacation satisfaction. This relationship has been demonstrated in both directions such that individuals with higher levels of vacation satisfaction enjoyed higher post-vacation life and job satisfaction and those who enjoyed pre-vacation life and job satisfaction were more satisfied with their vacations (e.g., Gilbert & Abdullah, 2002; Lounsbury & Hoopes, 1986). While highly satisfied vacationers enjoyed a stronger ameliorating effect on burnout, this effect was the quickest to fade among them (Westman & Eden, 1997).

Lounsbury and Hoopes (1986) found that vacation satisfaction moderated the pre-post vacation effect, better explaining variability in variables related to both work and non-work than any activity or other vacation feature (e.g., duration). Expectations regarding the vacation differ among individuals as does the discrepancy between ideal and actual vacation (Utechet & Aldag, 1989). When expectations are fulfilled, leaving a small discrepancy between expectations and experience, and where vacation or its activates are of importance to the individual, the outcome of that vacation would be beneficial to both the vacationing employee and the organization (Etzion, 2003; Hoeksma, Guy, Brown, & Brady, 1993; Ragneh & McKinney, 1993).

In our own studies, the effect of 13 leisure activities were tested, including reading, sports, hobbies, school education, community involvement, socialization with friends, socialization with family, entertainment, tours in nature, isolation and reflection, internet surfing, television watching, and listening to music. However, in one study, only sports had significant associations with burnout, so that increased time spent on workout decreased the extent of burnout. An insignificant association suggested that increased workout might reduce stress (Raviv, 2002). For Israeli flight attendants working less than a full-time job in shift format, we found that most were highly active in their leisure time, engaging by average in seven types of leisure activities. For the most part, these employees preferred passive types of leisure (social time with friends, movies, TV, reading, etc.). Working out, entertainment in the form of theater, and artistic inclination each attenuated burnout. Serious leisure (measured in time invested in the activity) was related to these three activities, as well as to school, excursions, reading, and dancing. The greater the number of leisure activities in which these flight attendants engaged, the greater the seriousness of their leisure. While serious leisure was not found to attenuate burnout, it strongly moderated it so that the more time invested in leisure activities, the lower the burnout (Ben-Moshe, 2006). In a different study, meeting with friends, tours in nature, and entertainment were found to be significantly and inversely related to stress, burnout, and exhaustion and positively related to vacation satisfaction, whereas pre-vacation planning of hobby activities had similar effects on post-vacation exhaustion and vacation satisfaction (Laski, 2002).

The vacation satisfaction index used in our own studies (e.g., Etzion, 2003) comprised items extracted from Lounsbury and Hoopes (1986) and translated into Hebrew (see Westman & Eden, 1997; see also Chapter 1). Our data nevertheless is inconclusive regarding the relationship between the level of satisfaction with the vacation and post-vacation job-related stress and burnout. For example, Sapir (1995) found no relationship between leisure satisfaction and post-vacation stress and burnout. Laski (2013), too, found no relationship between leisure satisfaction and post-vacation stress, although some leisure activities were shown to contribute to leisure satisfaction. Similarly, she found no relationship between vacation activities and respite outcomes in her later study, a finding that may be explained by the length of the summer vacation taken by her teacher sample (Laski, 2013; Laski, Etzion, & Westman, 2013). In contrast, Westman and Eden (1997) found that vacations had a greater effect of stress and burnout proportionate to vacation satisfaction. Similar findings emerged when the respite took the form of military reserve duty (Etzion et al., 1998) or teachers' summer break (Zeevi-Ohana, 2000). In the other direction, Yoffe (1997) found that the greater the satisfaction of business travelers from their trip, the greater the attenuation in stress and burnout. Among students, we found that the greater their pre-vacation stress, the less satisfaction they derive from their vacation, and the greater their vacation satisfaction, the lower their post-vacation stress and burnout. These relationships nevertheless did not moderate the influence of vacation over stress and burnout (Zvi, 2004). This finding support previous studies conducted elsewhere among students (see Sapir, 1995). Perhaps these findings indicate a spillover pattern (Wilensky, 1960) between pre-respite and within-respite psychological well-being, as well as between within-respite and post-respite psychological well-being. This is further supported our own findings (Shaviv, 1998) showing that 49–55% of the variance in post-vacation stress and 73% of the variance in post-vacation burnout may be explained by pre-vacation stress and burnout levels. Vacation satisfaction in this study moderated only the variance in post-vacation stress but not burnout. An additional explanation may be that vacationers at large work fewer hours upon their return to work, and satisfied vacationers work fewer additional hours immediately after returning from their vacation (although vacationers' absenteeism was also reduced) (Avni, 2007).

Ron (2003), studying the effect of leisure satisfaction among 70 employees of four Israeli hi-tech companies (mean age 29, 67% males), reported that vacation satisfaction was positively associated with stress originating in the lack of positive characteristics. Raviv (2002), too, studied the effect of leisure satisfaction among Israeli hi-tech employees (69 employees in five companies, 57% males). Her findings show that the higher the leisure satisfaction, the lower the stress that is associated with negative work characteristics, and the less burnout and exhaustion are experienced. Similar findings emerged in the study of Magen-Neuhaus (1993) among factory workers and in the study of Yoffe (1997) among business travelers (both during the business trip and after it), while similar (insignificant) trends were also found by Vincour-Hai (2002) among administrative workers in an Israeli academic institution. Israeli mothers on a lengthy maternity leave also experienced greater alleviation in burnout, the greater their vacation satisfaction was (Silverberg, 2007).

As with the latter study, our meta-analysis data (Etzion, 2003) show that the length of the vacation tends to moderate the relationship of satisfaction with the vacation and job stress and burnout. In the group of workers who took a short vacation the relationships between satisfaction with the vacation and stress and burnout tended to be negative while among individuals who took a long vacation, they tended to be positive. While these relationships were insignificant, the difference between the coefficients of satisfaction from vacation and job stress in the two groups was significant.

Additionally, certain populations, personal preferences, or circumstances may also yield different results. Post-vacation stress and burnout levels were found to decrease among active vacationers compared to passive vacationers (Sraya-Wallin, 1999). However, based on stimulation needs, individuals seeking lifestyle stimulation preferred invigorating and innovative vacations, whereas stimulation avoiders preferred more structured or enriching vacations (Wahlers & Etzel, 1985). By the same token, teachers who traveled during their summer break enjoyed their vacation more than those who had not traveled (Zeevi-Ohana, 2000). Presumably, then, congruence or dissonance with certain preferences or situational factors would affect one's respite experience and satisfaction thereby influencing the vacation's impact on burnout, stress, and well-being.

Degree of detachment

Regardless of an employee's respite experience, employees may still be engaged, physically or mentally, with their work and their workplace. Psychological detachment from work was found to be one of four recovery experiences by which individuals unwind and recuperate from work during leisure time; the others being relaxation, mastery, and control (Sonnentag & Fritz, 2007). Lacking the opportunity or ability to detach from work may lead to insufficient recovery experiences, which, in turn, may lead to different psychological and physical ailments (Etzion et al., 1998; Sonnentag & Bayer, 2005). In an Israeli sample of employees and their significant others, Potok and Littman-Ovadia (2014) found that attachment avoidance moderated the negative association between workload and self-reported psychological detachment, whereas over-commitment was a full mediator between workload and role conflict as job stressors and psychological detachment. Detachment may additionally enhance the respite experience. Psychological detachment reported by employees and their family member was found to be a factor of workload, job involvement, and recovery efficacy suggesting that by managing one's workload, one might increase recovery efficacy (Sonnentag & Kruel, 2006). In a diary-based study of daily respites at the end of the work day, after hours work-related activities negatively influence situational well-being of individuals by their bedtime, whereas low-effort activities, social activities, and physical activities positively influence their well-being before going to sleep (Sonnentag, 2001). Moreover, the effect of a vacation on employees' cholesterol levels varied according to high or low perceived domestic and job demands on vacationers with an effect that lasted longer than the effects of the vacation of burnout and well-being (Strauss-Blasche, Ekmekcioglu, & Marktl, 2002; Strauss-Blasche et al., 2003).

Our study, Etzion et al. (1998), was the first to have subjectively measured the psychological dimension of detachment from their work. Detachment determined whether individual benefited from a positive respite experience: the greater detachment, the greater the transfer of a positive off-job experience back to the job, whereas individuals who remained attached to their jobs missed out on the relief they could have otherwise enjoyed. Detachment from work during respite allows individuals to be more fully present in their respite experience and therefore more influenced by it, whereas continuous, active contact with the job buffers the respite experience, dulling its effects.

Our studies thereafter yielded inconsistent results regarding the effect of psychological detachment in ameliorating post-vacation burnout and well-being. Based on questionnaire items taken from Etzion et al. (1998), which measured how frequently employees contacted their workplace either by phone or by physically visiting the premises, a study of 102 employees yielded no significant relationship between post-vacation stress and burnout and the level of detachment (Etzion, 2003). In one study, we even found borderline interaction effect suggesting that post-vacation burnout increased among detached vacationers (Silverberg, 2007).

In contrast, a study conducted among 92 Israeli employees (mean age 40, 59% males) in technological fields indeed showed a negative correlation between the level of detachment from work and post-vacation job stress (Sraya-Wallin, 1999). Similarly, a study conducted among 58 workers (mean age 47, 48%

males, 87% managers) in the Israeli public sector showed that the greater the detachment from work throughout the vacation, the lower the burnout level after it (Radian, 1998). Among 103 hi-tech employees (59 male) in both junior and managerial roles, detachment was negatively related to burnout, exhaustion and stress originating in negative work characteristics. It was suggested in this study that perhaps the level of pre-vacation burnout determined the level of work detachment during the vacation (Moshkowitz, 2003). Post-vacation stress levels tended to decrease among detached vacationers who were satisfied and enjoyed low job responsibility workload expectations and who had no substitutes at work while on vacation (Shaviv, 1998). In this study, detachment from work explained 2% of the variability in post-vacation stress. We additionally found that the greater the burnout experienced by employees, the greater their need to detach and relax (Raviv, 2002). Detachment allowed students to enjoy the respite effect as stress, burnout, and exhaustion were all reduced for detached vacationers (Zvi, 2004). Detachment was also found to be related to vacation satisfaction among mothers on maternity leaves (Silverberg, 2007).

It nevertheless seems that high detachment from work during a vacation is an insufficient precursor to reduced burnout. Among 42 female military workers in the IDF on maternity leave who reported high detachment from work, burnout was higher for those women who were highly involved with non-work activities compared to the women who reported low involvement in non-work activities (Ben Yoseff, 2004). Moreover, one's vocation may dictate one's respite experience. For example, while vacation satisfaction reduced burnout among 156 Californian high school teachers, it was found that their engagement in vacation activities that were related to school-work nevertheless reduced post-vacation exhaustion (Zeevi-Ohana, 2000). Similar findings with respect to post-vacation stress were also found in a study of 86 Israeli high school teachers (Laski, 2002). Perhaps this is related to the absence of a clear boundary between work and non-work spheres as a great proportion of the teachers' work is conducted while outside the classroom and outside official working hours (Ho, 1996; Stanton-Rich & Iso-Ahola, 1998). This may also be the reason for combat soldiers to have not experienced the respite effect during their vacation.

Possibly, employees who remain attached to their workplace or job during their respite enjoy the effect of detachment from their regular, everyday work (and life) circumstances. Physical detachment from the workplace seems to offer sufficient relief from the organizational climate and perpetual job stressors for business travelers to enjoy the respite effect (Westman & Etzion, 2002). Similar to satisfied vacationers, detached vacationers, too, worked fewer additional hours immediately after returning from their vacation (Avni, 2007). The interaction between job detachment and other personal traits are further discussed in the next chapter.

The longer the vacation, the greater the detachment experienced by mothers on maternity leave; however, the longer the vacation the greater their physical exhaustion upon their return to work (Ben Yoseff, 2004). Detachment did not interact with work engagement to influence post-vacation burnout, but it did interact with engagement in non-work life domain to produce such influence. Women who were highly detached from work during their maternity leave experienced an increase in post-vacation burnout when they were highly engaged in non-work life domains. No such effect was found for low levels of detachment regarding of level of engagement in non-work life domains. Work engagement did not interact with detachment to produce any similar (or other) effects.

Kunda (1993) found that detachment was challenging to vacationing hi-tech employees. Zvi (2004) found positive relationships between detachment from work during one's vacation and their degree of vacation satisfaction, degree of conscientiousness, and age. Silverberg (2007), however, found a strong negative relationship between education and detachment from work during one's vacation. It is possible that physical detachment from work is insufficient by itself to allow for the vacation to influence burnout and that the degree of detachment, both physical and emotional are related to engagement in both work and non-work areas of life (Ben Yoseff, 2004).

Degree of change

Respite studies typically perceive the respite itself as a change from the ongoing encounter with job stressors at the workplace. This feature of the respite, in and of itself, is expected to have an ameliorative effect on strain (Etzion, 2003). The need for a respite as a means of escapism is almost equally explained as a desire to forget one's obligations, as well as escape from daily reality (Sapir, 1995). Indeed, in interviews with hi-tech male employees whose respite was in the form of military reserve duty, interviewees singled out the degree of change from their daily routine as one of the important traits of their reserve duty (Lapidot, 1992).

Special respite cases

Some types of respite can be considered special in their overall characteristics. These include: mandatory organizational vacations, business travel, sabbaticals, maternity leaves, and reserve service.

Voluntary vs. mandatory vacations

In the organizational context, respites are often studied when employees are collectively forced to take a leave (e.g., Etzion & Zvi, 2004; Westman & Etzion, 2001). Following Parker's (1971) model that emphasized the importance of choice in leisure time, it may be surmised that the power of the respite to ameliorate work-related burnout and stress would grow when the respite was planned voluntarily.

The differential effects of organizational and personal vacations were measured in a study of 90 female social workers (mean age 38) in a religious community in Israel. Results showed a decrease in burnout after the organizational vacation and an increase after the personal vacation. In contrast, job stress decreased after the personal vacation for employees working 80% or more of a full-time job, but increased for those working less than that.

Shifts

Work in the format of shift has been described as a stressor, mostly due to the impact of shifts on the physiological, psychological, and social rhythms (Cooper & Marshall, 1976; Monk, 1988). Consequently, most of the issues discussed in the literature with respect to shifts relate to night shifts, which require workers to adjust their sleeping and waking cycle. Night shifts also generate desynchronization and conflict between the different environmental rhythms and the altered circadian cycle, which impede adaptation of the physiological functions of individuals to the new time structure. Shift employees therefore often experience fatigue and frailty (Carpentier & Cazamian, 1977). In terms of resource conservation, night shift employees are working during times which are generally used for recovery and resource replenishment (Aschoff, Hoffmann, Pohl, & Wever, 1975).

Day shift employees report a greater fit between work and non-work aspects of their lives, whereas night shift employees experience greater work-non-work conflict. Evening and night shifts may require employees to forgo certain family events such as the family dinner or participation in child rearing and education. Social involvement beyond family relations is also influenced by the irregular working arrangements of these employees. Such concessions are experienced as particularly grave by those working on weekend shifts as well, and they lead to greater levels of stress and more health issues (Banks, 1956; Coleman, 1989; Drenth & Hoolwerf, 1976; Jamal & Baba, 1992). Rotating shifts, as opposed to fixed shifts, additionally lead to role overload and role ambiguity in addition to greater work–non-work conflict (Barton, 1994; Jamal & Badawi, 1995; Staines & O'Connor, 1980). Possibly, night shift employees attempt to reduce their work–non-work conflict at the expense of sleeping hours, which in turn, leads to fatigue and strain interpersonal relationships (Staines & Pleck, 1984). Shift employees often experience conflicting desires between their need to rest and their will to spend time with their families, leading to a greater degree of divorce compared to employees who participate in more common working arrangements (Overman, 1993).

Research in which shift work was not found to negatively influence the work–non-work conflict was explained through natural selection, where shift employees had learned to adapt to their work arrangement in such a way that refrained from straining the work–non-work conflict (Skipper, Junk, & Coffey, 1990). Some researchers have additionally claimed that shift work may facilitate the reduction of conflict and stress, especially when work and non-work facets of life are well managed. Perceived control in rotating shift work was also found to reduce the degree of work–non-work conflict as opposed to long hours in inflexible working arrangements (Bacharach, Bamberger, & Conley, 1991; Greenhous & Beutell, 1985). Each of these working arrangements is expected to lead to a different blend of conflict with different coping alternatives, especially when involving different vocations (Richter, 1992). Rotating shift work may even be advantageous in freeing time for social and family interaction.

We conducted one study that was dedicated to respite under the circumstances of shift work in which participated 88 shift employees comprising hospital nurses and hotel staff (Aviv, 1998). Burnout but not stress nor work–non-work conflict was higher during routine work than it was prior to the vacation. None of these variables had changed after the vacation compared to their levels just prior to it. Role conflict, however, was significantly lower after the vacation compared to routine times before it. Where engagement in the work–non-work conflict was low, the discrepancy in stress between the pre-vacation routine period and post-vacation period influenced burnout. The same was true for individuals just prior to their vacation compared to post-vacation measurements. No such effect was found for high engagement in the work–non-work conflict. Neither work engagement nor vacation satisfaction were found to influence burnout or job conflict.

Business travel

In the US, one in five adults have been reported to travel for business at least once in 2004 and this number is likely to be on the rise. While in the US travel generated an estimated $153.2 billion in 2004, in 2007, leading Israeli business organizations incurred $972 million in business trip-related expenses. Business trips, which are typically short, are defined as traveling for the organization for period of a week or so while crossing international borders (Westman, Etzion, & Chen, 2009b). As a form of respite, business travel both consumes and replenishes psychological resources. Travel itself may be strenuous, as experienced by travelers who were not obligated to work during their travel (Fisher & Cooper, 1990).

It may be hypothesized that characteristics of the business trip, including frequency, workload, *a priori* notice, and control over different trip characteristics, would be related to perceived stress and health. The more frequent the travel, the shorter the notice, the greater the workload, and the lower the perceived control, the greater the perceived stress and the lower the perceived health. Similarly, the more active the purpose of the business trip, the greater the perceived stress we may find (Kremer-Fuks, 1999). The time of travel notice may be important as shorter notices may be perceived as restricting one's control over travel characteristics while also increasing level of ambiguity and uncertainty and hence stress and burnout, too (see Koniak, 2003). Jetlag may be a detrimental outcome of business trips, as well (Striker, Dimberg, & Liese, 2000).

This is supported by a study conducted with the World Bank (Sharkey, 2000) in which was found a significantly greater number of medical insurance claims among frequent business travelers compared to sedentary peers. Frequent business travelers also complained more regarding difficulties in psychological adjustment including anxiety and depression. Travelers also reported higher stress levels and negative influence of the business travel on their physical and psychological health. Finally, business travel tends to influence the psychological strain of the traveler's family due to the traveler's absence of crossover between spouses. Research shows that business travel involves medical risk, evidenced in higher rates of medical insurance claims filed by travelers compared to non-traveling peers (Striker et al., 1999). At the same time, the business trip may have a positive effect on employees, who may enjoy the change from a routine and who have been found to have experience better physical and psychological well-being compared to non-traveling peers (Striker et al., 2000).

The work–non-work conflict may be experienced as particularly acute by business travelers who leave their families behind. Travelers experiencing homesickness may also experience distress rising from their preference to conceal such emotions, potentially leading to grief-like responses. Homesickness was found related to dissatisfaction (as well as lack of satisfaction) of the new, unfamiliar environment (Fisher & Cooper, 1990). Overload may also stem from job-related role conflict or a frequent requirement to travel. Additional strain may originate in the discrepancy between routine work and the work related to the business travel including preparation, accumulated work while absent, and conclusion of their travel (DeFrank et al., 2000). First-time business travelers may experience stress stemming from uncertainty and ambiguity. They may also feel a need to succeed in the mission which was entrusted to them by their employing organization (Yoffe, 1997). Moreover, business travels cannot detach themselves from their work although they are able to detach themselves of the workplace (Koniak, 2003). Another stressor specific to business travel is the purpose of travel. Active purposes require initiative (negotiations, closing deals, presenting at expositions, presenting a live demo, rendering support, or making a business presentation) on behalf of the traveler, who is largely responsible for the business outcome of the trip. A passive goal, in contrast, requires little initiative on behalf of the traveler as their potential to influence the outcome is limited. Passive travelers may be considered guests (in a conference, exposition, or course) rather than pivotal players in the business scenario (Koniak, 2003). In some locations, travelers may additionally be threatened by political turbulence or terror activity, thereby increasing the psychological load of the trip (Koniak, 2003).

Studying 35 business travelers of the cultures (American, Israeli, and Swedish; mean age 44) taking between three and 24 trips a year, Westman (2004) identified four phases in the business travel that require different coping strategies: (a) pre-trip, (b) journey, (c) stay, and (d) post-trip. In other words, the business trip can be perceived as an unfolding experience during which stress and burnout may be subject to change. Each phase in the journey is involved with different travel stressors that could potentially hinder the overall positive effect of the respite on work related burnout and stress.

The respite experience of a business trip can be enhanced based on the tenets of the job demands-resources model and Conservation of Resources theory. While some studies found that business travel has an adverse effect of psychological health (e.g., DeFrank et al., 2000), among 57 employees (mean age 34, 79% males) of hi-tech companies in Israel, Westman & Etzion (2002) found an ameliorative effect of business travel on work-related stress and burnout. Westman, Etzion, and Gattenio (2008) further determined the unique nature of business travel as respite, noting that not only burnout was lowest during the trip but so were work–family conflict and family–work conflict. However, this finding was moderated by gender. Their findings show that work–family conflict and family–work conflict among men remained rather constant throughout the travel experience, dipping lowest upon returning home, whereas for women they were lowest during the trip itself and highest upon returning home.

Westman, Etzion, and Chen (2009a) explored the influence of three specific resources on the respite effect of business travel, namely perceived control over travel schedules, cultural intelligence and the ability to adapt to new cultural settings, and self-efficacy. They found that perceived control had no significant impact on either of the measured outcome variables, exhaustion and vigor, possibly because it was not truly left to the control of the traveler but was largely determined by organizational policies. Similarly, family support did not contribute significantly to either outcome. However, cultural intelligence measured before traveling predicted vigor during the trip (but not after it) and exhaustion both during and after travel. Thus, enhancing cultural intelligence prior to travel may be an effective means for reducing exhaustion and increasing vigor, thereby improving business travelers' performance.

In our study of 100 business travelers of different seniority and roles working in large, leading Israeli hi-tech firms (63% males), travel frequency was found to be related to perceived work overload but not to stress, although work overload and stress were significantly related. Work overload was inversely related to control so that the greater the perceived workload, the lesser the perceived control and vice versa. Travel frequency was positively related to short notices, so that the more the employee traveled, the shorter were their travel notices. It was also inversely related to perceived control, so that the more frequent the travel, the lesser the perceived control. Perceived control, in turn, was inversely related to stress so that lesser perceived control was related to higher stress levels. Shorter notices were also positively related to perceived health, although perceived stress was not (Koniak, 2003).

In a different study of 57 (45 males) business travelers employed in different hi-tech companies, for whom the studied travel was not their first, burnout was related to work chores that amalgamated in the absence of the traveler. Stress was generally higher before the trip than after it. Stress originating in the presence of negative work characteristics was aligned with the general stress levels while stress originating in the absence of positive work characteristics showed the inverse direction being higher upon return from the trip. Stress levels during the trip were similar to their level prior to traveling (Yoffe, 1997).

A special class of business travel is that of flight attendants whose work is business travel. An additional characteristic of these employees is part-time work often involving shifts. Part-time jobs are a development of the last half of the twentieth century aimed at addressing individual needs and bettering the work environment for employer and employees combined. Both part-time and shift work formats affect the work–non-work relationship compared to the manner by which this relationship is experienced by full-time employees in day jobs. A flexible work arrangement is expected to enhance the quality of life for those employees who plan their non-work and leisure time, that is, for those individuals for whom leisure is highly valued (Ronen, 1984). We conducted a study among flight attendants in the Israeli airline of El-Al who had little control over their schedule, which would change from one month to the next and one season to the next. Rarely did these flight attendants work full time and monetary remuneration was in accordance to flight time. Burnout tended to decrease during the business trip and after it compared to its level before the trip. Here, too, control, perceived or actual, was not related to burnout. However, when gender and tenure were considered, women and veteran employees felt they had greater control over their working arrangement. Tenure was only slightly and weakly negatively related to burnout (Ben-Moshe, 2006).

Sabbaticals

Sabbaticals are a special form of respite in that they are particularly long and offer various opportunities for personal growth and professional advancement. Purportedly, they stimulate high quality employees, enrich knowledge and expertise in the organization, and reduce stress and burnout, thereby contributing to a more productive workforce that is more willing to serve until retirement (Toomey & Connor, 1988). Indeed, faculty members have reported an association between sabbaticals and career advancement based on their appreciation of role and status changes, formal professional opportunities, and general opportunities to learn new things and take on new challenges (Sorcinelli, 1986). Among 148 faculty members from ten universities in three countries, the respite promoted sense of control, positive affect, and life satisfaction (Westman, Eden, & Davidson, 2002). Similarly, among 85 faculty members at the 19 responding schools, 80% of sabbatical takers had a favorable experience, while an estimated 75% of sabbaticals yielded substantial accomplishments. Sabbatical satisfaction was ascribed to pre-sabbatical planning (Jarecky & Sandifer, 1986).

Sabbaticals may be considered more of a short-term relocation than a vacation, as they are paid time involving work, sometimes of different nature, and most typically at a different location. Sabbatical travelers, then, may be considered expatriates, required to cope with the adaptation to a foreign land, language, and culture and later – with their repatriation upon their return home (Davidson et al., 2010). Indeed, post-sabbatical depression appears to be quite common (Jarecky & Sandifer, 1986). Post-sabbatical challenges notwithstanding, Westman et al. (2002) demonstrated the ameliorative effect of sabbaticals as respite, reporting significant declines in perceived stress and burnout during the respite and thereby extending the known length of an effective respite from days or weeks to months and years. This effect seems to apply across gender (Hendel & Solberg, 1983).

Maternity leave

A rising trend in working arrangements, including time off, is the substantial growth in the number of countries, known as family-friendly countries, in which formal arrangements and legislation address the family needs of the working population (see also Chapter 4). The needs of female employees before and after childbirth have been recognized by the International Labour Organization as early as 1919 (www.ilo.org/dyn/normlex/en/f?p=NORMLEXPUB:12100:0:NO:P12100_ILO_CODE:C003). The revised Convention C103 from 1952 stipulates that upon giving birth, female employee should be entitled to a 12-week maternity leave or longer, during which protection would be conferred upon her from dismissal and she would be entitled to cash-paid wages from public funds, rather than the employer's funds, at a rate that would be sufficient "for the full and healthy maintenance of herself and her child in accordance with a suitable standard of living" (ibid.). Despite this early recognition, application in some countries is taking longer than in others. For example, in Israel, the Women's Labor Act was enacted in 1954, securing fully paid maternity leave due to childbirth or adoption for fourteen weeks, six of which may be taken before the event. In addition, employees may not dismiss women during their pregnancy or maternity leave and up to one year following child birth should the woman extend her maternity leave without pay. During this time, women are also entitled to work one hour less for their full wages (Richter, 1992). An amendment in 1997 allows men to take up to half of the maternity leave following childbirth should the mother opt to return to her workplace (Etzion & Dar, 2004). In the US, the Family and Medical Leave Act (FMLA) was enacted only in 1993, stipulating the organization of over 50 employees should allow unpaid leave of 12 weeks or longer with respect to family and health reason (including childbirth or adoption) to employees with at least one year of tenure and secure the same or equal job upon return from that vacation (Kelly & Dobbin, 1999). The European Union's Parental Leave Agreement and Directive was given legal force in June 1996, stipulating that all employees regardless of gender are entitled to a three-month or longer vacation in the case of childbirth or adoption. The directive allows to extend this entitlement until the child has reached the age of 8. Thus, Sweden, a country with a generous vacation policy, has set parental leave to a year during which one is paid 80% of the wages of the vacationing employee, whether the mother or the father (www.eurofound.europa.eu/observatories/eurwork/comparative-information/the-eu-parental-leave-agreement-and-directive-implications-for-national-law-and-practice).

Due to the extensive length of maternity leave, like sabbaticals, it is expected to play a dual role with respect to job-related, as well as general stress, both driving it and alleviating it. As a form of respite, maternity leave is a complex phenomenon due to the physical recovery involved in the early stages of the leave (Tulman & Fawcett, 1988), as well as the necessity for positive adaptation, bonding, and attachment between the mother and newborn (Zigler & Muenchow, 1983). Recently laboring mothers are additionally at risk for psychological complications, which appear to be ameliorated by the maternity leave itself (Hyde, Klein, Essex, & Clark, 1995). The length of the maternity leave further compounds its effect as respite.

The two studies we conducted regarding the respite effects of maternity leave produced conflicting results. We may assign these discrepancies to the small samples, as well as the different organizational settings from which participants were recruited. Studying 42 female military workers in the IDF, Ben Yoseff (2004) found little confirmation to any of the study hypotheses. Burnout among participants was not found to be related to maternity leave nor to vacation satisfaction or the extent to which the leave perceived as legitimate. The interaction between work detachment during leave and work engagement while at work was similarly not found to have influenced burnout. Only one significant interaction measure was found, but it operated in the direction opposite to the hypothesis: burnout of women who were highly involved in non-work activities and highly detached from work was higher than for women with low involvement in non-work activities and high detachment from work. Ben Yoseff (2004) concluded that with respect to burnout, the effects of maternity leave were different than those of a regular break from work.

In contrast, among a snowball sample of 86 women who returned to work up to six months following a maternity leave of up to one whole year, Silverberg (2007) found a strong positive relationship between full-time work and exhaustion, preventive coping style, and the strength of the work–non-work conflict an interaction effect on post-leave burnout between detachment from work and involvement with non-work activities. The length of the leave was strongly and positively related to perceived self-efficacy and negatively related to tenure. Longer tenure was inversely related to the length of vacation but positively related to mothers' intentions to return to the same organization. Mothers' intentions to return to the same organization were inversely related to the length of leave.

Military reserve duty

Active reserve duty is a special type of respite because the venue itself is highly stressful. The premises upon which it was assumed that reserve service might indeed offer an ameliorative respite effect was that work-related stress declined with any change from the job. Here, too, only one work is known in which the respite effects of reserve service were measured. Studying 81 Israeli employees (mean age 38) active reserve service of two to four weeks and 81 matched controls (mean age 39) who continued to work during the same period, Etzion et al. (1998) indeed showed a significant decline in job stress and in burnout among the reservists and no change among the controls. Aligned with other research described here, the quality of respite experience (service) and degree of detachment from work interacted in moderating the respite effects; the greater the detachment, the stronger the effect positive experience had in relieving reservists from work-related stress and burnout.

References

Aschoff, J., Hoffman, K., Pohl, H., & Weaver, R. A. (1975). Re-entrainment of circadian rhythms after phase-shifts of the Zeitgeber. *Chronobiologia, 2(1)*, 23–78.

Averill, J. R. (1973). Personal control over aversive stimuli and its relationship to stress. *Psychological Bulletin, 80(4)*, 286–303.

Aviv, N. (1998). The meaning of vacation among women working shifts: The effect of vacation on job stress, burnout and work-home conflict. M.Sc., Tel Aviv University, Israel. [Hebrew].

Avni, T. (2007). The impact of organizational vacation on the amount of regular working hours, overtime, sick days and absenteeism among the organization's employees. M.Sc., Tel Aviv University, Israel. [Hebrew].

Bacharach, S. B., Bamberger, P., & Conley, S. (1991). Work–home conflict among nurses and engineers: Mediating the impact of role stress on burnout and satisfaction at work. *Journal of Organizational Behavior, 12(1)*, 39–53.

Banks, S. O. (1956). Continuous shift work: The attitudes of wives. *Occupational Psychology, 30*(April), 69–84.

Barton, J. (1994). Choosing to work at night: A moderating influence on individual tolerance to shift work. *Journal of Applied Psychology, 79(3)*, 449–454.

Ben Yoseff, N. (2004). The effect of maternity leave on burnout among women serving in the I.D.F. M.Sc., Tel Aviv University, Israel. [Hebrew].

Ben-Moshe, L. (2006). Work features, "serious leisure" and burnout among flight attendants. M.Sc., Tel Aviv University, Israel. [Hebrew].

Byrne, B. M. (1993). Burnout: Testing for the validity, replication, and invariance of casual structure across elementary, intermediate, and secondary teachers. *American Educational Research Journal, 31(3)*, 645–673.

Caldwell, L. L. & Smith, E. A. (1988). Leisure: An overlooked component of health promotion. *Canadian Journal of Public Health, 79(2)*, (S44–S48).

Caplan, R. D. & Jones, K. W. (1975). Effects of work load, role ambiguity, and Type A personality on anxiety, depression and heart rate. *Journal of Applied Psychology, 60(6)*, 713–719.

Carpentier, J. & Cazamian, P. (1977). *Night work*. Geneva: International Labor Office.

Chen, C.-C. & Petrick, J. F. (2013). Health and wellness benefits of travel experiences: A literature review. *Journal of Travel Research, 52(6)*, 709–719.

Chen, S., Westman, M., Etzion, D., & Dach-Zahavi, A. (2009). The impact of a short respite on well-being and performance (Working Paper). Tel Aviv University, Henry Crown Institute of Business Research, Faculty of Management, Tel Aviv, Israel.

Cohen, S. & Hoberman, H. M. (1983). Positive events and social supports as buffers of life change stress. *Journal of Applied Social Psychology, 13(2)*, 99–125.

Coleman, P. M. (1989). Shift work scheduling for the 1990's. *Personnel Psychology, 66*(January), 10–15.

Conlin, L., Billings, A. C., & Averset, L. (2016). Time-shifting vs. appointment viewing: The role of fear of missing out within TV consumption behaviors. *Communication & Society, 29(4)*, 151–164.

Cooper, C. L. & Marshall, J. (1976). Occupational sources of stress. *Journal of Occupational Psychology, 49(1)*, 11–28.

Csikszentmihalyi, M. (2000). *Beyond boredom and anxiety: Experiencing flow in work and play.* San Francisco, CA: Jossey-Bass.

Davidson, O. B., Eden, D., Westman, M., Cohen-Charash, Y., Hammer, L. B., Kluger, A. N., et al. (2010). Sabbatical leave: Who gains and how much? *Journal of Applied Psychology, 95(5)*, 953–964.

Day, A. & Stevens, S. (2011). Developing and validating an expanded recovery experiences scale. Paper presented at the Work, Stress & Health Conference, Orlando, FL.

de Bloom, J., Compier, M., Guerts, S., de Weerth, C., Toon, T., & Sonnentag, S. (2009). Do we recover from vacation? Meta-analysis of vacation effects on health and well-being. *Journal of Occupational Health Psychology, 51(1)*, 13–25.

de Bloom, J., Geurts, S., & Kompier, M. A. (2013). Vacation (after-) effects on employee health and well-being, and the role of vacation activities, experiences and sleep. *Journal of Happiness Studies, 14(2)*, 613–633.

de Bloom, J., Geurts, S. A. E., Taris, T. W., Sonnentag, S., de Weerth, C., & Kompier, M. A. J. (2010). Effects of vacation from work on health and well-being: Lots of fun, quickly gone. *Work & Stress, 24(2)*, 196–216.

Deem, R. (1996). No time for a rest? An exploration of women's work, engendered leisure and holidays. *Time and Society, 5(1)*, 5–25.

DeFrank, R. S., Konopaske, R., & Ivancevich, J. M. (2000). Executive travel stress: Perils of the road warrior. *Academy of Management Executive, 14(2)*, 58–71.

Dhoest, A. & Simons, N. (2016). Still "watching" TV? The consumption of TV fiction by engaged audiences. *Media and Communication, 4(3)*, 176–184.

Diddams, M., Surdyk, L. K., & Daniels, D. (2004). Rediscovering models of Sabbath keeping: Implications for psychological well-being. *Journal of Psychology and Theology, 32(1)*, 3–11.

Drenth, P. J. D., Hoolwerf, G., & Thierry, H. (1976). Psychological aspects of shift work. In P. E. Warr (Ed.), *Personal goals and work design* (pp. 209–233). New York: Wiley.

Eden, D. (1990). Acute and chronic job stress, strain and vacation relief. *Organizational Behavior and Human Decision Processes, 45(2)*, 175–193.

Eden, D. (2001). Vacations and other respites: Studying stress on and off the job. In C. L. Cooper & I. T. Robertson (Eds.), *International review of industrial and organizational psychology* (pp. 121–146). Chichester, UK: Wiley.

Etzion, D. (2003). Annual vacation: Duration and relief from job stress and burnout. *Anxiety, Stress, and Coping, 16(2)*, 213–226.

Etzion, D. & Boklis, V. (2006). Meta-analytic study of respite and fade-out effects. Report by: Tel Aviv University, Henry Crown Institute of Business Research, Faculty of Management, Tel Aviv, Israel.

Etzion, D. & Dar, A. (2004). Patterns of leisure and vacation across cultures. *Panim- Quarterly of Society, Culture and Education, 27*, 23–38 [Hebrew].

Etzion, D., Eden, D., & Lapidot, Y. (1998). Relief from job stressors and burnout: Reserve service as respite. *Journal of Applied Psychology, 83(4)*, 577–585.

Etzion, D. & Ofek, A. (1998). The influence of two types of vacation on perceived job stressors and burnout among social workers. Paper presented at The 7th International Conference on Stress Research Budapest, Hungary.

Etzion, D. & Zvi, E. (2004). Stress and burnout among students before and after the Passover break: Vacation activities and personality variables as moderators. Report by: Tel Aviv University, Henry Crown Institute of Business Research, Faculty of Management, Tel Aviv, Israel.

Fisher, S. & Cooper, C. L. (1990). *On the move: The psychology of change and transition.* Chichester, UK: Wiley.

Folkman, S. & Moskowitz, J. T. (2000). Positive affect and the other side of coping. *American Psychologist, 55(6)*, 647–654.

Folkman, S., Moskowitz, J. T., Ozer, E. M., & Park, C. L. (1997). Positive meaningful events and coping in the context of HIV/AIDS. In B. H. Gottlieb (Ed.), *Coping with chronic stress* (pp. 293–314). New York: Plenum Press.

Frankenhaeuser, M., Lundberg, U., Fredrikson, M., Melin, B., Tuomisto, M., Myrsten, A.-L., et al. (1989). Stress on and off the job as related to sex and occupational status in white-collar workers. *Journal of Organizational Behavior, 10(4)*, 321–346.

Fredrickson, B. L. (2001). The role of positive emotions in positive psychology: The broaden-and-build theory of positive emotions. *American Psychologist, 56(3)*, 218–226.

Fritz, C. & Sonnentag, S. (2005). Recovery, health, and job performance: Effects of weekend experiences. *Journal of Occupational Health Psychology, 10(3)*, 187–199.

Fritz, C. & Sonnentag, S. (2006). Recovery, well-being, and performance related outcomes: The role of workload and related experiences. *Journal of Applied Psychology, 91 (4)*, 936–945.

Ganster, D. C. & Fusilier, M. R. (1989). Control in the workplace. In C. L. Cooper & I. Robertson (Eds.), *International review of industrial and organizational psychology* (pp. 235–280). Chichester, UK: Wiley.

Gilbert, D. & Abdullah, J. (2002). A study of the impact of the expectations of a holiday on an individual's sense of well-being. *Journal of Vacation Marketing, 8(4)*, 352–361.

Greenhous, J. H. & Beutell, N. J. (1985). Sources of conflict between work and family roles. *Academy of Management Review, 10(1)*, 76–88.

Hendel, D. D. & Solberg, J. (1983). Sabbatical and leave experiences of female and male faculty at a large research university. Paper presented at the Annual Meeting of the American Educational Research Association, Montreal, Canada.

Ho, J. T. S. (1996). Stress, health and leisure satisfaction: The case of teachers. *International Journal of Educational Management, 10(1)*, 41–48.

Hobfoll, S. E. (1989). Conservation of resource – A new attempt at conceptualizing stress. *American Psychologist, 44(3)*, 513–524.

Hobfoll, S. E. & Freedy, J. (1993). Conservation of resources: A general stress theory applied to burnout. In W. B. Schaufeli, C. Maslach, & T. Marek (Eds.), *Professional burnout: Recent developments in theory and research* (pp. 115–129). Washington, DC: Taylor & Francis.

Hoeksma, J. H., Guy, J. D., Brown, C., & Brady, J. (1993). The relationship between psychotherapist burnout and satisfaction with leisure activities. *Psychotherapy in Private Practice, 12(4)*, 51–57.

Horsman, P. A. & Kelloway, E. K. (2011). Does one day make a difference? Investigating the impact of weekend length on recovery experiences. Paper presented at the American Psychology Association (APA), Orlando, FL.

Hyde, J. S., Klein, H. M., Essex, M. J., & Clark, R. (1995). Maternity leave and women's mental health. *Psychology of Women Quarterly, 19(2)*, 257–285.

Iso-Ahola, S. E. (1984). Social psychological foundations of leisure and resultant implications for leisure counseling. In E. T. Dowd (Ed.), *Leisure counseling, concepts and applications*. Springfield, IL: Charles C Thomas.

Jamal, M. & Baba, V. V. (1992). Shiftwork and department-type related to job stress' work attitude and behavioral intentions: A study of nurses. *Journal of Organizational Behavior, 13(5)*, 449–464.

Jamal, M. & Badawi, J. (1995). Job stress, type 'A' behavior and employees' well-being among Muslim immigrants in North America: A study in workforce diversity. *International Journal of Commerce & Management, 5(4)*, 6–23.

Jarecky, R. K. & Sandifer, M. G. (1986). Faculty members' evaluations of sabbaticals. *Journal of Medical Education, 61(10)*, 803–807.

Kabanoff, B. & O'Brien, G. E. (1986). Stress and the leisure needs and activities of different occupations. *Human Relations, 39(10)*, 903–916.

Kelim Shluvim Research Ltd. (2010). Vacation days submitted to Coca Cola Israel. Report by: Kelim Shluvim Research Ltd [Hebrew].

Kelly, I. & Dobbin, F. (1999). Civil rights law at work: Sex discrimination and the rise of maternity leave policies. *American Journal of Sociology, 105(2)*, 455–492.

Kelly, J. R. (1982). *Leisure*. Englewood Cliffs, New Jersey: Prentice Hall.

Kelly, J. R. (1983). *Leisure identities and interactions*. London: George Allen & Unwin.

Kirkcaldy, B. D. & Cooper, C. L. (1993). The relationship between work stress and leisure style: British and German managers. *Human Relations, 46(5)*, 669–680.

Koniak, M. (2003). Overseas business trips: The impact of business trip purpose, frequency of business trip and warning time before business trip on stress and health. M.Sc., Tel Aviv University, Israel. [Hebrew].

Kremer-Fuks, S. (1999). The impact of vacation and slack work periods on job stress and burnout among newcomers and veteran Israelis. M.Sc., Tel Aviv University, Israel. [Hebrew].

Kunda, G. (1993). *Engineering culture*. Philadelphia, PA: Temple University Press.

Lapidot, Y. (1992). The effects of a military-reserve service on work stress and burnout. M. Sc., Tel Aviv University, Israel. [Hebrew].

Laski, S. (2002). The effect of summer break activities and personality variables on stress and burnout among high school teachers in Israel. M.Sc., Tel Aviv University, Israel. [Hebrew].

Laski, S. (2013). Recovery during vacation and its fadeout – A longitudinal study. Ph.D., Tel Aviv University, Israel.

Laski, S., Etzion, D., & Westman, M. (2013). Psychological and physiological recovery from work stressors during vacation and its fadeout: A longitudinal study. Paper presented at the 121st Annual Convention for the American Psychological Association (APA), Honolulu, Hawaii.

Levi, M. (2002). The impact of summer break and personal variables on job stress and burnout among teachers in a Kibbutz regional school. M.Sc., Tel Aviv University, Israel. [Hebrew].

Lounsbury, J. W. & Hoopes, L. L. (1986). A vacation from work: Changes in work and non-work outcomes. *Journal of Applied Psychology, 71(3)*, 392–401.

Magen-Neuhaus, T. (1993). The effect of vacation and feelings of stress at work on emotional burnout. M.Sc., Tel Aviv University, Israel. [Hebrew].

Mannfredo, M. J., Driver, B. L., & Tarrant, M. A. (1996). Measuring leisure motivation: A meta-analysis of the recreation experience preferences scales. *Journal of Leisure Research, 28(3)*, 188–213.

McCann, L. & Holmes, D. (1984). Influence of aerobic exercise on depression. *Journal of Personality and Social Psychology, 46(5)*, 1142–1147.

Mitchell, T. R., Thompson, L., Peterson, E., & Cronk, R. (1997). Temporal adjustments in the evaluation of events: The "rosy view". *Journal of Experimental Social Psychology, 33 (4)*, 421–448.

Moen, P., Dempster-McClain, D., & Williams, R. M. J. (1989). Social integration and longevity: An event history analysis of women's roles and resilience. *American Sociological Review, 54(4)*, 635–647.

Monk, T. H. (1988). Coping with stress of shift work. *Work & Stress, 2(2)*, 169–172.

Moshkowitz, S. (2003). The relationship between work-overload, leisure activities and burnout among high-tech employees. M.Sc., Tel Aviv University, Israel. [Hebrew].

Naegel, I. J. & Sonnentag, S. (2011). The relationship between leisure time sports activity and personal resources. Paper presented at the Annual APA Convention (August 4–7), Washington, DC.

Nawijn, J., de Bloom, J., & Geurts, S. (2013). Pre-vacation time: Blessing or burden? *Leisure Sciences, 35(1)*, 33–44.

Nawijn, J., Marchand, M., Veenhoven, R., & Vingerhoets, A. (2010). Vacationers happier, but most not happier after a holiday. *Applied Research in Quality of Life, 5(1)*, 35–47.

Neulinger, J. (1974). *The psychology of leisure: Research approaches to the study of leisure.* Springfield, IL: Charles C Thomas.

Overman, S. (1993). HR magazine. *Not the usual 9-to-5, 38(1)*, 47.

Parker, S. R. (1971). *The future of work and leisure.* New York: Praeger Publishers.

Pines, A. (1983). On burnout and the buffering effects of social support. In B. A. Farber (Ed.), *Stress and burnout in the human service professions* (pp. 155–173). Elmsford, NY: Pergamon Press.

Potok, Y. & Littman-Ovadia, H. (2014). Does personality regulate the work stressor-psychological detachment relationship? *Journal of Career Assessment, 22(1)*, 43–58.

Radian, R. (1998). The moderating effect of workaholism and Type A on the impact of vacation on job stress and burnout. M.Sc., Tel Aviv University, Israel. [Hebrew].

Ragneh, M. G. & McKinney, J. (1993). Campus recreation and perceived academic stress. *Journal of College Students Development, 34(1)*, 5–10.

Raviv, E. (2002). Effects of leisure activity on occupational stress and burnout among high-tech employees. M.Sc., Tel Aviv University, Israel. [Hebrew].

Reich, J. W. & Zautra, A. (1981). Life events and personal causation: Some relationships with satisfaction and distress. *Journal of Personality and Social Psychology, 41(5)*, 1002–1012.

Richter, J. (1992). Balancing work and family in Israel. In S. Zedeck (Ed.), *Work, families and organizations* (pp. 362–394). San Francisco, CA: Jossey-Bass.

Ron, A. (2003). Effects of organizational vacation on occupational stress and burnout among high-tech employees. M.Sc., Tel Aviv University, Israel. [Hebrew].

Ronen, S. (1984). *Alternative work schedules.* Homewood, IL: Dow Jones Irwin.

Rook, K. S. (1987). Social support versus companionship: Effects on life stress, loneliness, and evaluation by others. *Journal of Personality and Social Psychology, 52(6)*, 1132–1147.

Rook, K. S. (1990). Social relationships as a source of companionship: Implications for older adults' psychological well-being. In B. R. Sarason, I. G. Sarason, & G. R. Pierce (Eds.), *Social support: An interactional view*. New York: Wiley.

Rubinstein, C. (1980). Vacations, expectations, satisfactions, frustrations, fantasies. *Psychology Today, 14(5)*, 62–66, 61–76.

Ryan, R. M. & Deci, E. L. (2000). Self-determination theory and the facilitation of intrinsic motivation, social development, and well-being. *American Psychologist, 55 (1)*, 68–78.

Sapir, O. (1995). The effect of vacation on stress and burnout at work. M.Sc., Tel Aviv University, Israel. [Hebrew].

Sharkey, J. (2000, May 10). The World Bank gauges the toll travel takes on employees and looks for ways to soften the effect, *New York Times*, p. C00008.

Shaviv, T. (1998). The impact of an organized tour vacation on work stress and burnout among managerial and non-managerial vacationers. M.Sc., Tel Aviv University, Israel. [Hebrew].

Silverberg, S. (2007). Home/work conflict and burnout among working mothers after maternity leave – Situational and personal characteristics as moderators. M.Sc., Tel Aviv University, Israel. [Hebrew].

Skipper, J. K., Junk, F. D., & Coffey, L. (1990). Nurses and shift-work: Effects on physical health and mental depression. *Journal of Advanced Nursing, 15(7)*, 835–842.

Sluiter, J. K., Frings-Dresen, M. H. W., van der Beek, A. J., & Meijman, T. F. (2001). The relation between work-induced neuroendocrine reactivity and recovery, subjective need for recovery and health status. *Journal of Psychosomatic Research, 50(1)*, 29–37.

Sonnentag, S. (2001). Work, recovery activities and individual well-being: A diary study. *Journal of Occupational Health Psychology, 6(3)*, 196–210.

Sonnentag, S. (2004). Flight attendants' daily recovery from work: Is there no place like home? *International Journal of Stress Management, 11(4)*, 366–391.

Sonnentag, S. & Bayer, U. V. (2005). Switching off mentally: Predictors and consequences of psychological detachment from work during off-job time. *Journal of Occupational Health Psychology, 10(4)*, 393–414.

Sonnentag, S. & Fritz, C. (2007). The recovery experience questionnaire: Development and validation of a measure for assessing recuperation and unwinding from work. *Journal of Occupational Health Psychology, 12(3)*, 204–221.

Sonnentag, S. & Kruel, U. (2006). Psychological detachment from work during off-job time: The role of job stressors, job involvement, and recovery-related self-efficacy. *European Journal of Work and Organizational Psychology, 15(2)*, 197–217.

Sonnentag, S. & Niessen, C. (2008). Staying vigorous until work is over: The role of trait vigour, day-specific work experiences and recovery. *Journal of Occupational and Organizational Psychology, 81(3)*, 435–458.

Sorcinelli, M. D. (1986). Sabbaticals and leaves: Critical events in the career of faculty. Paper presented at the Annual Meeting of the American Educational Research Association, San Francisco, CA.

Sraya-Wallin, O. (1999). The effect of vacation characteristics and the fit between job and career-anchors on stress and burnout among employees in a hi-tech environment. M. Sc., Tel Aviv University, Israel. [Hebrew].

Staines, G. L. & O'Connor, P. (1980). Conflict among work, leisure and family roles. *Monthly Labor Review, 103(8)*, 35–39.

Staines, G. L. & Pleck, J. H. (1984). Nonstandard work schedules and family life. *Journal of Applied Psychology, 69(3)*, 515–523.

Stanton-Rich, H. M. & Iso-Ahola, S. E. (1998). Burnout and leisure. *Journal of Applied Social Psychology, 28(21)*, 1931–1950.

Strauss-Blasche, G., Ekmekcioglu, C., & Marktl, W. (2000). Does vacation enable recuperation? Changes in well-being associated with time away from work. *Occupational Medicine, 50*, 167–172.

Strauss-Blasche, G., Ekmekcioglu, C., & Marktl, W. (2002). Moderating effects of vacation on reactions to work and domestic stress. *Leisure Sciences, 24(2)*, 237–249.

Strauss-Blasche, G., Ekmekcioglu, C., & Marktl, W. (2003). Serum lipids responses to a respite from occupational and domestic demands in participants with varying levels of stress. *Journal of Psychosomatic Research, 55(6)*, 521–524.

Strauss-Blasche, G., Ekmekcioglu, C., Marktl, W., Riedmann, B., Riedmann, G., Humpeler, E., et al. (2004). Vacation at moderate and low altitude improves perceived health in individuals with metabolic syndrome. *Journal of Travel Medicine, 11(5)*, 300–307.

Strauss-Blasche, G., Muhry, F., Lehofer, M., Moser, M., & Marktl, W. (2004). Time course of wellbeing after a three-week resort-based respite from occupational and domestic demands: Carry-over, contrast and situation effects. *Journal of Leisure Research, 36*, 293–309.

Strauss-Blasche, G., Reithofer, B., Schobersberger, W., Ekmekcioglu, C., & Marktl, W. F. (2005). Effect of vacation on health: Moderating factors of vacation outcome. *Journal of Travel Medicine, 12(2)*, 94–101.

Striker, J., Dimberg, L., & Liese, B. H. (2000). Stress and business travel: Individual, managerial, and corporate concerns. *National Productivity Review, 20(1)*, 3–9.

Striker, J., Luippold, R. S., Nagy, L., Liese, B., Bigelow, C., & Mundt, K. A. (1999). Risk factors for psychological stress among international business travellers. *Occupational and Environmental Medicine, 56(4)*, 245–252.

Tinsley, H. E. A. & Tinsley, D. J. (1986). A theory of the attributes, benefits and causes of leisure experience. *Leisure Sciences, 8(1)*, 1–45.

Toomey, E. L. & Connor, J. M. (1988). Employee sabbaticals: Who benefits and why. *Personnel, 65(4)*, 81–84.

Tulman, L. & Fawcett, J. (1988). Recovery from childbirth: Looking back 6 months after delivery. *Health Care for Women International, 12(3)*, 341–350.

Utechet, K. M. & Aldag, R. J. (1989). Vacation discrepancy: Correlates of individual differences and outcomes. *Psychological Reports, 65(3)*, 867–882.

Vincour-Hai, V. (2002). Stress and burnout before and after vacation and their crossover between spouses. M.Sc., Tel Aviv University, Israel. [Hebrew].

Wahlers, R. G. & Etzel, M. J. (1985). Vacation preference as a manifestation of optimal stimulation and lifestyle experience. *Journal of Leisure Research, 17(4)*, 283–295.

Webb, D., Kohlbacher, F., & Prieler, M. (2017). Life satisfaction in Japan: The role of hedonic values, advertising and watching TV. *Social Business, 7(2)*, 155–175.

Westman, M. (1999). Applying Hobfoll's conservation of resources (COR) theory to respite research. Paper presented at The Academy of Management Symposium, Chicago, IL.

Westman, M. (2004). Strategies for coping with business trips: A qualitative exploratory study. *International Journal of Stress Management, 11(2)*, 167–176.

Westman, M. & Eden, D. (1997). Effects of a respite from work on burnout: Vacation relief and fade-out. *Journal of Applied Psychology, 82(4)*, 516–527.

Westman, M., Eden, D., & Davidson, O. B. (2002). Sabbatical leave as a "time out" from chronic academic job stress: A test of Conservation of Resources theory (Working Paper No. 18/02). Tel Aviv University, Henry Crown Institute of Business Research, Faculty of Management, Tel Aviv, Israel.

Westman, M. & Etzion, D. (2001). The impact of vacation and job stress on burnout and absenteeism. (Special issue on burnout). *Psychology & Health, 16(5)*, 595–606.
Westman, M. & Etzion, D. (2002). The impact of short overseas business trips on job stress and burnout. *Applied Psychology: An International Review, 51(4)*, 582–592.
Westman, M., Etzion, D., & Chen, S. (2009a). Are business trips a unique kind of respite? In P. Perrewé, D. Ganster, & S. Sonnentag (Eds.), *Research in occupational stress and well-being* (pp. 167–204). Oxford: JAI Press/Elsevier.
Westman, M., Etzion, D., & Chen, S. (2009b). Are business trips a unique kind of respite? In S. Sonnentag, P. L. Perrewé, & D. C. Ganster (Eds.), *Current perspectives on job-stress recovery* (pp. 167–204). Oxford: JAI Press/Emerald Group Publishing.
Westman, M., Etzion, D., & Gattenio, E. (2008). International business travels and the work-family interface – A longitudinal study. *Journal of Occupational and Organizational Psychology, 81(3)*, 459–480.
Westman, M., Etzion, D., & Yoffe, Y. (1998). The effect of overseas business trips on job stress and burnout (Working Paper). Tel Aviv University, Henry Crown Institute of Business Research, Faculty of Management, Tel Aviv, Israel.
Wilensky, H. (1960). Work careers, and social integration. *International Social Sciences Journal, 12(4)*, 543–560.
Yoffe, Y. (1997). The effects of overseas business trips on job stress and burnout. M.Sc., Tel Aviv University, Israel. [Hebrew].
Zeevi-Ohana, A. (2000). The impact of summer break activities and personality variables on job stress and burnout among high school teachers. M.A., The Hebrew University, Jerusalem, Israel.
Zigler, E. & Muenchow, S. (1983). Infant day care and infant-care leaves: A policy vacuum. *American Psychologist, 38(1)*, 91–94.
Zvi, E. (2004). Stress and burnout among students before and after the Passover break – Vacation activities and personality variables as moderators. M.Sc., Tel Aviv University, Israel.

3

PERSONAL CHARACTERISTICS

Respite features set aside, the vacation experience is subjective (Rubinstein, 1980), dependent on the interaction between environmental factors and personal characteristics. The latter may determine the manner by which individuals interpret demands, opportunities, and meaning of the respite, as well as the manner by which work stress and stressors are handled. In a cluster analysis of longitudinal data collected among Finnish teachers over a school autumn term, Kinnunen (1988) found four different stress-type groups that differed in terms of personality characteristics and coping strategies. That is, personality characteristics have the power to affect the experience of relief through respite, its timing and duration, and its degree. The study of Hoxter and Lester (1988) suggests that personal characteristics determine travel preferences so that, as noted in Chapter 2, alignment between personality and respite activities or location may serve as important factors in determining a vacation's potential to replenish individuals' resources.

Personal characteristics range from demographic background, to different classifications of personality, to work-related attitudes, to situational factors including employment and family relations. In this chapter are discussed the relationships between the respite effect and different personal characteristics.

Demographics

The quality of the leisure experience is influenced by both environmental and personal factors, such as geographical, physical, and financial accessibility, educational legacy, and general cultural attitude (Csikszentmihalyi, 2000). Different demographic variables have also been studied for their connection with burnout. Whereas some studies report no significant effects for any measured demographic variables (e.g., Burke & Greenglass, 1995a; Greenglass, Fiksenbaum, & Burke,

1994), other found significant relationships between demographic characteristics and burnout. For example, age was found to define the extent of burnout experienced by teachers – the age of 30 marked a significant increase in burnout among teachers (Fiske, 1982; Friedman, 1991). This, however, may be a derivative of tenure, which was found to be positively related to burnout. Similarly, in the case of teachers, education level (which might correlate with age and tenure) was positively related to burnout so that teachers with university degrees experienced greater burnout compared to teachers with only college degrees (Friedman, 1991).

Age and place of residence

Based on the model of person-environment fit, one's culture of origin may bear significance to variance in perceptions of stimuli and response patterns. Comparing American and Israeli employees holding similar organizational positions, Etzion (1988b) found differences in levels of burnout, which she attributed to the cultural origins of her study participants. Similarly, Dubin et al. (1976) attributed to cultural origins differences found between American and British employees with respect to their work involvement and the importance they attributed to different job components. Some studies (Krau, 1984; Toren, 1984) found various differences with respect to work attitudes, perceptions, and behaviors within the Israeli society between native Israeli employees and those who immigrated from ex-Soviet countries. The latter have been found to experience greater levels of stress and burnout compared with native Israelis at all times (Kremer-Fuks, 1999). Past studies are few exploring teachers' burnout as a function of their location of residence. Whereas Schwab and Lwanicki (1982) found no pattern of influence between location of residence and burnout among teachers, Farber (1983) found burnout among 10% of rural teachers compared to 22% of urban teachers.

Few of our studies found clear relationships between stress, burnout, or respite and the characteristics of age, place of residence, or socioeconomic background. In a study of occupational workloads, leisure activities, and burnout conducted among 103 (59 men) hi-tech programmers, team leaders, and project managers, a significant negative relationship was found between age and the exhaustion facet of burnout. Contrary to the hypothesis, the younger the employee, the greater their exhaustion. A similar relationship was found between age and stress originating in the lack of positive work characteristics (the younger the employee, the greater the reported level of stress). Age was also negatively related to leisure expectations, active leisure, and family leisure, so that greater leisure expectations and more active and family leisure were typical of older employees (Moshkowitz, 2003). In a study on the effect of organized vacation on work-related stress and burnout conducted among 72 engineers (43% men) of different companies, age was found to moderate the effect of the vacation over stress levels. The older the employees, the greater the decrease in post-vacation work-related stress (Shaviv, 1998).

Place of residence may be important in understanding stress and burnout, especially with respect to one's place of work. Studying the effect of the two-month summer break on job stress and burnout of 44 Israeli teachers (23% males) working in a kibbutz school, Levi (1972) found a moderating effect for type of residence on burnout. Her results show that kibbutz residents suffered the highest levels of burnout, followed by kibbutz members who experienced intermediate levels of burnout, whereas teachers living outside of the kibbutz were characterized by the lowest levels of burnout. Teachers living within the kibbutz experience no detachment from their pupils and their parent after work. In contrast, they enjoy greater social support compared to urban peers. These factors may have some influence on Levi's findings.

Etzion (1988a) also found that age was positively related to perceptions of job success and satisfaction thereby reducing burnout. The work–non-work conflict is additionally attenuated over the years.

Gender

Gender is also often found to be a marker of stress and burnout tendencies (Etzion, 1984; Maslach & Jackson, 1981; Pines & Aronson, 1988; Westman & Eden, 1997). Stress and burnout literature typically reports women with higher levels of either compared to men (e.g., Barnett & Baruch, 1987; Baum & Grunberg, 1991; Etzion, 1984; 1987; Etzion & Pines, 1986; Johnson & Hall, 1996; Jones, 1993; Pines et al., 1981; Pines & Kafry, 1981). As such, Etzion and Pines (1986) found greater burnout among women than among men in both the US and in Israel, a difference that preserved through different demographic variables. Pines and Aronson (1988) attributed the greater degree of burnout among women to gender-based role stereotypes in that women have a greater tendency to work in people-oriented occupations where the emotional relationship with other individuals may be lengthy and demanding thereby more prone to instill burnout. Schaufeli and Enzmann (1998) explained that women additionally react with greater emotional sensitivity and tend more to expose emotional difficulties and health problems compared with men. Their study showed greater exhaustion in women compared to greater de-personalization scores among men. In alignment with this, Etzion (1984) noted that women exhibit greater propensity toward burnout compared with men in any vocation. She found no gender-based difference in work stress although women experience greater stress than men in the non-work lives. Cherniss (1995) noted that women are more worried and preoccupied with home and family obligations, whereas men are more concerned about work and career issues. These concerns seem to subside later in life, allowing women to renew their pursuit of their careers (Bailyn, 1973; Blitz, 1988; Pines & Kafry, 1981).

The greater levels of stress women, compared with men, experience in their lives combined with greater commitment to the home and family realms suggests greater work–non-work conflict (Galin, 2002; Pines & Aronson, 1988). Consequently, women's success in work is associated with a reduced degree of success in non-work sphere of their life and vice versa (Etzion, 1988a). Indeed, correlations were found by Etzion and Bailyn (1994) between work–non-work conflict and burnout among women in different continents; by Grinker (1991) among women who had pursued technology-related careers; and by Zeevi-Ohana (2000) among high school teacher in the US; as well as by other researchers (e.g., Cooke & Rousseau, 1984; Etzion, 1987, 1988a; Freudenberger & North, 1985). Men were found to experience such burnout and stress more frequently than women typically in service and care professions, and particularly teachers.

In alignment with the above, Moshkowitz (2003) found greater overall burnout, as well as exhaustion among women compared with men. Similarly, in a study of 92 Israeli hi-tech employees, women were found to experience from burnout more than men before a vacation but no differences were found in post-vacation burnout levels. Women were also found to have greater levels of job stressors before, as well as after their vacation. This may be surprising, given that 79% of the men in the sample compared to 57% of the woman experienced career anchor misfit (Sraya-Wallin, 1999). Men were often found to react less to stress (Clark & Reiker, 1986; Mallinckrodt, Leong, & Kraji, 1989), perhaps as a result of socialization process teaching men that emotional expression is unmanly, reflecting weakness among men (Davidson-Katz, 1991). This result, however, may be vocation-dependent, as in the teaching profession male teachers have been found to burnout more (Burke, Greenglass, & Schwarzer, 1996; De Heus & Diekstra, 1999; Farber, 1991; Friedman, 1991; Van Horn, Schaufeli, Greenglass, & Burke, 1997). Compared to their female colleagues, male teachers have been found to experience greater burnout and stress from their work, they were less committed to their vocation, and less satisfied from it (Burke & Greenglass, 1995b; Schwab & Iwanicki, 1982). One explanation offered for this irregularity was the stress emerging from perceptual gender misalignment, when men work in a profession generally regarded as "female" (Etzion, 1984). Additionally, women find the vocational conditions to be in alignment with their needs as mothers and secondary breadwinners, whereas male teachers may feel pressured to seek additional income or change vocation due to the pressures of being their family's primary breadwinner (Farber, 1991). The differences found between male and female teachers may also be attributed to the relative ease by which women are able to procure social support outside of work (Etzion, 1984), on the one hand, and the tendency to assign male teachers to more problematic classes, rendering their job harder than that of female teachers accompanied with less social support (Farber, 1991). Among our studies, Kremer-Fuks (1999) also found greater burnout among women than men.

In that same study, men attributed greater importance to non-work life following their vacation, whereas women attributed a degradation in importance assigned to non-work life throughout the study. Vacation satisfaction was found to be greater among men, who also benefited more from the vacation compared to women (Kremer-Fuks, 1999). Similar results were found by Shaviv (1998) who documented greater work-related stress among men, as well as greater ability to detach from work during their vacation. Consequently, women experienced greater stress levels immediately after the vacation, whereas men experienced lower stress levels upon their return to work. These results, however, are controversial, as Westman and Eden (1997) found greater relief in post-vacation burnout among women rather than men, assuming that work–non-work conflict is diminished during vacation time thereby reducing burnout. Rubinstein (1980) had also found that men enjoy their vacation less than women, and that they had greater desire to return to work after their vacation. Zvi (2004) also found higher levels of stress and burnout among women just prior to a vacation in her student sample while post vacation stress was higher among women. In contrast, in this study, vacation worked to reduce burnout among women to a greater extent compared with men. The beneficial influence of vacation on both men and women may also be surmised from the study of Avni (2007), in which participated 850 full-time employees from a communication company. Men significantly reduced overtime upon their return from a vacation, whereas women reduced their extent of absences from work following their vacation. Possibly, women's distinctive experiences and time concepts within their social, cultural and economic contexts, and particularly with respect to holidays, serves to explain these gender differences (Deem, 1996). It is also possible that the fact that women often ascribe success in life to a number of factors, whereas men attribute it to success in both work and non-work spheres insinuates that compared to women, men experience fewer success components during their vacation in which they are less exposed to work components (Etzion, 1988a). In other words, we might expect vacation to positively influence women more than men, at least in terms of stress and burnout. Findings by Laski (2002) who studied 86 teachers from 13 high schools located in central Israel did not support this notion.

Gender often shapes roles perceptions with respect to the work–non-work balance. Women appear to bear a double-edged role conflict in this respect as focusing upon a career reduces their femininity in the eyes of men while choosing their family life over a career renders them incompetent and invalidated in the eyes of men. That is, the typical female conflict is stereotyped by stipulating that women may be skilled or successful but not both. Translating these perceptions to employees, we found that men and women exhibit different patterns of exploitation of vacation days as a function of the age of their children. Women use more vacation days when children are under the age of six, as well as during their adolescence (ages 13–18), whereas men use more vacation days when children are young (ages six–13) or adults (over 19 years old). Interviews with these employees suggest that women used their vacation days for different errands rather than vacationing (Braun (Cohen), 2007). One explanation for these findings is that the younger the children, the greater the family burden experienced and the greater the time demands on the parents (Greenhous & Kopelman,

1981). Life–family conflict was positively related to burnout among both Israeli and American woman, especially among women with children under six years old (Etzion & Bailyn, 1994) and across different vocations (e.g., hi-tech employees in Zeevi-Ohana, 2000; and school teachers in Grinker, 1991).

In the study of Moshkowitz (2003), men and women experienced similar rates of leisure time during their work, yet women tended to enjoy more leisure activities during work. This may be in compensation to the large difference found between the two groups in time dedicated to the leisure activities outside of their work experience, perhaps due to the dual role of women in their non-work spheres. These findings are striking especially given that men in this study work an average of two hours more daily compared with the women. Perhaps related to these data is women's report of more detrimental health conditions.

For the majority of female teachers (who comprise approximately 75% of teachers in Israel), work–non-work conflict was found to be a source of stress (Oliver-Lomerman, 1983). Stress at work leads to stress in non-work spheres, especially the family sphere (Paludi, 1986), whereas vocations characterized by stress tend to increase stress crossover within the workplace, as well as between the work and family spheres (Westman & Etzion, 2001). Although working women may be susceptible to a greater degree of stress, one study found that they were able to express a greater sense of self-worth and life satisfaction; they were able to better cope with their stresses, reported less work–non-work conflict rates and greater job satisfaction (Paludi, 1986).

Gender was positively related to two characteristics of long-term respites among 1495 (80% men) American university faculty members. The data showed more men recipients than female recipients (approximately 40% compared to approximately 30% respectively in both a one-year sabbatical at half salary and a one-quarter sabbatical at full pay), as well as gender-based differences regarding reasons for not taking such a respite (Hendel & Solberg, 1983). Notwithstanding, evidence has been accumulating to show that men and women experience leisure differently (Colley, 1984; Henderson, 1994).

Marital status, parenting, and age of children

Some studies have found significant effects of marital status so that non-married individuals were found to burn out more (Li, Wong, Law, Chiu, & Yuen, 1994). Similarly, Yoffe (1997) found greater burnout among non-married individuals before, during, and after their business trip a broad. Non-married individuals were also characterized by greater levels of stress in general, as well as the result of the absence of positive work characteristics. In contrast, Moshkowitz (2003) found greater levels of stress resulting from negative work characteristics among married individuals. Women appear to be more susceptible than men to stress crossover from other family members (Westman, 2001). Parenting contributes to family stress (Greenhous & Beutell, 1985).

Marital status, parenting, and age of children influence the vacation experience and one's perception of leisure and relaxation during the vacation. For example, young, single individuals with lower education levels and non-managerial jobs tended more to engage in active leisure (Moshkowitz, 2003). The addition of children to the family limits time, space, and financial resources that would have otherwise been available for leisure. Personal leisure is often converted to whole-family leisure, especially when children are young (Kelly, 1982). Additionally, vacation time in itself may operate as a source of stress both originating and further contributing to family conflicts and economic hardship (Etzion & Boklis, 2006; Holmes & Rahe, 1967). When children grow older, parenting allows for greater flexibility to parents who are no longer required to provide for all of their child's needs so that leisure may be reintroduced to the life of the parent. Flexibility increases yet again when children grow into adolescence and later into young adulthood. While parents may still divert some of their personal resources toward their children's needs, their autonomy concerning their leisure time increases so that they may address their own intrinsic needs rather than extrinsic ones (Kelly, 1982).

The age of children is negatively associated with the degree of family stress, so that the younger the children, the greater the stress experienced by parents (Beutell & Greenhouse, 1985; Greenhous & Kopelman, 1981). Working mothers with young children experience lack of time, stress, and fatigue (Paludi, 1986) and are at greater risk of experiencing psychological distress (Brown, 1975), whereas working mothers who experience parenting challenges are at greater risk of depression (Vanfassen, 1981). The younger the children and the greater their number, the greater the parents' burnout (Grinker, 1991). Our own study supported the association between the number of children and the extent of exhaustion and cynicism as facets of burnout and an inverse association between the age of the oldest child and the extent of burnout in general, as well as its facets of exhaustion and cynicism (Silverberg, 2007). This may be related to the growing demands that the passage of time incurs upon our societies as an earlier study found no such effect (Kariacou & Sutcliffe, 1978).

We accordingly found that men tended to accumulate unmaterialized vacation time when children are very young (0–6 years old), or rather mature (12–18 years old), whereas women tended to accumulate unmaterialized vacation time when children are in between these age groups or have matured to young adults (19 years old and older) (Braun (Cohen), 2007). These are important findings as employees who are able to balance their work and non-work commitments while availing time for leisure maintain lower levels of burnout (Cherniss, 1995). In fact, we found that teachers who were parents experienced an increase in burnout following the summer break compared with teachers who were not parents; that is to say that the break had a greater positive impact on childless individuals (Zeevi-Ohana, 2000). In a different study (Laski, 2002), vacation negatively affected stress levels originating in the absence of positive work characteristics among parents of pre-school children (under the age of six) while reducing burnout upon return from the vacation (although burnout levels rose again during the subsequent period of routine work). Parents of school children experienced a different effect: stress was reduced on return from the vacation (but rose again during routine work) and burnout remained unchanged throughout the study. Vacation, then, for parents of young children enhances the burden of their "second job" denying parents the opportunity to properly replenish their resources and stop the cycle of resource loss (Farber, 1991).

Parents report more pronounced work–non-work conflict compared with individuals who are not parents. Stress and burnout originating at work may spill over to family life, leading employees to seek support at home which is not available to them at their workplace. Burned out employees whose resources have diminished over time are inattentive to other family members, they assume a passive role in their family life and are unavailable emotionally (Repetti & Wood, 1997). Social support extended by a family member, and particularly a spouse, is thus particularly important in coping with stress and burnout (Brown, 1975). The greater the support, the more reduced the pathological symptoms of distress (Aneshensel, 1986).

Education

Education tends to moderate the effect of vacation over burnout so that individuals with greater education levels benefit more from their vacation in terms of ameliorating burnout (Shaviv, 1998). Education is positively related to the work–non-work conflict, as well as exhaustion as a facet of burnout, both of which grow with greater levels of education. More educated individuals also find it more difficult to detach from work. Adversely, proactive coping by strategic planning increases as does the level of education (Silverberg, 2007).

Personality types

Personality helps explain why individuals often react quite differently to similar situations at work. It has been argued that because individuals differ from each other in their degree of vulnerability to stress and their coping mechanisms, no discussion of burnout would be complete without reference to personal and personality factors. For example, a broad range of research reveals personality types as a factor in the experience of stress and burnout. In fact, personality variables seem to explain variance in burnout better than demographic, environmental, or organizational factors (Byrne, 1993; Huebner & Mills, 1994; Piedmont, 1993). For example, self-esteem and locus of control were found to moderate the relationship between stress and burnout (Byrne, 1993), whereas different types of thinking and emotional patterns were related to dwindling energy resources (Garden, 1989). A more limited range of research explored relationships between personality and leisure or respite, and hence their ameliorating effect on stress and burnout. Notwithstanding, family environment, encouragement from parents and teachers, relationships with peers and socioeconomic factors in childhood predict adult attitudes to and engagement in leisure (Cassidy, 2005).

Because leisure activities are a matter of choice based on one's tendencies, desires, perceived significance, and intrinsic motivation and satisfaction, the exploration of leisure activities with respect to personality may be worthwhile. Based on the notion of person-environment fit, an optimal fit between the person and their leisure activities would yield the greatest and best results for one's well-being. Csikszentmihalyi (1990; 1997; 2000) suggested the concept of *flow* to describe complete balance between the opportunity to act and the individual's skill and characteristics, which lead to an optimal experience. The balance is disrupted when one's skill are insufficient for carrying out the activities or when they exceed available opportunities leading to a negative experience of boredom. Additionally, a greater fit between one's psychological needs and benefits gained through any given leisure activity increases the probability of a significant leisure experience (Tinsley & Tinsley, 1986).

Research exploring the relationship between leisure activities and satisfaction emphasizes the significance of leisure to one's well-being by way of addressing needs of self-expression and self-actualization (Meir & Melamed, 1986; Ragheb, 1993; Schreyer & Driver, 1989; Tinsley & Tinsley, 1986). The Optimal Arousal Model (Iso-Ahola, 1980) suggests that an optimal level of stimuli would be reached through balancing between the need for stability and the need for change and diversity in leisure activities. The preference of leisure activities is nevertheless a derivative of the balance between the optimal reality and the present reality so that individuals would compensate for both deficient and excessive stimuli at the workplace by seeking the opposite degree of stimuli in their leisure activities (Wahlers & Etzel, 1985). Indeed, a series of studies conducted by Melamed and Meir (Meir & Melamed, 1986; Meir, Melmed, & Abu-Frecha, 1990; Melamed & Meir, 1981; Melamed et al., 1995) supported the notion of leisure-person fit (Holland, 1985), showing that individuals tend to elect leisure activities that are in line with their personality and that indeed such an alignment is beneficial to one's well-being, spilling over to work satisfaction and reduced burnout.

Eden (1990) suggested that vacation would be a source of stress for individuals characterized by certain personality traits while Westman and Eden (1997) suggested that variance in individuals' levels of stress and burnout in response to acute or chronic events attests to the moderating effect of personality. Individuals' response to stress, burnout, and respites have been studied primarily with respect to three classifications, which do not represent all of human variation but are nevertheless useful representations of it: The Big Five, Types A and B, and workaholics.

The Big Five

The Big Five, originally known as the Five Factor Model, first emerged in the 1930s through numerous studies (Allport, 1937; Lewin, 1935; Murray, 1938) that independently aimed to clarify personality psychology (John, 1990b; Wiggins & Trapnell, 1997). It was further developed by Cattell (1945), Tupes and Christal (1961), Norman (1963), Digman (1979) and Goldberg (1982), culminating in the most known of all – McCrae and John (1992) and Costa and McCrae (1985). The five personality factors of the model have been validated across many different backgrounds, cultures, ages, instruments, and observers, and they were found to remain stable throughout one's life (McCrae & Costa, 1990). In a nutshell, *extraversion* relates to traits such as being sociable, gregarious, assertive, talkative, and active; *emotional stability* (also called *neuroticism*) is associated with being anxious, depressed, angry, embarrassed, emotional, worried, or insecure; *agreeableness* refers to being courteous, flexible, trusting, good-natured, cooperative, forgiving, soft-hearted, and tolerant; *conscientiousness* (or *dependability*) is associated with being careful, thorough, responsible, organized, planned, and occasionally also associated with being hardworking, achievement-oriented, and persevering; *openness to experience* (also called *culture* or *intellectence*) refers to being imaginative, cultured, curious, original, broad-minded, intelligent, and artistically sensitive (Barrick & Mount, 1991; Digman, 1990; McCrae & John, 1992; Zellars, Perrewe, & Hochwarter, 2000).

The Big Five personality factors have been associated with different work-related variables and outcomes. For example, conscientiousness has been established as a predictor of job performance, whereas extraversion, as well as openness to experience, were associated with job performance in occupations involving a high degree of social interaction, such as those of managers and sales people (e.g., Barrick & Mount, 1991; Hurtz & Donovan, 1998). Similarly, neuroticism was positively related to burnout (Piedmont, 1993) but negatively related with conscientiousness (Huebner & Mills, 1994; Piedmont, 1993; Wilson & Mutero, 1989).

Gallagher (1990) found that individuals characterized by neuroticism interpret stress as threat, whereas individuals characterized by extroversion interpret stress as a reward-bearing challenge. Innes and Kitto (1989) established a relationship between neuroticism and occupational stress, as well as health issues. Wilson and Mutero (1989), however, established relationships between stress and all personality types other than openness to experience. Piedmont (1993) and Zellars et al. (2000) both found a positive relationship between neuroticism and emotional exhaustion and a negative relationship between agreeableness and emotional exhaustion, as well as depersonalization. Piedmont (1993) additionally found a negative relationship between extraversion and emotional exhaustion. A study of Finnish teachers (Kinnunen, 1988) supports this observation, offering a classification of teachers based on their stress-related experiences. In contrast to teachers whose stress levels lowered after the weekend, teachers experiencing chronic stress and those experiencing no stress could be associated with personality types. Chronically stressed individuals tended to score highly on the scale of neuroticism and exhibited a coping style of somatic complaints and depression, whereas those experiencing no stress scored lower on the scale of neuroticism and exhibited a rational coping style. Other studies have found personality traits captured in the five-factor structure to be related to burnout (e.g., John, 1990a). Two meta-analyses (Barrick & Mount, 1991; Hurtz & Donovan, 1998) revealed burnout to be negatively correlated to conscientiousness, extraversion, low neuroticism, and agreeableness (see also Huebner & Mills, 1994; Piedmont, 1993). Explanations have been offered for these relationships between neuroticism, stress and burnout suggesting the individuals characterized by neuroticism are emotionally unstable, and as such are also highly sensitive to environmental stimuli, and therefore are more susceptible than other personality types to experience psychological distress (Innes & Kitto, 1989; Maslach et al., 2001).

In our own studies, Zeevi-Ohana (2000) found negative relationships between agreeableness, low neuroticism, and conscientiousness and all facets of stress and burnout. Extraversion and low neuroticism significantly predicted burnout prior to the vacation and extraversion was also negatively related to lack of self-efficacy prior to the vacation. Openness, in contrast, was positively related to exhaustion. Zvi (2004) also found a negative relationship between conscientiousness and burnout, as well as its facet of exhaustion.

Investigating the respite effect of the two-month long summer break of 86 high school teachers, Laski (2002) found moderating effects were for vacation activities, demographic variables, and Big Five personality traits. Neuroticism was positively related to stress originating both from the presence of negative work characteristics, as well as the absence of positive work characteristics; it was also positively related to burnout. The higher an individual scored on the neuroticism scale, the more stress and burnout they experienced at their workplace. Additionally, neuroticism tended to moderate the effect of vacation over the exhaustion and cynicism facets of burnout. The presence of negative work characteristics was more acutely felt prior to the vacation, whereas the absence of positive work characteristics was more acutely felt upon individuals' return to their workplace, as well as later on, during their routine period. Neuroticism was negatively related to activities during the summer break that were related to school and the preparation of the next school year so that the higher the individual scored on this scale, the less time was diverted to work during their break.

Agreeableness was negatively related to the absence of positive work characteristics, as well as to burnout, so that individuals scoring higher on the scale experienced less stress from absence of positive work characteristics, as well as less burnout. During their break, individuals scoring high on this scale diverted less time to family tasks and more time to entertainment (Laski, 2002).

Conscientiousness was negatively related to burnout after the vacation and during the routine period while individuals scoring high on this scale diverted more time to work during their break (Laski, 2002). Zvi (2004) found conscientiousness to moderate the influence of a vacation over exhaustion but not over general burnout or stress. In this study, individuals differed in additional aspects when characterized by low compared to high conscientiousness such as the degree to which involvement in one's main occupation during the vacation was related positively to stress and exhaustion and negatively to vacation satisfaction. For individuals characterized by low conscientiousness, active leisure served to ameliorate post-vacation stress, whereas post-vacation stress increased among highly conscientious individuals who partook inactive leisure. Conscientiousness also moderated the degree to which excursions in nature ameliorated post-vacation stress, as well as the contribution of family time during the vacation to individuals' vacation satisfaction.

Openness to experience was negatively related to burnout only the routine period. Extraversion was related negatively with the absence of positive work characteristics only prior to the vacation; it was also negatively related to burnout, but only during routine periods (that is, not prior to or immediately after the vacation). All personality factors except neuroticism were negatively related to the exhaustion or cynicism facets of burnout and positively related to self-efficacy (Laski, 2002).

Etzion and Sapir (1996), as well as Sraya-Wallin (1999) suggest that personality types that react differently to stress and burnout are the factors influencing activities chosen during leisure or ordinary times. Flaxman, Ménard, Bond, and Kinman (2012) found a divergent pattern of respite to post-respite effects among academics, who were classified as being high versus low in a self-critical form of perfectionism (doubts about actions). Deterioration in well-being experienced by perfectionist academics when first returning to work was mediated by their tendency for perseverative cognition (i.e., worry and rumination) about work during the respite itself.

With respect to respite, personality determines baseline levels of fear and sadness thereby moderating change in disgust across individuals' vacations. It may thus be suggested that it is possible to influence the effect of vacation over vacationing individuals should the role of personality and its association with such emotional transformation in tourists' experience be acknowledged by vacation-related agents, as well as by researchers (Lin, Kerstetter, Nawijn, & Mitas, 2014). In their study of the personality and tourism, Hoxter and Lester (1988) found differences in travel preferences correlating with different personality profiles, so that *psychocentrics* preferred travel in distance and culture closer to home, whereas *allocentrics* chose more exotic destinations, sought contact with indigenous peoples, and expressed interest in unusual experiences. In a Taiwanese study of personality, leisure motivation, and leisure satisfaction, Chen (2006) found significant relationships between the five-factor model of personality, leisure motivation, and leisure satisfaction. The author also found significant differences between dimensions of the five-factor model of personality and select demographics, and accordingly, components of leisure motivation and leisure satisfaction were also related to select demographics. Similarly, Wilkinson and Hansen (2006) identified personality rather than mood or affect variables as significant predictors of leisure.

In a different study of personality and vacation effects conducted at the entrance and checkout points of an all-inclusive vacation village, Besser and

Priel (2006) found that lower negative mood during vacation, higher confirmation of positive affective expectations, and greater satisfaction with vacation leisure activities were all related to a positive model of other. Confirmation of positive affective expectations mediated the effect of the positive model of other on low negative mood during vacation so that vacationers who reported having a high positive model of others felt that they had fulfilled their positive affective expectations, which, in turn, lowered their levels of negative mood during the vacation and promoted higher levels of satisfaction with various vacation leisure activities. In a similar study, Besser and Shackelford (2007) were able to establish relationships between higher conscientiousness and lower negative mood during vacation, as well as higher confirmation of positive affective expectations, and lower perceived stress. However, perceived stress mediated the effect of conscientiousness on low negative mood during vacation and on the confirmation of positive affective expectations. The authors noted that individuals scoring high on conscientiousness may enjoy planned vacation to achieve freedom from work demands, especially if they are low in neuroticism, resulting in decreased stress, positive affect, and fulfillment of positive expectations.

Zeevi-Ohana (2000) found that low neuroticism, extraversion, conscientiousness, and agreeableness correlated negatively with most stressors and burnout measures, whereas vacation satisfaction correlated strongly and positively with extraversion, conscientiousness, and openness. Zvi (2004) found neuroticism moderated the effect of vacation over stress and exhaustion but not over other facets of burnout or burnout in general. For individuals who scored low on the scale of neuroticism, a relationship was found between work activities during the vacation and vacation satisfaction so that the more they worked during their vacation, the less satisfied they were with it. For these individuals a relationship was also established between leisure activities involving entertainment and burnout as well vacation satisfaction. The more these individuals consumed entertainment, the greater their vacation satisfaction and the lesser their burnout. Findings were slightly different for individuals who scored highly on the scale of neuroticism: the more they worked during their vacation, not only were they less satisfied but they were also more stressed after the vacation. Vacation satisfaction as a function of leisure activities for these individuals involved excursions in nature and time spent with friends.

Types A and behavior patterns

Freudenberger (1974) suggested that the primary victims of burnout are individuals who are dynamic, charismatic, and task-oriented, who hold unrealistic expectations and persist in their efforts to realize their expectations. Maslach (1982) pointed out the in order to burn-out one must have first been fired up, or enthusiastic, suggesting that passionate individuals tend to burnout. Etzion (1987) also suggested the highly ambitious individuals are at high risk of burnout. Conversely, it has also been suggested that individuals prone to burnout are sensitive, anxious, humanitarian, and people-oriented (Edelwich & Brodesky, 1980; Farber, 1983).

American cardiologists Rosenman and Friedman first coined the term *Type A behavior pattern (TABP)* to identify individuals experiencing various cardiovascular conditions (Rosenman et al., 1964). The term typically refers to individuals who are highly competitive, ambitious, achievement-oriented, work-driven, time-conscious, impatient, intolerant, aggressive, fast-talkers, and who tend to work independently rather than delegate under stress (Lee, Early, & Hanson, 1988). In contrast, individuals characterized by *Type B behavior pattern (TBBP)* are considered more serene and lax, less influenced by an internal drive for the type of intensity characterizing individuals of Type A (Friedman, 1977).

Burnout has been associated with Type A personality (Nagy & Davis, 1985) whose burnout was significantly more acute compared to individuals characterized by Type B behavior patterns (Radian, 1998). The association between subjective workload and anxiety was found greatest for individuals characterized by Type A behavior (Caplan & Jones, 1975), who in a different study also reported greater perceived workload and sense of control compared to individuals characterized by Type B behavior (Kushnir & Melamed, 1991). Stress, however, was found unrelated to Type characteristics (Radian, 1998).

Smith and Anderson (1986) considered Type A behavior not as a reaction to strenuous conditions but as an instigator of challenges and demands. They noted that individuals characterized by Type A behavior patterns choose to partake in circumstances that are objectively more strenuous. They additionally subjectively perceive those circumstances to pose greater challenge and demands than may be objectively true. Their expression of aggressiveness further triggers challenging and demanding responses from other individuals with whom they interact. Finally, they are only selectively open to feedback and underestimate their achievements, which heightens the needs to improve and meets self-imposed demands. In a forced workplace shutdown that was perceived to be stressful, individuals characterized by Type A behavior were more strongly affected, possibly because of their high job involvement (Caplan, 1983). Their high job involvement might also explain why individuals characterized by Type A behavior seem to enjoy greater perceived intrinsic job satisfaction (Day & Bedeian, 1991). Lee et al. (1988) identified that individuals characterized by Type A behavior performed better than those characterized by Type B behavior in assignments involving multiple, simultaneous tasks compared to assignments involving focus and cooperation. The authors suggested that Type A characteristics of impatience, a sense of urgency, and strenuous timelines may hinder performance where collaboration, patience, and focus are required. Accordingly, it may be surmised that conditions involving a decreased sense of control, lax workload, and absence of complex tasks, as may be experienced during a vacation, would be interpreted as strenuous for individuals characterized by Type A behavior (Kushnir & Melamed, 1991). Indeed, the tendency toward Type A behavior explained variance among employees in taking vacation days based separately on both exhaustion and cynicism (Braun (Cohen), 2007). Radian (1998) found that women were more likely than men to be characterized by Type A behavior.

Research seems to indicate that future studies may benefit from looking at different aspects of Type A behavior separately (Bluen, Barling, & Burns, 1990). For example, when dividing Type A characteristics into *achievement striving* (comprising the extent to which individuals take their work seriously, are active, and are hardworking) and *impatience irritability* (impatience, anger, hostility, and time-related obsession), performance was tied to the former while health complaints were associated with the latter (Spence, Helmreich, & Pred, 1987). Similarly, Bluen et al. (1990) found that achievement striving was positively related to sales performance and job satisfaction, whereas impatience irritability was positively related to depression and negatively related to job satisfaction. These results might explain studies yielding contradicting results in which the quality of work was both related and unrelated to employees characterized by Type A behavior (Jamal, 1985; Radian, 1998).

Workaholism

Workaholics are individuals who work excessively long hours. They have an irresistible inner drive to work compulsively and excessively, allocating a rather stable and considerable amount of time to work even in the absence of an extrinsic compulsion (Schaufeli, Bakker, van der Heijden, & Prins, 2009). They tend to feel restless and guilty when not working (Machlowits, 1980; Spence & Robbins, 1992); their dependency upon work fulfills their need of self-worth, so that they think incessantly about their work, even when not at the workplace (Schwartz, 1982). Retirement might seem as threatening as a death verdict to workaholics (Machlowits, 1980; Rubinstein, 1980). Workaholics under the age of 40 are prone to be men rather than women, suggesting perhaps that women allow themselves to immerse themselves in their work after their children have grown older (Doerfler & Kammer, 1986). Studying workaholism as discretionary investment of considerable time at work, Snir and Zohar (2008) found workaholism to be associated with continuous work engagement, a preference for work over leisure activity, and higher positive affect during work activities than during leisure activities. Not all employees working long hours are workaholics. Different scales of work enthusiasts have been identified of individuals who are motivated to invest considerably in their work efforts in order to derive satisfaction based on income, leisure, prerequisites, or other results (e.g., Douglas & Morris, 2006; Spence & Robbins, 1992). Workaholics, in contrast, derive satisfaction from the work *per se*. The authors also found that work enthusiasts enjoy high job satisfaction, whereas workaholics enjoy only low job satisfaction. Spence and Robbins (1992) further corroborated that workaholics are characterized by higher measures of perfectionism, non-delegation of responsibility, and job stress compared with work enthusiasts and other groups of employees.

Interviewing and observing workaholics, Machlowits (1980) found that excessive work is workaholics' means to overcome their perceived shortcomings and doubts concerning their aptitude. The line between work and pleasure is a blurry one for workaholics, and they are therefore – despite limiting personality traits – generally happy people as they have the opportunity to do what brings them joy – work (Burke, 1999; Machlowits, 1980).

Workaholism is detrimental to both the employee and the workplace. Excessive working hours are not necessarily productive; mistakes, failure, and occupational accidents are, among others, the result of long working hours combined with little rest (Fassel & Wilson-Schaef, 1989). Workaholism may also be an attempt to overcome insufficient productivity during regular working hours. Workaholism is positively related to a sense of responsibility and seniority (Rivyon-Ablass, 1989). Whether workaholism is related to stress and burnout is yet to be determined. While some studies found such a relationship (Nagy & Davis, 1985), others have not (Radian, 1998). Radian (1998) found a strong correlation between workaholism and Type A behavior patterns. These traits, however, might pose difficulties in any attempt workaholics make to recover from work (Van Wijhe, Peeters, Schaufeli, & Ouweneel, 2013). Indeed, some researchers have found that workaholics often do not benefit from their vacation (Westman, Eden, & Magen-Neuhaus, 1994). Others, however, found no effect of the vacation over workaholics compared with non-workaholics. It is thus perhaps unsurprising that detachment from work was found to be a moderating factor in burnout among workaholics immediately after their return from their vacation (Radian, 1998). Additionally, workaholics strong motivation to work inhibits their drive to invest in non-work. In her study of 26 employees based on their accumulation of time off at their workplace Braun (Cohen) (2007) found a significant positive relationship between one's drive to work and their balance of vacation days. That is, the greater the drive of the employee to work, the less time off they take. When they finally do take their respite, workaholics are often incapable of enjoying a vacation and cutting themselves off psychologically from their work. Moreover, vacation bears a negative psychological influence on workaholics, who regard it as an inappropriate waste of time (Machlowits, 1980; Rubinstein, 1980). Magen-Neuhaus (1993) found workaholics enjoyed only slight relief through their vacation when they were unable to detach from work. Workaholics not only exhibited lower levels of leisure satisfaction compared to non-workaholics, but they also experienced greater work-family conflict (Brady, Vodanovich, & Rotunda, 2008).

Studying employees at an industrial plant following an annual break, Magen-Neuhaus (1993) found that the ameliorating effect of vacation over stress and burnout was lower among workaholics compared to non-workaholic peers. Van Wijhe et al. (2013) found that negative emotions at the end of the workday hampered recovery during the evening of workaholic employees more than of non-workaholics. A cycle ensued, in which workaholics spent more time on work-related activities during the evening than non-workaholics when feeling negative emotions at the end of the workday, and, in turn, the lack of recovery experiences during the evening had influenced negative and positive emotions the following morning. Negative affect, however, seems to be a precursor of workaholism as does perfectionism, whereas boredom and loneliness, which is also predicted by perfectionism, mediate the effect of workaholism on life satisfaction (Bovornusvakool, Vodanovich, Ariyabuddhiphongs, & Ngamake, 2012). Bakker, Demerouti, Oerlemans, and Sonnentag (2013) also found that work-related activities during the evening affect more negatively the daily well-being of workaholics than non-workaholics. However, physical activities during the evening affected the daily well-being of workaholics more positively than their effect over non-workaholics, suggesting that physical activities during the evening may promote daily recovery from work for such individuals.

Vacation objectors

The literature clearly indicates that certain individuals do not benefit from their vacations and perhaps even suffer from it. For example, a study conducted by Hilton Hotels reported in 1995 that 38% of interviewed employees had not taken any vacation in over a year (Bailey Reinhold, 1997). Our own studies also show that individuals differ in their rates of unused vacation days based on demographic characteristics (gender, age, age of children), as well as on personality traits. This accumulative evidence seems to point at a new, potentially beneficial, research trajectory to better understand *vacation objectors*, individuals who refrain from using their vacation days, and the phenomenon of *vacation avoidance*.

We conducted a study among people who rarely take vacations to explore associations between burnout, enjoyment, personality characteristics (workaholism and Type A), and attitudes toward work (organizational commitment and job satisfaction). Among approximately 1,000 men and women working in two centers of a public organization in Israel, 144 employees had accumulated more than 55 unused vacation days. Of these, a total of 26 agreed to take an active part in the study by filling a questionnaire, and eight of these also agreed to be interviewed. The data was gathered during a 12-month period where each participant was measured only once on seven variables: burnout, Type A, work addiction, organizational commitment, job satisfaction, and personal and organizational background data.

Our quantitative data revealed a significant positive correlation between the burnout indices and the number of unused vacation days moderated by gender (men tended to take fewer vacation days than women) when considering family status and age of young children. The qualitative data asserted that the power of a vacation to attenuate stress and burnout is dependent on a plethora of moderating factors including personality, environmental conditions, and phase in one's lifecycle. The interviews also shed light on vacation avoidance. For some individuals, the vacation was a source of stress, others shun their vacation to avoid negative affect resulting from burdening colleagues with additional workloads while they themselves were resting and enjoying themselves. Vacation avoidance also ensued monetary and other ambitions and work addiction, which could, in turn, be attributed to the personal characteristics of employees. The Hilton study revealed similar apprehensions albeit among vacationers: 7% were nervous about their work while on vacation, 19% couldn't stop thinking about it, and 13% took their work along with them. Most vacationers had not realized how exhausted they truly were or to what degree they were shortchanging their organization by not taking time to replenish their resources (Bailey Reinhold, 1997).

Results of our study also imply that both the macro and micro levels of the corporate culture in which employees are embedded may contribute to the phenomenon of vacation avoidance, for example by way of organizational commitment and approval of the legitimacy of the vacation in the eyes of the organization at large. Alternatively, organizational cultures that endorse flexible working arrangement might offer sufficient leisure time so that employees do not feel the need to take any additional time off.

Leisure sickness

While individuals have been reported in different disciplines to suffer from various degrees of illness symptoms during or just prior to leisure times, the first systematic report on *leisure sickness*, also known as *weekend maladies* was produced only in Vingerhoets et al. (2002). The authors noted that about 3% of the Dutch population whom they surveyed reported that they had been suffering from such symptoms constantly for lengthy periods (ten years or more). Respondents reported experiencing fatigue and suffering headaches, migraines, dizziness, and muscular pains, as well as cold-driving viral infections, fever, and flu-like symptoms experienced over the weekend or during vacations. The authors related these data to several risk factors, including: workload, personal characteristics, difficulties in adjusting to non-work conditions, a high need for achievement, a sense of responsibility with respect to work. Some similarities may be drawn between causes suggested by Vingerhoets et al. (2002) and notions delineated by Machlowits (1980), who noted that workaholics tend to develop leisure sickness, feeling physically or psychologically unwell at the face of vacation. Kop, Vingerhoets, Kruithof, and Gottdiener (2003) found that the first two days of the vacation may be particularly detrimental as diagnosis of myocardial infarction occurred disproportionately higher during these days.

Explaining why people develop such symptoms during their time off may be easier than explaining why these symptoms develop before leaving the workplace. For example, Vingerhoets et al. (2002) suggest that exposure to sickness inducing environmental factors (pathogenic factors in the non-work environment) or different life style during leisure time contribute to leisure sickness. At the psychological level, the authors suggest physiological problems of adaptation the development of health problem that occur during leisure time and low appreciation of leisure activities and secondary gains from leisure periods translate to psychological gains of leisure sickness. Additionally, the prospects of re-stressing after de-stressing during leisure time may be stressing enough to cause leisure sickness. We may add that other employment considerations may be at work, as shown by our finding that younger employees tend to hospitalize themselves at the emergency room on non-work days due to work-related considerations (Tuval, 2001).

Drawing from data linking increases in death rates to religious holidays, Van Heck and Vingerhoets (2007) suggested the possibility that individuals are unconsciously able to time their illnesses to a time that is more convenient for them. Alternatively, it is possible that individuals do not become sick during leisure time but merely become more aware of the symptoms during this time (Pennebaker, 1994; 2000; Van Heck & Vingerhoets, 2007). While our own research showed that asthma and headaches drive patients to the emergency room more often on non-work days, we generally found a decrease in hospital emergency check-ins during non-work days (Tuval, 2001). Similarly, some studies of weekend migraines seem to disprove the essence of the term (e.g., Alstadhaug, Salvesen, & Bekkelund, 2007; Linnman, Maleki, Becerra, & Borsook, 2013).

Employment

The literature addresses several characteristics of the workplace and one's employment as being material to the influence of the workplace over stress and burnout, as well as the influence of the vacation over the employee. First and foremost, are various work stressors: the presence of negative work characteristics, the absence of positive work characteristics, workload, and stress intensity. Other relevant employment characteristics are tenure, seniority, and vocation.

Vocation, seniority, and tenure

The literature suggests that burnout among service and care professionals may be identified early in one's career and that the highest levels of burnout are prevalent among employees with less than six years of experience. These findings cross vocations, typifying teachers, nurses, lawyers, social workers, and psychological therapists, who were sometimes found burned out within their first working year (Cherniss, 1995; Dworkin, 1987; Gray & Freeman, 1987; Maslach, 1982; Maslach et al., 2001; Pines & Aronson, 1988). With greater work experience, these inclinations alter. Schaufeli and Enzmann (1998) found burnout was more characteristic of employees under the age of 30 compared to older employees. Specifically, depersonalization and emotional exhaustion were negatively related to tenure and age. Woods (1999) explained this relationship in terms of coping strategies, suggesting that employees who survived the system had better survival, acceptance, or adjustment strategies. That is, they were better able to adjust their psychological contrast to the system, its values, and changes occurring within it, thereby attenuating the value-based conflict between the employee and the system so that burnout was reduced. Cherniss (1995) found that teachers who arrived late into their vocation because they held a different job prior, experienced less burnout that those for whom teaching was a first vocation.

Hennefrund (1987) found that 71% of responding executives had never taken a four-week vacation. Most of them vacated for up to a week or during the weekend. Approximately a third of the respondents reported taking work with them on their vacation while 41% described taking a suitcase of work with them. Only 25% of the respondents reported to have never taken any work with them on their vacations. Etzion (1987, 1988a) suggested that employees with greater motivation who hold high expectations of themselves feel less at liberty to break from work.

Our studies found that the longer one's tenure at the current workplace and the longer the hours they spend at work, the less exhausted they feel (Moshkowitz, 2003). This may be related to a different finding of ours that tenure is positively related to preventative styles of coping (Silverberg, 2007). Our studies also found a relationship between the level of detachment and tenure. Inexperienced personnel, who are characterized by greater motivation idealism and expectations, as well as greater stress and burnout, seem to find it more difficult to detach from work during their vacation. Perhaps more experienced employees have learned that detachment is not only possible but also desirable and would not compromise the quality of future work (Etzion et al., 1998; Sraya-Wallin, 1999). Following Wood's (1999) assertion that surviving teachers have psychologically adapted to their workplace, such experienced employees are better able to prepare themselves for the summer break in order to extract the most of it.

Managers and non-managers

Studies on manager burnout are particularly important for their potential to address trickle down or spillover effect to non-managers in the organization (Golembiewski & Munzenrider, 1986; Golembiewski, Munzenrider, & Stevenson, 1986; Harvey & Raider, 1984; Kadushim, 1985; Seltzer & Numerof, 1988). Moshkowitz (2003) explored differences between software engineers, team leaders, and project managers with respect to work and vacation characteristics and behaviors. The groups differ with respect to the diversity of their tasks and the number of interfaces required on the job. Software engineers have relatively few job interfaces as the majority of their tasks are completed independently and require no coordination with other teams. They additionally enjoy a flexible timetable as they are typically free to set their own schedule within the given timeline. The work of team leaders comprises both independent tasks and managerial duties *vis-à-vis* their subordinate software engineers. The degree (typically little or moderate) to which they are exposed to interfaces within their own organization, as well as the client organization is a function of the size of their team. Project managers, in contrast, must interface with many different teams within the organization, as well as at the client organization. Their days are characterized by meetings and coordination of numerous efforts to this effect, and their outputs are expected by multiple parties, so that their schedules are rather inflexible. It might this be expected that the greater the schedule flexibility, the greater the extent of leisure experienced during the work hours. Indeed, software engineers reported more leisure during work than both other groups although differences between team leaders and project managers were not significant. Interestingly, despite the differences in time dedicated to leisure during work hours, the three groups hold similar perceptions regarding their investment in leisure during work hours and outside of them.

Sraya-Wallin (1999) found that 93% of the managers in her study exhibited a job-career anchor misfit compared with 58% of non-managers. Employees experiencing such a misfit showed a moderate decrease in post-vacation burnout, whereas as those experiencing a job-career anchor fit showed a moderate increase in post-vacation burnout. Among mothers who had taken a maternity leave within six months prior to the study, Silverberg (2007) found managers experienced greater exhaustion. In a different study of ours, Raviv (2002) found self-efficacy differentiated managers from non-managers (higher among the former), while other burnout measures did not demarcate the two groups. Leisure satisfaction also differed between the two groups. While in both groups leisure satisfaction was negatively related to stress originating in the presence of negative work stressors, among non-managers leisure satisfaction was additionally negatively related to general burnout, as well as exhaustion. Among non-managers, the degree to which leisure was needed for detachment was positively related to their burnout level, whereas the degree of perceived freedom was negatively related to exhaustion. Interaction effects were found for cynicism between managerial role and leisure satisfaction, as well as for general burnout between managerial role and the need for leisure for the purpose of detachment.

Among managers, leisure satisfaction was negatively related to stress originating in the presence of negative work stressors while the degree of sport activities was positively related to the level of cynicism.

The hierarchy of employees may influence their given vacation time so that high-ranking executives may enjoy more vacation time and special vacation benefits compared to low-ranking managers (Lounsbury & Hoopes, 1986). Boone, Kurtz, and Fleenor (1989) found among senior executives among both large industries and service industries that approximately two-thirds take an average of three to four weeks of vacation time annually, while about a third take an average of two weeks or less and only a small group take five weeks off work annually.

In our studies, however, Shaviv (1998) found that one's sense of responsibility influences the effect of vacation over work stress. Employees with a low sense of responsibility experienced lower perceptions of job stress after their vacation, whereas employees with a high sense of responsibility experienced a high rise in their post-vacation job stress. While the vacation itself had not influenced employees' level of burnout, the presence of a replacement at work worsened post-vacation burnout while the absence of such a replacement alleviated post-vacation burnout. Sapir (1995), too, found that knowledge of a replacement at work increased post-vacation stress levels.

Moshkowitz (2003) studied three hierarchical levels of employees: software engineers, team leaders, and project managers. The software engineers experienced significantly higher overload – both quantitative and qualitative – compared with the other groups, who experienced similar levels of overload of both types. This is in line with earlier findings suggesting that overload is a primary stressor for managers in addition to ambiguity and the multiplicity and conflicting nature of job demands (Buttenr, 1992; Edwards & Cooper, 1990; Smith & Tisak, 1993).

Teachers

Stress factors specific to teachers include issues of discipline and violence among pupils, the degree of heterogeneity in the classroom, scholarly issues, the need to meet educational goals set by the board or management of the school and the need to meet administrative demands occasionally at the expense of educational goals, inconsideration on behalf of parents, and reciprocal relationships between teachers and their role partners at the school (Cherniss, 1981; Friedman & Lotan, 1985; Kariacou & Sutcliffe, 1975; Oliver-Lomerman, 1983; Schwab & Lwanicki, 1982; Woods, 1999). As care and service professionals, teachers are expected to experience burnout (Farber, 1983; 1991; Kyriacou, 2001). In fact, it has been reported that 25–35% of teachers may be experiencing burnout (Kariacou & Sutcliffe, 1975). This may be attributed to certain personalities, such as *feelers* compared with *thinkers*, who may be more prone to be drawn to such professions (Garden, 1985). In countries where teachers work less than what would be considered a full-time job, teachers may additionally be susceptible to variables that are related to part-time work; for example, women working part-time were found to experience a greater degree of work–non-work conflict (Greenhous & Beutell, 1985). Burned out teachers may develop fatigue and a desire to stay at home, negative attitudes toward pupils, and a diminished sense of self-actualization. Professionally, they may exhibit poor teaching and execution performance levels, lengthy absenteeism behavior, and early retirement. Burned out teachers may neglect preparation of their classes, act harshly or sternly toward pupils, hold minimal expectations from their pupils, develop intolerance toward their frustrations, develop exhaustion, and loosen their sense of commitment (Cherniss, 1981; Farber, 1983; Maslach, 1976; Schwab, 1983).

Among American teachers, Dworkin (1987) found burnout rose in the first five years of work and then dropped incrementally with every additional year of tenure. In contrast to studies suggesting burnout is experienced in early ages, Friedman and Lotan (1985) found a positive relationship between burnout and tenure among elementary school teachers, so that burnout increased with tenure. Graham (1999) suggested that teachers who left the vocation within five years had not experienced burnout. One explanation for greater burnout among experienced teachers may be a shift in their job demands, creating greater overload as students are more desiring of passing thresholds of higher education institutions. While all teachers experience a similar workload, veteran teachers additionally had to adjust to the evolution and transformation of values within the system. In our own studies, Zeevi-Ohana (2000) found that teachers experiencing the greatest degree of burnout were those with five to ten years of tenure, whereas those with lesser or greater tenure experienced less burnout. Similarly, Leiter (1999) found cynicism and exhaustion were highest among employees of three to 15 years of tenure. Possibly, younger teachers, closer in age to their students, can identify with their pupils more easily, thereby increasing burnout. As teachers grow slightly older and more distant from their students, their level of identification, and hence tolerance of students' behaviors decreases. When they grow older still and have children of their own, once again teachers may relate to their students but not necessarily identify with them, thereby experiencing less burnout (Gray & Freeman, 1987). A decrease in burnout after the age of 40 or 15 years of tenure was also found by Schwab and Iwanicki (1982) and Oliver-Lomerman (1983). Among our studies, seniority among teachers was positively related to exhaustion and cynicism (Laski, 2002).

Findings related to tenure and burnout are typically sectional rather longitudinal. They are thus based on burnout levels of teachers who boast different length of tenure at a single point in time (Shirom, 1989). Maslach et al. (2001) referred to this limitation as the *survival bias* in the sense that employees who experienced burnout earlier in their career might leave their vocation earlier. Thus, cross-sectional studies, by their very nature, expose trends among surviving employees who have experienced lesser degrees of burnout to begin with. In other words, lower burnout among teacher with greater length of tenure could be the result of attrition of teachers who experienced greater levels of burnout. The least experienced teachers typically boast the greatest levels of motivation and enthusiasm (Cherniss, 1995; Farber, 1991).

Among teachers, we found that vacation satisfaction negatively correlated with burnout and that engagement in vacation activities related to school work was negatively connected to post-vacation exhaustion (Zeevi-Ohana, 2000). Etzion (1998) found that higher levels of education are positively related to expectations of self-actualization and career advancement. Vocations such as teaching may impede the realization of such expectations, which in turn, is likely to increase burnout (Cherniss, 1980; 1995; Farber, 1991). Teachers are in a unique position with respect to time off which is derivative of their industry. For example, teachers are forced to take a lengthy summer break. In Jewish communities, as well as in Israel, teachers are on school breaks for approximately 30% of the year due to holidays (Laski, 2002). Laski (2002) found differences in stress and the effect of vacation over stress between home room teachers and professional teachers to the point that some effects took opposing directions. Not all finding, however, were significant. Variance among teachers was also found with respect to the grades they taught: teachers of higher grades were found to burn out more than those in elementary schools (Farber, 1991).

Values, beliefs, and attitudes

Beliefs

Schlossberg (1991) found that among vacationing managers 78% believed vacations play an important role in preventing burnout among managers and 81% believed their vacation facilitates greater work productivity. In a different survey, 90% believed that relaxation, sleep, and taking breaks is a helpful technique for coping with stress (Crampton, Hodge, Mishra, & Price, 1995).

Values

We each carry a different series of values. For example, following the Protestant ethics, some might consider work to represent the essence of life, conferring upon individuals a sense of empowerment and self-confidence. Others, in contrast, may consider work as merely a means to an end. Perceptual variance may lead to variance in post-vacation effects and their degrees (Braun (Cohen), 2007).

Organizational commitment

Organizational commitment has been defined as a subjection relationship between the employee and their organization, a psychological attachment that contributes to the explanation of employees' behavior over time. Among these is employees' decision to perpetuate their membership in the organization and exert effort toward achieving the best interest of the organization (Allen & Meyer, 1990). The literature discerns between *attitudinal commitment* that reflects the employee's perceptions regarding the organization and *behavioral commitment* that represents the behavior aspects of attitudinal commitment (Mowday, Porter, & Steers, 1982). Organizational commitment has also been divided into *affective commitment, normative commitment*, and *continuance commitment*. Whereas the latter construct reflects extrinsic motivation to sustain organizational membership, the first two constructs reflect intrinsic motivation to do so. We investigated the relationship between these constructs in only one study, the sample of which was rather small. Given the limitations of this study, we found that affective organizational commitment was negatively related to burnout so that the more employees are affectively committed to their organization, the less burnout they experience, whereas the greater burnout they experience, the less affectively committed they are to the organization. In contrast to this hypothesis, a negative tendency was found in affective organizational commitment and unused vacation days (Braun (Cohen), 2007).

Klein, Hyde, Essex, and Clark (1998) found that the effect of the length of a maternity leave on the well-being of mothers a year after birth was dependent upon personal characteristics that express their relative commitment to work and family. Role incongruence in the form of work salience as an expression of commitment enhanced depression the longer mothers stayed away from work, whereas women whose choices regarding work and family were congruent with their commitments were less distressed (Klein et al., 1998).

Job satisfaction

Job satisfaction is considered an affective expression of the employee toward their job conditions and content; it is a subjective feeling of the perceived gap between desired and current conditions (Locke, 1984). Braun (Cohen) (2007) asserted the negative relationship between job satisfaction and burnout (the greater the burnout, the lesser the satisfaction derived from the job), particularly with respect to extrinsic motivators for job satisfaction. The greater one's joy at their workplace and the greater their self-efficacy, the greater their job satisfaction, in particular with respect to intrinsic motivators for job satisfaction. Job satisfaction did not, however, influence the number of unused vacation days of employees (Braun (Cohen), 2007). Laski (2002) found that job satisfaction was greater immediately after the vacation but decreased later on. Her study yielded differences in life satisfaction as a function of the vacation.

Self-efficacy

Bandura (1997) suggested that self-efficacy reflects one's belief in one's own capabilities to orchestrate and execute actions that will facilitate the management of given circumstances. Individuals characterized by high self-efficacy perceive a positive atmosphere and some degree of control over their circumstances, which in turn leads them to successful performance. In contrast, individuals expressing low self-efficacy perceive their influence to be limited, leading them to low performance. Among students, for example, self-efficacy was positively related to grades (Schunk & Zimmerman, 1997).

Saks (1994) found self-efficacy was negatively related to anxiety, which was positively related to stress, suggesting a negative relationship between self-efficacy and stress. In a different study, individual self-efficacy was found to be only moderately related to anxiety and weakly related to frustration. Group self-efficacy, however, was found to be strongly related to anxiety, frustration, dissatisfaction, and propensity for attrition. Group self-efficacy was additionally found to moderate the relationship between the number of working hours and satisfaction, anxiety, and propensity for attrition. It also moderated frustration and mediated the relationship between constraints and job satisfaction, as well as propensity for attrition (Jex & Gudanowski, 1992). In our own studies, we were unable to establish moderation of self-efficacy over the relationship between stress and burnout (Morad, 2009).

Perceptions of legitimacy

Perception of vacation legitimacy influence the effect of the vacation over vacationing employees. Legitimacy may be culturally driven, explaining the phenomenon of *karoshi* discussed earlier or the propensity of French employees to utilize their vacation days in full (Etzion & Dar, 2004). Findings of Etzion et al. (1998) showing positive post-leisure effects of reserve duty among Israeli employees may also be explained by the legitimacy of such time off work perceived by the Israeli society. Forced company vacation may bear similar legitimacy, whereas the personally tailored vacation timing may be perceived as selfish by other co-workers thereby diminishing the potential of a post-vacation alleviation effect (Magen-Neuhaus, 1993).

The nature of the maternity leave, being relatively long and for reasons that may be perceived as selfish desires, may negatively affect its potential for any attenuation in post-leisure stress and burnout (Ben Yoseff, 2004). Indeed, laboring employees who perceived the organizational culture of their workplace as family-friendly were found to show greater organizational commitment and planned to return earlier than mothers who did not perceive the organizational culture at their workplace as supportive. Returning early from a maternity leave may, however, be the result of quite different perceptions. Mothers who had working long hours prior to their maternity leave due to perceiving the organizational culture at their workplace as highly demanding returned early from their leave in order to meet organizational demands and standards (Lyness, Thompson, Francesco, & Judiesch, 1999). In her study of maternity leave, Ben Yoseff (2004) found that legitimacy did not play a role in post-vacation burnout and trends were in opposition to hypotheses. Mothers who perceived their leave to be illegitimate experienced less burnout upon their return to work compared with mothers who perceived their leave to be legitimate and experienced greater burnout upon their return to work. Legitimacy nevertheless demarcated the two groups in the sense that those who perceived their leave to be illegitimate experienced greater levels of burnout both before and after their leave compared with those who perceived their leave to be legitimate. Pre-leisure burnout levels were also positively associated with greater engagement in non-work life domains (and negatively associated with vacation satisfaction). Greater engagement in work domains was positively related to perceptions of legitimacy of the maternity leave, whereas greater engagement in work domains was also found related to seniority. It may be then, that an engaging attitude toward life in general, in both work and non-work domains, is related to feelings of legitimacy and lower baseline burnout levels. However, the feeling of legitimacy and engagement perhaps operate to create expectations that cause the vacation to influence burnout in opposition to hypotheses.

Work engagement

Sonnentag (2003) found a positive relationship between day-level recovery through leisure was positively related to day-level work engagement and day-level proactive behavior during the subsequent work day such as personal initiative and pursuit of learning.

Negative and positive stressors

Research has long since shown that the presence of negative job stressors and the absence of positive job characteristics are different types of strains that operate differently and independently of each other (Etzion et al., 1998; Hackman & Oldham, 1975; Kanner et al., 1978); for example, job dissatisfaction emerged in the absence of positive job characteristics but not in the presence of negative job stressors (Kanner et al., 1978). Negative job stressors are considered to be extrinsic and hence have the power to influence job dissatisfaction but lack the ability to influence job satisfaction (Herzberg, 1996). The literature specifies the following work characteristics as being negative: role conflict, role ambiguity, and overload. Managers may additionally experience decision-making, responsibility for subordinates, and meeting deadlines as negative work characteristics (see Moshkowitz, 2003). Overload may be perceived as a form of conflict; quantitative overload is experienced when individuals perceive there to be too many tasks to be accomplished with their given resources while qualitative overload is experienced when individuals are expected to complete tasks at a quality greater than they might be able to achieve with the given resources (Kahn, 1974; Kahn et al., 1964). Job control and job demands (working hours, decision latitude, break control/autonomy, and mental, emotional, and physical demands) moderated health problems while the need for recovery measured based on fatigue and other related symptoms was a major predictor of psychosomatic complaints, sleep complaints, and emotional exhaustion (Sluiter, Van der Beek, & Frings-Dresen, 1999).

The presence or absence of positive job characteristics influences one's extent of job satisfaction. Hackman and Oldham (1975) had identified the following factors as essential for feelings of satisfaction: variance of tasks and skills, role identity (the extent to which the job reflects the full cycle of the work), perceived significance of work, role autonomy (freedom, independence, and discretion in prioritization and execution), and feedback regarding performance efficiency and the outcomes of different activities. The absence of these factors may lead to a sense of failure, alienation, disappointment, and distress, and thus to perceived job stress (Kanner et al., 1978). Our own studies revealed distinct patterns between stressors and burnout: the absence of positive work characteristics correlated strongly and positively with cynicism and lack of professional efficacy, while the strain of the work–non-work conflict correlated strongly and positively with exhaustion. While stress originating in negative work characteristics correlated strongly and positively with all types of burnout (Zeevi-Ohana, 2000), it was most strongly (Zeevi-Ohana, 2000) or exclusively (Yoffe, 1997) related to exhaustion. The absence of positive job characteristics was related positively with cynicism but negatively with self-efficacy (Yoffe, 1997).

Similarly, Levi (1972) suggested that in the absence of a certain level of psychosocial stimuli, one's subjective quality of life, and specifically levels of burnout, may be detrimentally affected.

Seniority was found to demarcate the effect of the two types of stress. Senior employees reported greater stress originating in the absence of positive work characteristics (Laski, 2002). These findings are in line with those of Westman (1992) who found that seniority confers greater coping resources, as well as richer jobs. Moshkowitz (2003) found that both managers and non-managers perceived their financial and health well-being similarly although managers enjoyed better financial well-being. Non-managers experienced more stress than managers regardless of its origin. While she found no difference between the groups with respect to general burnout, non-managers ranked higher on cynicism. These findings are in line with those of Westman (1992) who suggested that the higher employees ranked in the organizational hierarchy, the more resources were available to the in coping with stress and burnout. Such resources include, among others, power, prestige, salary, authority, and autonomy. Employees not privy to such resources are more vulnerable to stress and burnout. Moshkowitz (2003) found no differences between managers and non-managers with respect to motivation for leisure, leisure expectations, leisure satisfaction and degree of detachment during leisure. The preference of non-managers to engage in active leisure compared with the preference of managers to engage in passive leisure may be a function of age.

The greater the stress originating in the presence of negative work characteristics, the greater the need for detachment and relaxation; no such relationship was found for the absence of positive job characteristics (Raviv, 2002). Raviv (2002) found that stress originating in the absence of positive job characteristics was more characteristic of non-managers, whereas stress originating in the presence of negative job stressors was more characteristic of managers. Nevertheless, leisure satisfaction was negatively related to stress originating in the presence of negative work stressors among both managers and non-managers. No significant relationship was found in either group between leisure characteristics and stress originating in the absence of positive job characteristics. Raviv (2002) also found that greater perceived freedom concerning leisure activities was related to lower stress originating in the presence of negative work characteristics. However, no relationship was found between this leisure characteristic and the absence of positive work characteristics. Leisure engagement yielded no relationship with either types of stress. Zeevi-Ohana (2000) found that the summer break facilitated a significant decrease in teachers' burnout and stress originating in the presence of negative work characteristics. However, following the break a significant increase was found in stress originating in the absence of positive work characteristics. Morad (2009) found a strengthening in the relationships between burnout and stress of both types following a short three-day vacation of Israeli nurses.

References

Allen, N. J. & Meyer, J. P. (1990). The measurement and antecedents of affective, continuance and normative commitment to the organization. *Journal of Occupational Psychology, 63(1)*, 1–18.

Allport, G. W. (1937). *Personality: A psychological interpretation*. Oxford: Holt.

Alstadhaug, K. B., Salvesen, R., & Bekkelund, S. (2007). Weekend migraine. *Cephalalgia, 27(4)*, 343–346.

Avni, T. (2007). The impact of organizational vacation on the amount of regular working hours, overtime, sick days and absenteeism among the organization's employees. M.Sc., Tel Aviv University, Israel. [Hebrew].

Bailey Reinhold, B. (1997). *Toxic work*. New York: Plume.

Bailyn, L. (1973). Family constraints on women's work. *Annals of The New-York Academy of Science, 208(1)*, 82–90.

Bakker, A. B., Demerouti, E., Oerlemans, W., & Sonnentag, S. (2013). Workaholism and daily recovery: A day reconstruction study of leisure activities. *Journal of Organizational Behavior, 34(1)*, 87–107.

Bandura, A. (1997). *Self-efficacy: The exercise of control*. New York: Freeman.

Barnett, R. C. & Baruch, G. K. (1987). Social roles, gender and psychological distress. In R. C. Barnett, L. Biener, & G. K. Baruch (Eds.), *Gender and stress* (pp. 122–143). New York: Free Press.

Barrick, M. R. & Mount, M. K. (1991). The big five personality dimensions and job performance: A meta-analysis. *Personality Psychology, 44(1)*, 1–26.

Baum, A. & Grunberg, N. G. (1991). Gender, stress and health. *Health Psychology, 10(2)*, 80–85.

Ben Yoseff, N. (2004). The effect of maternity leave on burnout among women serving in the I.D.F. M.Sc., Tel Aviv University, Israel. [Hebrew].

Besser, A. & Priel, B. (2006). Models of attachment, confirmation of positive affective expectations, and satisfaction with vacation activities: A pre-post panel design study of leisure. *Personality and Individual Differences, 41(6)*, 1055–1065.

Besser, A. & Shackelford, T. K. (2007). Mediation of the effects of the big five personality dimensions on negative mood and confirmed affective expectations by perceived situational stress: A quasi-field study of vacationers. *Personality and Individual Differences, 42 (7)*, 1333–1346.

Beutell, N. J. & Greenhouse, J. H. (1985). Sources of conflict between work and family roles. *Academy of Management Review, 10(1)*, 76–88.

Blitz, T. (1988). Burnout among women managers in Israel. M.Sc., Tel Aviv University, Israel.

Bluen, S. D., Barling, J., & Burns, W. (1990). Predicting sales performance, job satisfaction, and depression by using the Achievement Striving and Impatience-Irritability dimensions of Type A behavior. *Journal of Applied Psychology, 75(2)*, 212–216.

Boone, L. E., Kurtz, D. L., & Fleenor, P. C. (1989). Games CEOs play – most CEOs make time for leisure in their lives. *American Demographics, 11(1)*, 13–15, 776 778.

Bovornusvakool, W., Vodanovich, S. J., Ariyabuddhiphongs, K., & Ngamake, S. T. (2012). Examining the antecedents and consequences of workaholism. *The Psychologist-Manager Journal, 15(1)*, 56–70.

Brady, B. R., Vodanovich, S. J., & Rotunda, R. (2008). The impact of workaholism on work-family conflict, job satisfaction, and perception of leisure activities. *The Psychologist-Manager Journal, 11(2)*, 241–263.

Braun (Cohen), A. (2007). "Vacation objectors" – The factors behind vacation avoidance in a public organization. M.Sc., Tel Aviv University, Israel. [Hebrew].

Brown, G. (1975). Social class and psychiatric disturbance among women in an urban population. *Sociology, 9(2)*, 225–253.

Burke, R. J. & Greenglass, E. A. (1995a). Longitudinal study of psychological burnout in teachers. *Human Relations, 48(2)*, 187–202.

Burke, R. J. & Greenglass, E. A. (1995b). Work stress, role conflict, social support, and psychological burnout among teachers. *Human Relations, 48(2)*, 187–202.

Burke, R. J. (1999). Workaholism and extra-work satisfactions. *International Journal of Organizational Analysis*, 7(*4*), 352–365.

Burke, R. J., Greenglass, E. A., & Schwarzer, R. (1996). Predicting teacher burnout over time: Effects of work stress, social support, and self-doubts on burnout and its consequences. *Anxiety, Stress, and Coping: An International Journal, 9(3)*, 261–275.

Buttenr, E. M. (1992). Entrepreneurial stress: Is it hazardous to your health? *Journal of Managerial Issues, 4(2)*, 223–240.

Byrne, B. M. (1993). Burnout: Testing for the validity, replication, and invariance of casual structure across elementary, intermediate, and secondary teachers. *American Educational Research Journal, 31(3)*, 645–673.

Caplan, R. D. (1983). Person-environment fit: Past, present, and future. In C. L. Cooper (Ed.), *Stress research* (pp. 35–78). New York: Wiley.

Caplan, R. D. & Jones, K. W. (1975). Effects of work load, role ambiguity, and Type A personality on anxiety, depression and heart rate. *Journal of Applied Psychology, 60(6)*, 713–719.

Cassidy, T. (2005). Leisure, coping and health: The role of social, family, school and peer relationship factors. *British Journal of Guidance & Counselling, 33(1)*, 51–66.

Cattell, R. B. (1945). The principal trait clusters for describing personality. *Psychological Bulletin, 42(3)*, 129–169.

Chen, C. (2006). Personality traits and their relationship to leisure motivation and leisure satisfaction in Southern Taiwan university students (China). *Dissertation Abstracts International: Section B: The Sciences and Engineering, 66*(*7-B)*,3988.

Cherniss, C. (1980). *Staff burnout: Job stress in the human services*. Beverly Hills, CA: Sage Publications.

Cherniss, C. (1981). Preventing burnout: From theory to practice. In J. W. Jones (Ed.), *The burnout syndrome* (pp. 172–176). London: London House Management Press.

Cherniss, C. (1995). *Beyond burnout: Helping teachers, nurses, therapists and lawyers recover from stress and disillusionment*. New York: Routledge.

Clark, E. J. & Rieker, P. P. (1986). Gender differences in relationships and stress of medical and law students. *Journal of Medical Education, 61(1)*, 32–40.

Colley, A. (1984). Sex roles and explanations of leisure behavior. *Leisure Studies, 3(3)*, 335–341.

Cooke, R. A. & Rousseau, D. M. (1984). Stress and strain from family roles and work-role expectations. *Journal of Applied Psychology, 69(2)*, 250–260.

Costa, P. T., Jr. & McCrae, R. R. (1985). *The NEO personality inventory manual.* Odessa, FL: Psychological Assessment Resource.

Crampton, S. M., Hodge, J. W., Mishra, J. M., & Price, S. (1995). Stress and stress management. *SAM – Advanced Management Journal, 60(3)*, 10–18.

Csikszentmihalyi, M. (1990). *Flow, the psychology of optimal experience.* New York: Harper Perennial.

Csikszentmihalyi, M. (1997). *Finding flow: The psychology of engagement with everyday life.* New York: Basic Books.

Csikszentmihalyi, M. (2000). *Beyond boredom and anxiety: Experiencing flow in work and play.* San Francisco, CA: Jossey-Bass.

Davidson-Katz, K. (1991). Gender roles and health. In C. R. Snyder & R. D. Donelson (Eds.), *Handbook of social and clinical psychology: The health perspectives* (pp. 246–266). New York: Pergamon Press.

Day, D. V. & Bedeian, A. G. (1991). Work climate and Type A status as predictors of job satisfaction: A test of the interactional perspective. *Journal of Vocational Behavior, 38(1)*, 39–52.

de Heus, P. & Diekstra, R. F. W. (1999). Do teachers burn out more easily? A comparison of teachers with other social professions on work stress and burnout symptoms. In R. Vandenberghe & A. M. Huberman (Eds.), *Understanding and preventing teacher burnout: A sourcebook of international research and practice* (pp. 269–284).Cambridge: Cambridge University Press.

Deem, R. (1996). No time for a rest? An exploration of women's work, engendered leisure and holidays. *Time and Society, 5(1)*, 5–25.

Digman, J. M. (1979). The five major domains of personality variables: Analysis of questionnaire data in light of the five robust factors emerging from studies of rated characteristics. Paper presented at the Annual Convention of the Society of Multivariate Experimental Psychology, Los Angeles, CA.

Digman, J. M. (1990). Personality structure: Emergence of the five-factor model. *Annual Review of Psychology, 41(1)*, 417–440.

Doerfler, M. C. & Kammer, P. P. (1986). Workaholism, sex, and sex role stereotyping among female professionals. *Sex Roles, 14(9–10)*, 551–560.

Douglas, E. J. & Morris, R. J. (2006). Workaholic, or just hard worker? *The Career Development International, 11(5)*, 394–417.

Dubin, R., Hedley, R. A., & Taveggia, T. (1976). Attachment to work. In R. Dubin (Ed.), *Handbook of Work, Organization and Society* (pp. 281–341). Chicago, IL: Rand McNally.

Dworkin, A. G. (1987). *Teacher burnout in the public schools: Structural causes and consequences for children.* Albany, NY: State University of New York Press.

Edelwich, J. & Brodesky, A. (1980). *Burnout: Stages of disillusionment in helping professionals.* New York: Human Science Press.

Eden, D. (1990). Acute and chronic job stress, strain and vacation relief. *Organizational Behavior and Human Decision Processes, 45(2)*, 175–193.

Edwards, J. R. & Cooper, C. L. (1990). The person – environment fit approach to stress. Researching problems and some suggested solutions. *Journal of Organizational Behavior, 11(4)*, 293–307.

Etzion, D. (1984). The moderating effect of social support on the relationship of stress and burnout. *Journal of Applied Psychology, 69(4)*, 615–622.

Etzion, D. (1987). Burnout: The hidden agenda of human distress (Working Paper No. 930/87). Tel Aviv, Israel: Henry Crown Institute of Business Research, Faculty of Management, Tel Aviv University.

Etzion, D. (1988a). The experience of burnout and work/non-work success in man and female engineers: A match-pairs comparison. *Human Resource Management, 27(2)*, 163 179.
Etzion, D. (1988b). Understanding the dynamics of burnout through cross-cultural research: Reflection on the comparison between Israelis and Americans (Working Paper No. 965/88). Tel Aviv, Israel: Henry Crown Institute of Business Research, Faculty of Management, Tel Aviv University.
Etzion, D. & Bailyn, L. (1994). Patterns of adjustment to the career/family conflict of technologically trained women in the U.S. and Israel. *Journal of Applied Social Psychology, 24(17)*, 1520–1549.
Etzion, D. & Boklis, V. (2006). Meta-analytic study of respite and fade-out effects. Report by: Tel Aviv University, Henry Crown Institute of Business Research, Faculty of Management, Tel Aviv, Israel.
Etzion, D. & Dar, A. (2004). Patterns of leisure and vacation across cultures. *Panim- Quarterly of Society, Culture and Education, 27*, 23–38 [Hebrew].
Etzion, D., Eden, D., & Lapidot, Y. (1998). Relief from job stressors and burnout: Reserve service as respite. *Journal of Applied Psychology, 83(4)*, 577–585.
Etzion, D. & Pines, A. (1986). Sex and culture in burnout and coping among human service professionals. *Journal of Cross – Cultural Psychology, 17(2)*, 191–209.
Etzion, D. & Sapir, O. (1996). The effect of an annual vacation in relieving stress and burnout at work (a controlled field study at an industrial enterprise in Israel). Report by: Tel Aviv University, Henry Crown Institute of Business Research, Faculty of Management, Tel Aviv, Israel.
Farber, B. A. (1983). Current trend, future directions. In B. A. Farber (Ed.), *Stress and burnout in the human service professions* (pp. 242–248). New York: Pergamon Press.
Farber, B. A. (1991). *Crisis in education. Stress and burnout in the American teacher.* San Francisco, CA: Jossey-Bass.
Fassel, D. & Wilson-Schaef, A. (1989). The high cost of workaholism. *Business and Health, 1*, 38–42.
Fiske, E. B. (1982 Sept. 19). Survey of teachers reveals morale problems. *New York Times*, p. A1, A52.
Flaxman, P., Ménard, J., Bond, F. W., & Kinman, G. (2012). Academics' experiences of a respite from work: Effects of self-critical perfectionism and perseverative cognition on postrespite well-being. *Journal of Applied Psychology, 97(4)*, 854–865.
Freudenberger, H. J. (1974). Staff burnout. *Journal of Social Issues, 30(1)*, 159–165.
Freudenberger, H. J. & North, G. (1985). *Women's burnout.* Garden City, New York: Anchor Press/Doubleday.
Friedman, M. (1977). Type A behavior pattern: Some of its pathophysiological components. *Bulletin of the New York Academy of Medicine, 53(7)*, 593–604.
Friedman, I. A. (1991). High- and low-burnout schools: School culture aspects of teacher burnout. *Journal of Educational Research, 84(6)*, 325–333.
Friedman, I. A. & Lotan, A. (1985). *The psychological burnout of the teacher in Israel in elementary school.* Jerusalem, Israel: Henrietta Szold Institute.
Galin, A. (2002). Flexible work patterns: Implication for the employment of women – the case of Israel (Working Paper No. 3/2002). Tel Aviv University, Henry Crown Institute of Business Research, Faculty of Management, Tel Aviv, Israel.
Gallagher, D. J. (1990). Extraversion, neuroticism and appraisal of stressful academic events. *Personality and Individual Differences, 11(10)*, 1053–1057.
Garden, A. M. (1985). Burnout: The effect of personality. Ph.D., Massachusetts Institute of Technology, Cambridge, MA.

Garden, A. M. (1989). Burnout: The effect of psychological type on research findings. *Journal of Occupational Psychology, 62(3)*, 223–224.
Goldberg, L. R. (1982). From ace to zombie: Some explorations in the language of personality. In C. D. Spielberger & J. N. Butcher (Eds.), *Advances in personality assessment* (Vol. 1, pp. 203–234). Hillside, NJ: Erlbaum.
Golembiewski, R. T. & Munzenrider, R. (1988). *Phases of burnout: Developments in concepts and applications.* New York: Praeger.
Golembiewski, R. T., Munzenrider, R. F., & Stevenson, J. G. (1986). *Stress in organizations: Toward a phase model of burnout.* New York: Praeger.
Gray, H. & Freeman, A. (1987). *Teaching without stress.* London: Paul Chapman Publishing.
Greenglass, E. R., Fiksenbaum, L., & Burke, R. J. (1994). The relationship between social support and burnout over time in teachers. *Journal of Social Behavior and Personality, 9(2)*, 219–230.
Greenhous, J. H. & Beutell, N. J. (1985). Sources of conflict between work and family roles. *Academy of Management Review, 10(1)*, 76–88.
Greenhous, J. H. & Kopelman, R. E. (1981). Conflict between work and family roles: Implications for the career planning process. *Human Resource Planning, 4(1)*, 1–10.
Grinker, R. (1991). Life phases as burnout factors in women with technological careers. M. A., Tel Aviv University, Israel.
Hackman, J. R. & Oldham, G. R. (1975). Development of the job diagnostic survey. *Journal of Applied Psychology, 60(2)*, 159–170.
Harvey, S. H. & Raider, M. C. (1984). Administrator burnout. *Administration in Social Work, 8(2)*, 81–89.
Hendel, D. D. & Solberg, J. (1983). Sabbatical and leave experiences of female and male faculty at a large research university. Paper presented at the Annual Meeting of the American Educational Research Association, Montreal, Canada.
Henderson, K. A. (1994). Broadening an understanding of women, gender, and leisure. *Journal of Leisure Research, 26(1)*, 1–7.
Hennefrund, W. (1987). Breaking away. *Association Management, 39*(August), 65–68.
Herzberg, F. (1996). *Work and the nature of man.* Cleveland, OH: World Publishing.
Holland, J. L. (1985). *Making vocational choices: A theory of vocational choices and work environments.* Englewood Cliffs, NJ: Prentice Hall.
Holmes, R. M. & Rahe, R. H. (1967). The social readjustment rating scale. *Journal of Psychosomatic Research, 11(2)*, 213–218.
Hoxter, A. L. & Lester, D. (1988). Tourist behavior and personality. *Personality and Individual Differences, 9(1)*, 177–178.
Huebner, E. S. & Mills, L. B. (1994). Burnout in school psychology: The contribution of personality characteristics and role expectations. *Special Services in the School, 8(2)*, 53–67.
Hurtz, G. M. & Donovan, J. J. (1998). Personality and job performance: Will the real "big five" please stand up? Paper presented at the 13th Annual Conference of the Society for Industrial and Organizational Psychology, Dallas, Texas.
Innes, J. M. & Kitto, S. (1989). Neuroticism, self-consciousness and coping strategies and occupational stress in high school teachers. *Personality and Individual Differences, 10(3)*, 303–312.
Iso-Ahola, S. E. (1980). *The social psychology of leisure and recreation.* Dubuque, LA: William C. Brown.
Jamal, M. (1985). Type A behavior and job performance: Some suggestive findings. *Journal of Human Stress, 11(2)*, 6–68.

Jex, M. S. & Gudanowski, M. D. (1992). Efficacy beliefs and work stress: An exploratory study. *Journal of Organizational Behavior, 13(5)*, 509–517.

John, O. P. (1990a). The "big five" factor taxonomy: Dimensions of personality in the natural language and questionnaires. In L. Pervin (Ed.), *Handbook of personality theory and research* (pp. 66–100). New York: Guilford.

John, O. P. (1990b). The search for basic dimensions of personality: A review and critique. In P. McReynolds, J. C. Rosen, & G. L. Chelune (Eds.), *Advances in psychological assessment* (Vol.7, pp. 1–37). New York: Plenum.

Johnson, J. V. & Hall, E. M. (1996). Dialectic between conceptual and causal inquiry in Psychosocial work-environment research. *Journal of Occupational Health Psychology, 1(4)*, 362–374.

Jones, R. W. (1993). Gender-specific differences in the perceived antecedents of academic stress. *Psychological Reports, 72(3)*, 739–743.

Kadushim, A. (1985). *Supervision in social work* (2nd ed.). New York: Columbia University Press.

Kahn, R. L. (1974). Conflict, ambiguity and overload, three elements in job stress. In A. Mclean (Ed.), *Occupational stress* (pp. 47–61). Springfield, IL: Charles C Thomas.

Kahn, R. L., Wolfe, D. M., Snoek, J. E., Quinn, R. P., & Rosenthal, R. A. (1964). *Organizational stress: Studies in role conflict and ambiguity* (Vol. 2). New York: Wiley.

Kanner, A. D., Kafry, D., & Pines, A. (1978). Conspicuous in its absence: The lack of positive condition as a source of stress. *Journal of Human Stress, 4(4)*, 33–39.

Kyriacou, C. & Sutcliffe, J. (1975). Teachers stress and satisfaction. *Educational Research, 21 (2)*, 89–96.

Kyriacou, C. & Sutcliffe, J. (1978). A model of teacher stress. *Educational studies, 4(1)*, 1–6.

Kelly, J. R. (1982). *Leisure*. Englewood Cliffs, New Jersey: Prentice Hall.

Kinnunen, U. (1988). Teacher stress during an autumn term in Finland: Four types of stress processes. *Work and Stress, 2(4)*, 333–340.

Klein, H. M., Hyde, J. S., Essex, M. J., & Clark, R. (1998). Maternity leave, role quality, work involvement, and mental health one year after delivery. *Psychology of Women Quarterly, 22(2)*, 239–266.

Kop, W. J. P., Vingerhoets, A. P., Kruithof, G. M., & Gottdiener, J. S. M. (2003). Risk factors for myocardial infarction during vacation travel. *Psychosomatic Medicine, 65(3)*, 396–401.

Krau, E. (1984). Commitment to work in immigrants: Its function and peculiarities. *Journal of Vocational Behavior, 24(3)*, 329–339.

Kremer-Fuks, S. (1999). The impact of vacation and slack work periods on job stress and burnout among newcomers and veteran Israelis. M.Sc., Tel Aviv University, Israel. [Hebrew].

Kushnir, T. & Melamed, S. (1991). Work-load, perceived control and psychological distress in Type A/B industrial workers. *Journal of Organizational Behavior, 12(2)*, 155–168.

Kyriacou, C. (2001). Teacher stress: Directions for future research. *Educational Review, 53 (1)*, 27–35.

Laski, S. (2002). The effect of summer break activities and personality variables on stress and burnout among high school teachers in Israel. M.Sc., Tel Aviv. M.Sc. University, Israel. [Hebrew].

Lee, C., Early, P. C., & Hanson, L. A. (1988). Are Type A better performers? *Journal of Organizational Behavior, 9(3)*, 263–269.

Leiter, M. P. (1999). Burnout among teachers as a crisis in psychological contracts. In R. Vandenberghe & A. M. Huberman (Eds.), *Understanding and preventing teacher burnout: A sourcebook of international research and practice*. Cambridge: Cambridge University Press.

Levi, L. (1972). *Stress and distress in response to psychosocial stimuli*. Oxford: Pergamon Press.

Lewin, K. (1935). *A dynamic theory of personality*. New York: McGraw-Hill.

Li, Q. F., Wong, K. F., Law, K. H., Chiu, L. S., & Yuen, S. W. (1994). A study of primary school teacher burnout: Its relationship with teacher characteristics, professionalism and social support. *Journal of Primary Education, 4(2)*, 17–26.

Lin, Y., Kerstetter, D., Nawijn, J., & Mitas, O. (2014). Changes in emotions and their interactions with personality in a vacation context. *Tourism Management, 40*(February), 416–424.

Linnman, C., Maleki, N., Becerra, L., & Borsook, D. (2013). Migraine tweets—What can online behavior tell us about disease? *Cephalalgia, 33(1)*, 68–69.

Locke, E. A. (1984). Job satisfaction. In M. Gruneberg & T. Wall (Eds.), *Social psychology and organizational behavior* (pp. 93–117). New York: Wiley.

Lounsbury, J. W. & Hoopes, L. L. (1986). A vacation from work: Changes in work and non-work outcomes. *Journal of Applied Psychology, 71(3)*, 392–401.

Lyness, K. S., Thompson, C. A., Francesco, A. M., & Judiesch, M. K. (1999). Work and pregnancy: Individual and organizational factors influencing organizational commitment, timing of maternity leave, and return to work. *Sex Roles, 41(7/8)*, 485–508.

Machlowits, M. (1980). *Workaholics, living with them, working with them*. Massachusetts: Adison Wesley Publishing.

Magen-Neuhaus, T. (1993). The effect of vacation and feelings of stress at work on emotional burnout. M.Sc., Tel Aviv University, Israel. [Hebrew].

Mallinckrodt, B., Leong, F. T., & Kraji, M. M. (1989). Sex differences in graduate students – Life change stress and stress symptoms. *Journal of College Student Development, 30 (4)*, 332–338.

Maslach, C. (1976). Burnout. *Human Behaviour, 5(9)*, 16–22.

Maslach, C. (1982). *Burnout: The cost of caring*. Englewood Cliffs, NJ: Prentice Hall.

Maslach, C. & Jackson, S. E. (1981). The measurement of experienced burnout. *Journal of Occupational Behavior, 2(2)*, 99–113.

Maslach, C., Schaufeli, W. B., & Leiter, M. P. (2001). Job burnout. *Annual Review of Psychology, 52*(February), 397–422.

McCrae, R. R. & Costa, P. P. (1990). *Personality in adulthood*. New York: Guilford.

McCrae, R. R. & John, O. P. (1992). An introduction to the five-factor model and its applications. *Journal of Personality, 60(2)*, 175–215.

Melamed, S. & Meir, E. I. (1981). The relationship between interest-job incongruity and selection of avocational activity. *Journal of Vocational Behavior, 18(3)*, 310–325.

Melamed, S., Meir, E. I., & Samson, A. (1995). The benefits of personality-leisure congruence: Evidence and implications. *Journal of Leisure Research, 27(1)*, 25–40.

Meir, E. I. & Melamed, S. (1986). The accumulation of person-environment congruencies and well-being. *Journal of Occupational Behavior, 7(4)*, 315–323.

Meir, E. I., Melamed, S., & Abu, F. A. (1990). Vocational, avocational, and skill utilization congruencies and their relationship with well-being in two cultures. *Journal of Vocational Behavior, 36(2)*, 153–165.

Morad, N. (2009). The effect of a weekend respite on burnout and stress among hospital nurses. M.Sc., Tel Aviv University, Israel. [Hebrew].

Moshkowitz, S. (2003). The relationship between work-overload, leisure activities and burnout among high-tech employees. M.Sc., Tel Aviv University, Israel. [Hebrew].

Mowday, R., Porter, L., & Steers, R. (1982). Employee–organization linkages. In P. Warr (Ed.), *Organizational and occupational psychology* (pp. 219–229). New York: Academic Press.

Murray, H. A. (1938). *Explorations in personality*. New York: Oxford University Press.

Nagy, S. & Davis, L. G. (1985). Burnout: A comparative analysis of personality and environmental variables. *Psychological Reports, 57(3)*, 1319–1326.

Norman, W. T. (1963). Toward and adequate taxonomy of personality attributes: Replicated factor structure in peer nomination personality ratings. *Journal of Abnormal and Social Psychology, 66(6)*, 574–583.

Oliver-Lomerman, A. (1983). Stress and strain among Israeli teachers – The mediating effect of social support and facilitating information flow. M.A., Tel Aviv University, Israel.

Paludi, M. A. (1986). Working mothers and the family context, predicting positive coping. *Journal of Vocational Behavior, 28(3)*, 211–253.

Pennebaker, J. W. (1994). Psychological bases of symptom reporting: Perceptual and emotional aspects of chemical sensitivity. *Toxicology and Industrial Health, 10(4/5)*, 497–511.

Pennebaker, J. W. (2000). Psychological factors influencing the reporting of physical symptoms. In A. A. Stone & J. S. Turkkan (Eds.), *The science of self-report: Implications for research and practice* (pp. 299–315). Mahwah, NJ: Erlbaum.

Piedmont, R. L. (1993). A longitudinal analysis of burnout in the health care setting: The role of personal disposition. *Journal of Personality Assessment, 61(3)*, 457–473.

Pines, A. & Aronson, E. (1988). *Career burnout: Causes and cures*. New York: Free press.

Pines, A., Aronson, E., & Kafry, D. (1981). *Burnout: From tedium to personal growth*. New York: Free Press.

Pines, A. & Kafry, D. (1981). Coping with burnout. In J. W. Jones (Ed.), *The burnout syndrome* (pp. 139–150). London: House Press.

Radian, R. (1998). The moderating effect of workaholism and Type A on the impact of vacation on job stress and burnout. M.Sc., Tel Aviv University, Israel. [Hebrew].

Ragheb, M. R. (1993). Leisure and perceived wellness: A field investigation. *Leisure Sciences, 15(1)*, 13–24.

Raviv, E. (2002). Effects of leisure activity on occupational stress and burnout among high-tech employees. M.Sc., Tel Aviv University, Israel. [Hebrew].

Repetti, R. L. & Wood, Z. (1997). Effects of daily stress at work on mothers' interactions with preschoolers. *Journal of Family Psychology, 11(1)*, 90–108.

Rivyon-Ablass, L. (1989). Workaholism in the city and the kibbutz. M.A., Tel Aviv University, Israel.

Rosenman, R. H., Friedman, M., Straus, R., Wurm, M., Kositchek, R., Hann, W., et al. (1964). A predictive study of coronary heart disease. *Journal of The American Medical Association, 189*, 103–110.

Rubinstein, C. (1980). Vacations, expectations, satisfactions, frustrations, fantasies. *Psychology Today, 14(5)*, 62–66, 61–76.

Saks, M. A. (1994). Moderating effects of self-efficacy for the relationship between training method and anxiety and stress reaction of newcomers. *Journal of Organizational Behavior, 15*, 639–654.

Sapir, O. (1995). The effect of vacation on stress and burnout at work. M.Sc., Tel Aviv University, Israel. [Hebrew].

Schaufeli, W. B., Bakker, A. B., van der Heijden, F. M. M. A., & Prins, J. T. (2009). Workaholism among medical residents: It is the combination of working excessively and compulsively that counts. *International Journal of Stress Management, 16(4)*, 249–272.

Schaufeli, W. & Enzmann, D. (1998). *The burnout companion to study and practice*. London: Taylor & Francis.

Schlossberg, H. (1991). Hyatt Survey: If you take a vacation, do it the right way. *Marketing News, 25*, 21–22.

Schreyer, S. R. & Driver, B. L. (1989). The benefits of leisure. In E. L. Jackson & T. L. Burton (Eds.), *Understanding leisure and recreation: Mapping the past, charting the future*. State College, PA: Venture Publishing.

Schunk, D. H. & Zimmerman, B. J. (1997). Developing self-efficacious readers and writers: The role of social and self-regulatory processes. In J. T. Guthrie & A. Wigfield (Eds.), *Reading engagement: Motivating readers through integrated instruction* (pp. 34–50). Newark, DE: International Reading Association.

Schwab, D. P. & Iwanicki, E. F. (1982). Perceived role conflict role ambiguity and teacher burnout. *Educational Administration Quarterly, 18(1)*, 60–74.

Schwab, R. L. (1983). Teachers burnout: Moving beyond psychobabble. *Theory into Practice, 22(1)*, 21–26.

Schwartz, S. H. (1982). Job involvement as obsession-compulsion. *Academy of Management Review, 7(3)*, 429–432.

Seltzer, J. & Numerof, R. E. (1988). Supervisory leadership and subordinate burnout. *Academy of Management Journal, 31(2)*, 439–446.

Shaviv, T. (1998). The impact of an organized tour vacation on work stress and burnout among managerial and non-managerial vacationers. M.Sc., Tel Aviv University, Israel. [Hebrew].

Shirom, A. (1989). Burnout in work organizations. In L. Cooper & I. Robertson (Eds.), *International review of industrial and organizational psychology* (Vol.*2*, pp. 25–48). New York: Wiley.

Silverberg, S. (2007). Home/work conflict and burnout among working mothers after maternity leave – Situational and personal characteristics as moderators. M.Sc., Tel Aviv University, Israel. [Hebrew].

Sluiter, J. K., van der Beek, A. J., & Frings-Dresen, M. H. W. (1999). The influence of work characteristics on the need for recovery and experienced health: A study on coach drivers. *Ergonomics, 42(4)*, 573–583.

Smith, C. S. & Tisak, J. (1993). Discrepancy measures of role stress revisited: New perspectives on old issues. *Organizational Behavior and Human Decision Processes, 56 (2)*, 285–307.

Smith, T. W. & Anderson, N. B. (1986). Models of personality and disease: An interactional approach to Type A behavior and cardiovascular risk. *Journal of Personality and Social Psychology, 50(6)*, 1166–1173.

Snir, R. & Zohar, D. (2008). Workaholism as discretionary time investment at work: An experience-sampling study. *Applied Psychology: An International Review, 57(1)*, 109–127.

Sonnentag, S. (2003). A new look at the interface between non-work and work recovery, work engagement, and proactive behavior recovery, work engagement, and proactive behavior. *JAP, 88(3)*, 518–528.

Spence, J. T., Helmreich, R. L., & Pred, R. S. (1987). Impatience versus achievement striving in Type A pattern: Differential effects on student's health and academic achievement. *Journal of Applied Psychology, 72(4)*, 522–528.

Spence, J. T. & Robbins, A. S. (1992). Workaholism: Definition, Measurement, and Preliminary Results. *Journal of Personality Assessment, 58(1)*, 160–178.

Sraya-Wallin, O. (1999). The effect of vacation characteristics and the fit between job and career-anchors on stress and burnout among employees in a hi-tech environment. M. Sc., Tel Aviv University, Israel. [Hebrew].

Tinsley, H. E. A. & Tinsley, D. J. (1986). A theory of the attributes, benefits and causes of leisure experience. *Leisure Sciences, 8(1)*, 1–45.

Toren, N. (1984). National cultures of science: A study of Soviet and American immigrant scientists in Israel. *Science and Public Policy, 11(3)*, 125–160.

Tupes, E. C. & Christal, R. E. (1961). Recurrent personality factors based on trait ratings. *Journal of Personality, 60(2)*, 225–251.

Tuval, A. (2001). Patterns of stress-related ER hospitalization at Sheba Medical Center between January 1997 and September 2000. M.Sc., Tel Aviv University, Israel. [Hebrew].

van Heck, G. L. & Vingerhoets, A. J. J. M. (2007). Leisure sickness: A biopsychosocial perspective. *Psihologijske Teme, 16(2)*, 187–200.

van Horn, J. E., Schaufeli, W. B., Greenglass, E. R., & Burke, R. J. (1997). A Canadian-Dutch comparison of teachers' burnout. *Psychological Reports, 81(2)*, 371–382.

van Wijhe, C., Peeters, M., Schaufeli, W., & Ouweneel, E. (2013). Rise and shine: Recovery experiences of workaholic and nonworkaholic employees. *European Journal of Work and Organizational Psychology, 22(4)*, 476–489.

Vanfassen, B. F. (1981). Sex differences in the mental health effects of spouse support and equity. *Journal of Health and Social Behavior, 72(2)*, 130–143.

Vingerhoets, A. J. J. M., van Huijgevoort, M., & van Heck, G. L. (2002). Leisure sickness: A pilot study on its prevalence, phenomenology, and background. *Psychotherapy and Psychosomatics, 71(6)*, 311–317.

Wahlers, R. G. & Etzel, M. J. (1985). Vacation preference as a manifestation of optimal stimulation and lifestyle experience. *Journal of Leisure Research, 17(4)*, 283–295.

Westman, M. (1992). Moderating effects of decision latitude on stress-strain relationship: Does organizational level matter? *Journal of Organizational Behavior, 13(7)*, 713–722.

Westman, M. (2001). Stress and strain crossover. *Human Relations, 54(6)*, 717–751.

Westman, M. & Eden, D. (1997). Effects of a respite from work on burnout: Vacation relief and fade-out. *Journal of Applied Psychology, 82(4)*, 516–527.

Westman, M. & Etzion, D. (2001). The impact of vacation and job stress on burnout and absenteeism. (Special issue on burnout). *Psychology & Health, 16(5)*, 595–606.

Westman, M., Eden, D., & Magen-Neuhaus, T. (1994). Effects of vacation on job stress and burnout: Relief and fade-out. Paper presented at The 15th International Conference on Social Stress Research, Honolulu, HI.

Wiggins, J. S. & Trapnell, P. D. (1997). The return of the Big Five. In R. Hogan, J. Johnson, & S. Briggs (Eds.), *Handbook of personality psychology* (pp. 737–765). San Diego, CA: Academic Press.

Wilkinson, T. J. & Hansen, J. C. (2006). The relationship among leisure interests, personality traits, affect, and mood. *Measurement and Evaluation in Counseling and Development, 39 (1)*, 31.

Wilson, D. & Mutero, C. (1989). Personality concomitants of teachers' stress in Zimbabwe. *Personality and Individual Differences, 11(10)*, 1195–1198.

Woods, P. (1999). Intensification and stress in teaching. In R. Vandenberghe & A. M. Huberman (Eds.), *Understanding and preventing teacher burnout: A sourcebook of international research and practice*. Cambridge: Cambridge University Press.

Yoffe, Y. (1997). The effects of overseas business trips on job stress and burnout. M.Sc., Tel Aviv University, Israel. [Hebrew].
Zeevi-Ohana, A. (2000). The impact of summer break activities and personality variables on job stress and burnout among high school teachers. M.A., The Hebrew University, Jerusalem, Israel.
Zellars, K. L., Perrewé, L. P., & Hochwarter, W. A. (2000). Burnout in health care: The role of the five factor of personality. *Journal of Applied Social Psychology, 30(8)*, 1570–1598.
Zvi, E. (2004). Stress and burnout among students before and after the Passover break – Vacation activities and personality variables as moderators. M.Sc., Tel Aviv University, Israel.

4

POLICIES AND NORMS

Organizational and environmental conditions

The need to balance work and non-work, or resting, periods has been acknowledged since antiquity, where a weekly resting day was combined with holidays that served as resting days. Modernity, however, caused shifts in ancient traditions, which evolved differently around the world. Despite wide agreement regarding the need for leave, there are vital differences among countries with respect to the number of working hours and leave days that are assigned to employees. The issue of working arrangements is often raised in public and academic discourses *vis-à-vis* differing economic, social, and demographic needs of organizations and their employees. In the past, the eight-hour day was a central issue in the development and consolidation of trade unions worldwide. Today, still, the regulation of working hours and arrangements is on the political and social agenda of many countries. The decrease in working hours and their reallocation are the concern of organizations and workers' unions, as well as certain social groups, particularly those championing social equality. In our studies, we have investigated extensively vacation-related policies and norms in four locations: the US, the European community, Japan, and Israel (Dar, 2002; Etzion & Dar, 2004). A general trend worldwide can be discerned in which societies and interested bodies within (such as labor organizations or feminist movements) attempt to improve working conditions of the labor force. The central vehicle for such improvement is considered the overall working hours of employees given time off, including paid holidays and paid leave.

Annual leave

Working hours and leave are typically standardized by law.

US

The US has been known as the "no-vacation nation" (Report of the Center for Economic and Policy Research – CEPR, cited by Ray, Sanes, & Schmitt, 2013), and its employees have been said to suffer from "vacation deficit disorder" (Robinson, 2003). As paid annual leave is not mandated by federal law, it is contingent on labor legislation in individual states or on employer-initiated policies and collective labor agreements. Uniformity in the leave provided by American employers ensues from the competition among them (Rosenberg, 2001). American companies imitate one another's leave policy in order to remain attractive to potential employees. According to a survey of the Central Bureau of Statistics, in the years 1993 and 1994, 76% of American employees received paid annual leave, 72% received paid holiday leave and only 18% received paid personal need leave (US Bureau of Labor Statistics, 2000). According to the National Compensation Survey of the US Department of Labor and Bureau of Statistics of 2012, 77% of private sector workers had access to paid vacations and holidays. The data reveal that paid vacations and holidays provided to employees in the private sector are unequally distributed. For instance, only half of low-wage workers (bottom fourth of earners) have any paid vacation compared to 90% of high wage workers.

Data collected via the National Compensation Survey: Employee Benefits in the United States (US Bureau of Labor Statistics, 2012) clearly indicates the fact that for American employees, vacation time depends on and is calculated according to length of service. For instance, in 2012, American private industry workers with one year of service received ten days of paid vacation on average, those with five years of service received 14 days on average, those with ten years of service were granted 17 days and those 20 years on the job received an average of 19 days of paid vacation in 2012.

European community

The EU regulations (Council Directive 93/104/EC) set out a framework of four weeks or 20 days per year, for all EU member countries. Leave is subject to the terms determined by law or common practice (generally collective labor agreements) in each member state. In reality, the number of days of paid leave is determined by law in individual EU members and in accordance with local collective labor agreements. These, however, may only improve employees' rights as laid down in European Economic Community (EEC) regulations and may not detract from them. Thus, in many countries, such as Austria, Denmark, Sweden, France, Luxemburg, and Finland, tenure of one working year merits a vacation of over a month (Mattes, 1992; Schor, 1992). EU regulations determine that annual leave cannot be converted into pay except in the case of employment termination. This was decided in order to encourage employees to take the leave due to them. In many countries, additional stipulations ensure that workers take their allotted leave each year. In some countries, employers are forbidden to offer extra pay for forfeiting vacation, Austria specifically offers "cashing out" the vacation while others provide deadlines for vacation taking. Several countries mandate more paid leave for specific populations such as younger or older workers. For example, Austria, Germany, Italy, and Switzerland offer more vacation time for young workers, while Norway offers workers over the age of 60 an additional one week of paid leave a year. Some European countries have regulations to guarantee that workers take some of their leave during the summer. Several nations offer additional leave for specific purposes such as jury service, house move, or marriage.

Many OECD (Organisation for Economic Co-operation and Development) countries guarantee paid holidays, including New Year's Day, Good Friday, Easter Monday, and Christmas. Other commonly paid holidays are Labor Day, Ascension Thursday, and All Saints' Day. In most countries, employers have flexibility around paid holidays and often have the option to schedule workers on holidays, provided that they pay higher rates than usual or offer paid days off on other occasions.

Japan

Japanese labor legislation (Labor Standards Law 1947, Amended in 2004) establishes the right of the employee to receive ten days of paid leave annually. No paid holidays are provided, and neither does Japanese legislation provide for sick leave or emergency leave, driving employees to accumulate leave for future emergencies. The number of days of leave received increases with seniority in the same place of work up to a maximum of 20 days. Although the law mandates paid leave, various surveys show that the Japanese in general do not take all the leave due to them. According to the results of the Vacation Deprivation Study 2012, Asian workers in general work the longest weeks (44 hours a week) and take the fewest vacation days, with Japanese workers breaking a record of taking only five days a year.

According to the *General Survey on Working Conditions* conducted by the Japanese Ministry of Health, Labor and Welfare (2012), in 2002 the number of paid annual leave actually used was 9.0 days, representing 48.1% of the amount due, both showing an increase compared to the previous year, when the average regular employee took only 8.8 paid holidays, representing 48.4% of the amount due. Although the average number of paid holidays given by firms was a record high, 18.1 days, the actual number of days taken had fallen for six consecutive years, revealing a widening gap between the system and reality (see Working conditions and the labor market, 2001a, 2001b). Inquiring into the reasons for why Japanese employees were not taking their leave, 90% of respondents noted that they could only take four consecutive days leave and only 5% said that they could take more than ten days; 36% ascribed not taking their leave to the pressure of work and their unwillingness to inconvenience fellow employees;

28% pointed to the attitude of their supervisors; 12% blamed the corporate atmosphere; and the remaining employees listed reasons such as a desire for promotion, the difficulty involved in requesting leave and the fact that they found work more interesting (see Imamura, 1998).

Israel

Israeli legislation (The Hours of Work and Rest Law of 1951; Paid Annual Leave Law Amendment, 1966) mandates paid annual leave, beginning with 14 days per year in the first four years of employment up to a maximum of 28 days increasing with tenure. The law restricts accumulation of annual leave or the possibility of cashing it in, presumably in support of employees' well-being by means of replenishing (see Shaviv, 1998). Employers are at liberty to offer more leave time to employees and often do so in their attempt to appear lucrative (Radian, 1998). Additionally, collective agreements tend to increase annual leave for the first few years.

Family-friendly policies

An obvious trend in working and leisure frameworks is the significant increase in the number of countries making arrangements and enacting legislation on leave for family needs. Such family-friendly policies reflect, to a large extent, the cultural values of a society when it comes to the personal needs of its members. It is the result of social and cultural changes expressing themselves in the considerable growth of the percentage of women in the workforce, in changing perceptions of gender roles and in the growing need of employees to find a balance between their working life and their family (Figart & Mutari, 1998). Notwithstanding, what is considered to be corrective legislation for the purpose of diminishing inequality may actually result in deepening the chasm of gender inequality (Vogel, 1993). Family-friendly policies can be regarded as evolving over three stages: the harmonization of work and family only with respect to working mothers and nursing care; acknowledgment of organizations' responsibility in adopting family-friendly policies that go beyond childcare; and finally, a central national policy that considers employees' lifestyles as a legitimate need (Galinsky, Freidman, & Hernandez, 1991).

US

The Family and Medical Leave Act (FMLA) legislated in 1993 (see United States Code. FMLA [On-line]) was a milestone in American history since it was the first time that the need for family leave received recognition and was given legal standing. The law addressed increasing pressure over the previous 25 years, during which the US underwent dramatic economic and social change leading to significant changes in the American labor market, including a significant increase in the number of women under employment. The intent of the FMLA was to facilitate better work–non-work balance in employees' lives by giving them the opportunity to take unpaid leave for family or medical purposes while considering employers' interests and minimizing the potential for gender discrimination at the workplace.

The FMLA marked a new era in US federal family and medical leave. Prior to it, there was no legal requirement to provide job-protective leave, except by the Pregnancy Discrimination Act (enacted in 1978) and other collective agreements that guaranteed unpaid leave for temporarily disabled employees. Employees at worksites that did not voluntarily endorse job-protective leave policies could be fired for taking leave for the birth of a child or for other familial medical situations. Some states adopted maternity leave statutes during the 1990s, which gave new mothers a right to job reinstatement after maternity leave. Notwithstanding, even prior to the FMLA, many employers had instigated leave programs for pregnant women and mothers. Kelly and Dobbin (1999) suggested that employers did so in response to the increase of women in the American workforce. The authors also claim that the law against sexual discrimination played a key role in the development of a leave policy for women and mothers since it forced employers to provide an answer to the needs of pregnant women.

FMLA allows workers to take up to 12 weeks of job-protected, unpaid leave for different familial needs such as maternity leave, care for an ill child, spouse or parent, and care for personal health conditions. Subsequent amendments allow family members of military personnel to take up to 26 weeks away from work to attend to situations arising from foreign deployment, including care for a service member with a serious injury or illness. The FMLA, however, is far from being universal, as its application is limited to certain terms of worksite and employee characteristics, for example, the firm must include at least 50 workers and the employee should have 12 months of tenure. In 2012, according to a survey conducted by the US Department of Labor, approximately half of the US workforce met FMLA standards. The survey also revealed that 55% of respondents took leave due to personal illness while only 21% took maternity leave. Most of the periods of leave were short, with 42% of the beneficiaries taking ten days and only 17% taking more than 60 days. Only 13% of responding employees reported taking leave for an FMLA reason in the preceding 12 months – a number unchanged since the earlier 2000 survey.

The benefits of the FMLA are thus limited. While the US Department of Labor concluded in 2012 that the law seems to have had a positive effect on workers and their families, only 2–3% of the leave allowance was taken *de facto*, because many employees could not afford the luxury of the legal right conferred to them by the FMLA. Many employees, especially the weak and needy, are unable to take leave for family or medical purposes. Employees' knowledge of their rights under the act was also found to be limited. These facts raise doubts regarding the extent to which the FLMA contributes, ultimately, to the welfare of the American employee in terms of bettering the balance between work and non-work obligations. On the positive side, the FMLA did not appear to impose an undue burden on employers. The vast majority (91%) of responding employers reported that the FMLA had either no effect or a positive effect on factors such as absenteeism, turnover, and morale. All in all, it seems that the FMLA has had a beneficial effect on most employees as it successfully changed the attitude toward leave and transformed it from a benefit determined by the employer and local legislation, into an entitlement under federal law (Dorman, 2000). Additionally, the law is encouraging the development of more uniform standards concerning family leave.

In 1999, the US Senate passed a motion enabling employees to use unemployment insurance (UI) in order to finance leave under the FMLA during the year following childbirth or adoption. Dorman (2000) suggested that despite recent improvements, the US is still not "family friendly" compared to other industrialized countries because American employees are not paid for the leave they take. In her view, the conflict between the FMLA and US workplace culture accounts for the paucity of employees, and particularly men, who actually exercise their legal right. American employers do not take a favorable view of male employees taking family leave, thus reinforcing traditional cultural attitudes concerning gender.

In the decades that have passed since the FMLA enactment, labor markets have continued to evolve, the number of women in the workforce has further increased, unionization has declined, and changes in the distribution of employment across sectors have transpired. Family and medical leave is still an active area of policy discussion. Some issues are still being debated, including its application for smaller worksites and the need to enact legislation for making leave paid.

European community

Unlike US legislation, EU legislation does not address general or personal family leave; rather, it offers provision for parental leave only, in particular for childcare. In further contrast to American law, such leave defined by European law is paid, at least in part. EU legislation nevertheless does not stipulate the extent of compensation but leaves that stipulation to the provision of the individual EU countries.

EU regulations regarding parental leave (Council Directive 96/34/EC) came into effect in 1996 following a European convention focused on gender equality and a social European convention aimed at advancing the idea of a balance between work and family areas of life. The regulations established the right of every employee regardless of gender to a minimum of three months' leave in the case of childbirth or adoption. The initial 1996 agreement was superseded by the new Framework Agreement enacted by the Council Directive 2010 (Council Directive 2010/18/EU), which extended the period of parental leave to a minimum of 14 weeks. Under the 2010 directive, the maximum period of entitlement is reached when the child reaches the age of eight and is determined by individual countries. The directive applies equally to all workers irrespective of their employment contract. EU regulations additionally obligate member states to provide employees with the legal right of an employee of paid parental leave due to *force majeure* or urgent family reasons, such as sickness or accident (Hall, 1997).

EU members are allowed to extend these legal rights as they see fit. Maternity and parental leaves vary between countries from 14 to 80 weeks for one or both parents. Sweden is particularly generous with respect to its leave policy. Swedish law establishes the right of employees to take leave for personal reasons, for personal needs or in order to take care of a sick relative. The Swedish law also confers the right of employees to receive financial compensation during these types of leave. This legal generosity was twofold: first, it aimed to increase the birthrate in the aftermath of World War II, and second, it was part of an effort to build a more egalitarian society and increase the participation of fathers in childrearing (Haas, 1993). Maternity leave was redesigned in the 1970s and turned into parental leave, in accordance with the political institutionalization of gender equality in the Nordic countries, with Sweden being the first country to introduce leave for both parents in 1974 (Haas & Rostgaard, 2011). Swedish couples are entitled to 480 days of parental leave. For the first 390 days (about 13 months) parents receive 80% of their usual wage while for the last three months they receive a flat rate benefit (Ellingsæter, 2009). One parent also has the right to work at only 75% of a full job scope until their children reach the age of eight. Men exercised their right to parental leave mostly in Sweden (78%), reflecting the influence of legislation on cultural perceptions and as a result, the norm – as opposed to Holland where only 9% of working men exercised this right or 1–3% in remaining EU countries (Hall, 1997).

Japan

The Childcare Leave and Family Leave Law was enacted in 1992 (and amended in 1995 and 2005) by the Japanese Ministry of Health, Labor and Welfare in response to the increase in two-income families and the ensuing need to support child and family care. A survey carried out by the Japanese Ministry of Health, Labor, and Welfare in 1993 revealed that 50.8% of all employers put into effect regulations relating to childcare and family leave even before the law came into force and that most of the employees who requested such leave were women. In 1999, 56.4% of the women who were mothers to children under the age of one took leave, while only 0.4% of the men who were fathers to children under the age of one did so. The reasons given by employees for not taking family leave, especially the men, were mainly: the atmosphere at the workplace or financial difficulties.

In 1999, the Japanese Ministry of Health, Labor, and Welfare established a system of financial compensation (grants) encouraging family-friendly companies. Many

Japanese companies have since come to understand that balancing work and private life is important from the perspective of both productivity and organizational efficiency. As a result, they have begun to advance appropriate plans, not only with regard to leave but also with regard to shorter and flexible working hours, in-house childcare facilities, and the re-employment of ex-employees who left in order to raise their children. Sato (2000) claims that most Japanese companies have a more generous leave policy than that required by law.

The 2005 amendment extended the coverage for family care leave and also introduced nursing care leave for up to three months allowing parents to tend to sick children. A survey conducted by the Japanese Ministry of Health, Labor, and Welfare showed a constant rise in the acceptance rate for childcare leave among women, amounting to 90.6% in 2008. The percentage of companies that had established rules for the application of childcare leave also rose, amounting to 66.4%.

Israel

Two main laws contribute to fashioning a family-friendly policy in Israel: The Women's Employment Law (1954) and the Employment (Equal Opportunities) Law (1988). By law, women to receive up to 26 weeks of maternity leave, 14 of which are fully paid, with no detriment to her rights in the workplace. The Israeli family-friendly legislation helped promote family-friendly policies in the workplace. Various sources have indicated that more than half of the women in the workforce continue working after marriage and childbirth (Richter, 1992; OECD, 2012). An amendment from 1997 allows men to take half the maternity leave, seven weeks after the birth, should the mother choose to return to work. The law further allowed fathers to be absent from work in order to care for a sick child. In the first three years after the 1998 enactment, men made little use of their right to maternity leave; only some 650 fathers made use of the opportunity afforded them to receive maternity leave. Despite what might be considered insufficient effort to inform parents of their opportunity to share the maternity leave, in most cases, fathers who took this opportunity expressed great satisfaction at being able to do so.

Cultural and social settings of leave and leisure

The cultural framework

A number of extensive comparative surveys have been conducted pointing to cultural and national differences concerning work and its significance as an expression of more general life values (Hofstede, 1980; 1998; Schwartz, 1995; 1999; Super & Sverko, 1995). In one of our studies, we examined national legislation in light of such comparative cultural dimensions identified by Hofstede (1980; 1994) and Schwartz (1994). The following is a brief summary of dimensions that we found related to national legislation.

Individualism versus collectivism. Hofstede (1980) described the relationship of the individual with the group reflecting the degree to which the society reinforces individual achievement and interpersonal relationships. In individualistic cultures, individuality and individual rights are paramount within the society so that individuals within such societies tend to form a larger number of looser relationships. In contrast, collectivistic cultures are characterized by close interpersonal relationships and extended families or collectives, where all members feel responsible for fellow group members. This dimension was examined by Hofstede according to four principal measures: the time devoted to personal and family life, good working conditions, job security, and job satisfaction. Very similar to this dimension is the dimension of **affective autonomy versus conservatism** (Schwartz, 1992; 1994), which describes the contrast between values that emphasize individuals and their right to act in their own best interest and values that emphasize the superiority of communal interests. Conservatism emphasizes the supremacy of the community's objectives and welfare while also favoring tradition and family. In individualistic or autonomous societies, we might expect legislation to be more permissive than protective of the individual's welfare, so that working hours and leave stipulations would be mostly determined by market forces and custom. Collective or conservative legislation, adversely, would be expected to protect the rights of employees and therefore we might expect it to be more favorable toward employees, that is, it would be expected to encourage more liberal leave and working-hours policies.

Masculinity versus femininity. Hofstede (1980) described the degree to which society differentiates between genders with respect to social functions. In masculine cultures this demarcation is highly notable as men are expected to be assertive, tough, and focused on material success while women are required to be more modest, gentle, and focused on the quality of life. The more feminine a society, the more similar are the social function of the genders. That is to say, both men and women are required to be modest, gentle and focused on the quality of life. According to Hofstede (1980), the more feminine a society, the greater the public sympathy with economic and social legislation benefiting the weak. We may assume that this dimension will express itself in the degree of family-friendly leave policies.

Hierarchy and mastery versus egalitarianism and harmony (Schwartz, 1992; 1994) represent the degree to which personal interests are fulfilled at the expense of others versus the degree to which personal interests are sacrificed in order to preserve the social and material environment. We may assume that work will have more importance in the life of society members where hierarchy and mastery predominate compared to societies where egalitarianism and harmony are more prevalent. Accordingly, we might expect more liberal leave policies to be found in societies characterized by egalitarianism and harmony. Since cultures characterized by these dimensions emphasize aspects of life such as recreation, community, and family over work. Societies characterized by hierarchy and mastery, in contrast, would be expected to yield less liberal leave and working-hours policies.

Legislation as a reflection of cultural values

Table 4.1 summarizes national legislation regarding working arrangements and leave. The differences between countries, even within European countries, is notable.

The legal arrangements for working hours, rest, and annual leave and the gap between law and custom vary significantly between the US, the European countries, Japan, and Israel. Countries vary with respect to all types of leave from no protection of rights to very generous leave packages. Whereas weekly working hours also vary considerably between the countries, in all these are determined by law. Notwithstanding, as detailed earlier, the countries additionally vary in the difference between the number of working hours stipulated by law and the average number of hours actually worked.

Table 4.2 details the cultural scores of countries based on the dimensions identified by Hofstede (1980) and Schwartz (1994). All scores were taken at original values presented in the original studies. We averaged the scores of Schwartz (1994) in two groups according to the degree of liberality we expected to find in each dimension. From a conceptual standpoint, we expected affective autonomy, conservatism, harmony, and egalitarianism to produce legislation that favored employees' welfare while we expected mastery and hierarchy to yield legislation that supported a free market.

Table 4.3 translates the data presented in Table 4.2 into a country ranking.

European countries with high averages on the dimensions of affective autonomy, harmony, and egalitarianism are generally those with a more benevolent policy toward employees regarding annual leave, particularly France and Denmark. Schwartz (1992; 1994) did not offer data for Sweden but estimated that Sweden's cultural profile would be similar to that of Denmark. In contrast, the US, Japan, and Israel display lower values on these dimensions and relatively high scores on the dimensions of hierarchy and mastery and therefore expected to yield less generous leave policy.

American policy regarding working hours and leave is limited in scope and relatively stringent. Annual leave is not statutory and family leave is unpaid, although the 40-hour working week is quite similar to the European average (42 hours). These data appear to be in line with Schwartz's findings (Schwartz, 1994; 1999) regarding dimensions that might indicate the centrality of work. When translating scores into ranks, the US, Japan, and Israel show exactly the same position in relation to other countries in the sample with respect to the average scores on the group of dimensions including hierarchy and mastery and those on the dimensions of affective autonomy, harmony, and egalitarianism. On the European front, the UK ranks second the US on the dimensions of individualism and masculinity. Indeed, similarities may be drawn between the two countries as the UK had been, up to application of the EU regulations, the only European country where annual leave was not mandated by law but by collective labor agreements. Schwartz (1994) found Ireland to have a similar cultural profile as the UK.

TABLE 4.1 Legalized working arrangements and leave in studied countries

Country	Weekly working hours	Paid annual leave (days)	Maternity and parental leave	
			Duration	Pay
EU – Austria	40	30–35 days	Maternity: 16 weeks Parental: 2 years	100% Partial
EU – Belgium	39	4 weeks	Maternity: 15 weeks Parental: 3 months each	75–82% Low flat rate
EU – Bulgaria			Parental: 68 weeks	90%
EU – Denmark	By EU legislation	25–30 days	Maternity: 18 weeks Parental: 32 weeks	100%, income dependent
EU – Finland	40	24–25 days	Maternity: 17.5 weeks Parental: 26 weeks	90% Flat rate
EU – France	35	5 weeks	Maternity: 16 weeks or more[1]	100% Flat rate
EU – Germany	48	20–29 days	Maternity: 14 weeks Parental: 3 years	100% Two years at flat rate, one year unpaid
EU – Ireland	48	20 days	Maternity: 42 weeks[1]	Unpaid
EU – Italy	40	28 days	Maternity: 20 weeks Parental: 6 months	80% 30%
EU – Norway		21 days		
EU – Portugal	48	22 days	Maternity: 6 months Parental: 2 years	100% Unpaid (except civil servants)
EU – Spain	40	30 days	Maternity: 16 weeks Parental: 3 years	100% Unpaid
EU – Sweden	40	25 days	Maternity and parental:~80 weeks (480 days)Additional 2 months	80% or flat rateUnpaid
EU – The Netherlands	40	20–31 days	Maternity: 16 weeks Parental: 26 x No. of weekly working hours	100% Unpaid
EU – UK	48	20–25 days	Maternity: 52 weeks Parental: 13 weeks	Partially paid at 90% and flat rate Unpaid
Israel	45	14–28 days	Maternity: 26 weeks Parental:	
Japan	40	10 days	Maternity: Parental: 3 months	

(Continued)

TABLE 4.1 (Cont.)

Country	Weekly working hours	Paid annual leave (days)	Maternity and parental leave	
			Duration	*Pay*
US	40	No legal provision for annual leave; duration and compensation left to individual states and companies. Maternity and parental: up to 12 weeks of job-protected, unpaid leave but dependent on worksite conditions		

1 Includes a provision for leave before birth.

The average scores of Japan on the dimensions of affective autonomy, harmony, and egalitarianism are relatively low compared with other countries, and they are very high for the opposing dimensions of mastery and hierarchy. We should nevertheless consider the changes that have occurred in Japanese society over the past decades. Japanese employees, especially the younger generation, aspire to a better balance between their working and private lives. As a result, more companies have been offering plans that aim to make the workplace friendlier for employees who have families of their own. More and more Japanese employees, both men and women, attach importance to a better balance between their private lives and their work. However, the common view among employees remains that creating this balance is the personal responsibility of the individual. As a result, many employees feel under pressure and experience stress in attempting to strike a balance between these two aspects of their lives (Makita & Ida, 2001).

Legislation in Israel regarding annual leave is not as generous toward employees as that which is found in Europe, although it does mandate paid annual leave. In practice, collective labor agreements increase the number of days of leave that Israeli employees enjoy. In Israel, employees take more leave than their American or Japanese counterparts but less than their European counterparts. With respect to working hours, Israel is closer to Japan, with a mandated working week of 45 hours, although in practice the working week amounts to 42.5 hours due to collective labor agreements.

Labor law in Israel today has its roots in Israel's origin as a welfare state with a socialist orientation and a very powerful trade union movement. Family plays a rather central in the life of Israeli women, for whom family and childrearing are considered the default choice. Israeli women thus adjust their careers and the effort they invest in work according to their family status and tend to compromise their work during early childrearing years (Etzion & Bailyn, 1994). In a report for the Israeli Ministry of Industry, Trade and Labor, Gal and Frenkel (2011) explained that Israel is a unique case study. The country's history and the political and demographic states before and since its establishment have, in fact,

TABLE 4.2 Country scores on the dimensions of Hofstede (1980) and Schwartz (1994)[1]

Country	*Hofstede's dimensions*		*Schwartz's dimensions*[2]							
Country	*Individualism*	*Masculinity*	*Affective autonomy*	*Conservatism*	*Harmony*	*Egalitarianism*	*Average*[3]	*Mastery*	*Hierarchy*	*Average*[4]
EU – Austria	55	79								
EU – Belgium	75	54								
EU – Denmark	74	16	4.0	3.6	4.2	5.5	4.33	4.0	2.2	3.01
EU – France	71	43	4.4	3.4	4.3	5.5	4.40	3.9	2.2	3.05
EU – Germany	67	66	4.0	3.4	4.4	5.4	4.30	4.1	2.3	3.20
EU – Ireland	70	68								
EU – Italy	76	70	3.0	3.8	4.8	5.6	4.30	4.1	1.7	2.90
EU – Norway	69	8								
EU – Sweden	71	5								
EU – The Netherlands	80	14	3.5	3.7	4.0	5.4	4.15	4.0	2.3	3.15
EU – UK	89	66								
Israel	54	47	3.6	4.1	3.0	4.8	3.88	4.1	2.7	3.40
Japan	46	95	3.5	3.9	4.1	4.7	4.05	4.3	2.9	3.60
USA	91	62	3.7	3.9	3.7	5.0	4.08	4.3	2.4	3.35

1 All scores as published in the original studies with the exception of averaged scores.
2 Countries with no scores were not represented in this study.
3 Average of affective autonomy, conservatism, harmony, and egalitarianism.
4 Average of mastery and hierarchy.

TABLE 4.3 Country ranking on the dimensions of Hofstede (1980) and Schwartz (1994)

Individualism		*Masculinity*		*Average of mastery and hierarchy*		*Average of affective autonomy, conservatism, harmony, and egalitarianism*[1]	
1	USA	1	Japan	1	EU – France	8	EU – Italy
2	EU – UK	2	EU – Austria	2	EU – Denmark	7	EU – France
3	EU – The Netherlands	3	EU – Italy	3	EU – Germany	6	EU – Denmark
4	EU – Italy	4	EU – Ireland	4	EU – Italy	5	EU – The Netherlands
5	EU – Belgium	5	EU – Germany	5	EU – The Netherlands	4	EU – Germany
6	EU – Denmark	5	EU – UK	6	USA	3	USA
7	EU – France	7	USA	7	Japan	2	Israel
7	EU – Sweden	8	EU – Belgium	8	Israel	1	Japan
9	EU – Ireland	9	Israel				
10	EU – Norway	10	EU – France				
11	EU – Germany	11	EU – Denmark				
12	EU – Austria	12	EU – The Netherlands				
13	Israel	13	EU – Norway				
14	Japan	14	EU – Sweden				

1 Order reversed to match conceptually with other dimensions.

contributed to it being generous in its welfare policy. Thus, the need for women's work on the one hand and an increased birth rate on the other, have led to investments that support and protect working mothers reinforced by legislation and workers' unions, altogether anchoring women's work as a desired norm. These processes may have led Israel to become the country with both the highest birth rate and the highest percentage of working mothers among industrialized countries.

The notion of gender equality, which played a key role in the socialist ideology of the founders of the State of Israel, stood in sharp contrast to the Jewish tradition, which puts emphasis on the family and on a traditional division of labor between men and women. These conflicting forces have been continuously at work in Israel with regard to gender equality generally and the family-friendly aspects of labor legislation in particular. Israel thus produced rather progressive legislation, albeit weakly enforced, side by side with informal support systems for working women.

The participation of women in the Israeli labor market has been consistently rising in recent decades amounting in 2008 to 80% of the Jewish (non-ultraorthodox) women of prime working age, while in 2011 almost 70% of Israeli women (ages 18–67) took part in the labor market. Men are nevertheless dominant in the in the labor market accounting for 88% of the workforce. Due to these proportions, Israel was rated tenth among OECD countries in its men-to-women labor ratio, and its male dominance among workers has been higher than the average among OECD countries. Further gender inequality in the labor market is expressed in the fact that less women are employed in higher positions, their wages are generally lower, and their representation in typically feminine jobs, characterized by lower wages and prestige, is very high (Ben David, 2014; Ministry of Labor, Social Affairs, and Social Services, 2018).

The many changes that Israeli society and the labor market have been undergoing are a result of both global technological and economic developments not specifically associated with Israeli society and nation-specific sociopolitical developments, such as the introduction of National Medical Insurance, which significantly diminished the power of the General Federation of Labor. These changes have, among others, influenced leave policy and working hours reflecting in the structure of the labor market, the weakening of the General Federation of Labor, and the replacement of collective labor agreements by personal contracts. Israeli society, according to Harpaz (1999), has become more individualistic and materialistic, and in practice, a trend may be discerned of more workers not taking full leave time owed to them.

Israeli legislation may be rated in generosity somewhere between the generous EU legislation and the less generous legislations found in Japan and the US. The country's average score on the dimensions of mastery and hierarchy are extremely high, second only to Japan's average and immediately followed by the average of the US. The country's average score on the dimensions of affective autonomy, harmony and egalitarianism are the lowest among the countries reviewed here, a result which one would expect to be associated with an even less generous leave policy than those of Japan and America.

Watanabe, Takahashi, and Minami (1997) claim that in Israel, gender roles are less traditional than in Japan, the UK, or in the US and that Israeli legislation is, in fact, more advanced. The problem in all countries, however, is not so much with legislation as with its implementation. In this respect, it is interesting to make a comparison with the Swedish legislation and the normative influence that could result from amending national legislation regarding the division of maternity leave between men and women. Possibly, as in Sweden, effective legislation guaranteeing payment for parental leave for men, as well as for women will lead to greater gender equality in the realization of parental leave.

National scores on the dimension of conservatism were not found to be directly linked to leave policy. As this dimension implies a devotion to family, the author assumed that the higher the score on this dimension, the less central will work be to country members, resulting in greater priority given to other aspects of life (Schwartz, 1999). We, however, found that countries with high scores on this dimension, such as Israel, the US, and Japan (in this order) enacted more restrictive leave policies and longer working hours compared to countries that scored low on this dimension, such as France, Denmark, and Germany that boast more generous leave policies.

A comparison of country scores on Hofstede (1980) cultural dimensions suggests that the dimension of individualism alone cannot explain the degree of generosity in national leave policies. Japan is the polar opposite of the US on this dimension (not only in our small sample but among all nationalities studied by Hofstede), showing the greatest tendency for collectivism. It was assumed that high collectivism would encourage policies of leave and working hours that are more beneficial to employees. From the viewpoint of legislation, Japan is indeed more generous than the US in that annual leave is mandatory and governmental agencies make a genuine effort to ensure compliance with the law. However, given our entire dataset, working hours and leave in the US and in Japan are quite similar and rather restrictive in their endorsement of employees' right.

This unexpected result may be explained by the fact that members of individualistic cultures tend to emphasize the rights of the individuals, their wishes, and their objectives, whereas in collectivist cultures, the individual is defined primarily in terms of membership in a group or a family, either nuclear or extended. Triandis (1994) noted that individualism in Japan is defined together with others and is mutually dependent on others. The Japanese perceive absolute lifelong loyalty to the group as a duty, and in the Japanese society the workplace is regarded as an organization toward which an employee has an absolute obligation. This obligation also appears to influence Japanese practices where employees take less leave than is awarded to them by law. With respect to leave and working hours, then, Japanese legislation alone has been found ineffective and must be accompanied by an educational effort toward a change in cultural values in order for it legislation to be implemented in full. This indeed has been encouraged by the Japanese government in recent years.

With respect to dimension of masculinity, low scores are characteristic of societies where work is less central in people's lives, where there is a greater belief in gender equality, and where there is a preference for leisure over pay. Japan and the UK boast high scores on this dimension, whereas low scores characterize the US and Israel. Sweden and Denmark, the Scandinavian countries with the lowest scores on this dimension, are also those countries with the most generous parental and family leave policies. The medium-high score of the US on this dimension could explain, to a certain extent, its restricted family leave policy, in particular the lacking mandated remuneration of employees, which stands in contrast to all other countries. Israel's low score on this dimension seems to correspond well the country's family-friendly social policy at work while equating parental rights.

Attempts to influence policy so as to increase flexibility in working hours and generosity in leave policy, not merely with respect to family leave, often originate in women's organizations or organizations calling for a balance between work and family. Working women with families stand, not unexpectedly, at the forefront of efforts to develop more flexible policies of working hours and leave that are more agreeable to employees. The change, however, cannot be induced by legislation alone. Culturally endorsed values greatly influence employees' ability to balance between work and family. Thus, informal social support systems play a vital role in employees' ability to handle the conflicting demands of work and family (Etzion & Rechnitz-Elpeleg, 2002).

It might be said that most societies are currently undergoing a fundamental change, part of which involves the ways in which time is organized. The rigid way in which we organize our lives (education, work, retirement) that evolved over the past century is slowly disintegrating, and the boundaries between the separate stages of life and various other activities have become blurred, fluid, and more reversible. The relationships between culture and the law and between law and norm are complex. They are influenced by a variety of political, cultural, and economic forces specific to each country. Our conclusion is that those organizations and groups striving to introduce more amenable working hours and leave for employees must work not only for legislative change but for a cultural change and a change in values as well. Examination of practice *vis-à-vis* legislation in each country further leads to the conclusion that a change in policy regarding working time cannot be affected by legislation alone. It can only result from a change in cultural values as each country has its own typical characteristics arising from cultural perceptions, political and legislative history, and its own unique labor relations. Some researchers (e.g., Hattiangadi, 2000) suggest that shorter work weeks and increased flexible working arrangements may be more beneficial and practical as a means to better balance work and non-work life domains than leave days.

Assuming that labor legislation and customs reflect cultural values, we examined the extent to which a relationship exists between the cultural dimensions, as defined by Hofstede (1980) and Schwartz (1994) and the extent to which the policies regarding various kinds of leave and working hours were liberal or restrictive. We found, as predicted, that Schwartz's dimensions – affective autonomy, harmony and egalitarianism – were associated with more liberal policies, whereas the values represented by the dimensions of hierarchy and mastery were associated with more restrictive policies. No direct connection was found between the dimension of conservatism and leave policies. The cultural framework suggests that the centrality of work nationally is associated with greater generosity in countries' leave policies while work values pointing to low centrality of work are associated with more liberal leave policies. The predicted relationship between the dimension of individualism and leave policy was not confirmed while a relationship between the dimension of masculinity family-friendly policies was indeed established. The lower the score on the masculinity dimension, the more family-friendly national policies were found to be.

References

Ben David, D. (Ed.). (2014). *State of the Nation Report. The Herbert M. Singer Annual Report Series*. Taub Center for Social Policy Studies in Israel. Jerusalem, Israel. Retrieved from: http://taubcenter.org.il/wp-content/files_mf/stateofthenation report2014english24.pdf.

Dar, A. (2002). Law and custom regarding leave and working hours in Israel, Japan and the EU – A cross-cultural comparison. M.Sc., Tel Aviv University, Israel. [Hebrew].

Dorman, P. (2000). Maternity and family leave policies: A comparative analysis. *The Social Science Journal, 38(2)*, 189–201.

Ellingsæter, A. L. (2009). *Fiscal system and female employment in Norway*. Report by: European Commission Directorate-General for Employment, Social Affairs and Equal Opportunities, Unit G1 'Equality between women and men'.

Etzion, D. & Bailyn, L. (1994). Patterns of adjustment to the career/family conflict of technologically trained women in the U.S. and Israel. *Journal of Applied Social Psychology, 24(17)*, 1520–1549.

Etzion, D. & Dar, A. (2004). Leave and leisure worldwide – A cross-cultural comparison among the USA, the EU, Japan and Israel. Report by: Tel Aviv University, Henry Crown Institute of Business Research, Faculty of Management, Tel Aviv, Israel.

Etzion, D. & Rechnitz-Elpeleg, O. (2002). Formal and informal support systems for women managers: Israel versus Sweden. In L. Nicolaou-Smokoviti (Ed.), *Scientific yearbook: Essays in honor of Professor Litsa Nicolaou-Smokoviti* (pp. 429–443). Piraeus, Greece: University of Piraeus.

Figart, D. M. & Mutari, E. (1998). Degendering work time in comparative perspective: Alternative policy frameworks. *Review of Social Economy, 56(4)*, 460–480.

Frenkel, M. (2011). Policy on work-family balance: Findings from global organizations, countries, and employers. Literature review. Report by: Ministry of Industry, Trade, and Labor, Israel.

Galinsky, E., Freidman, D. E., & Hernandez, C. A. (1991). *The corporate reference guide to work-family programs*. New York: Families and Work Institute.

Haas, L. (1993). Nurturing fathers and working mothers: Changing gender roles in Sweden. In J. C. Hood (Ed.), *Men, work and family* (pp. 238–261). Thousand Oaks, CA: Sage Publications.

Haas, L. & Rostgaard, T. (2011). Fathers' rights to paid parental leave in the Nordic countries: Consequences for the gendered division of leave. *Community, Work & Family, 14 (2)*, 177–195.

Hall, M. H. (1997). The EU parental leave agreement and Directive: Implications for national law and practice, Eironline. Retrieved from: http://eiro.eurofound.ie/1998/01/study/TN9801201S.html.

Harpaz, I. (1999). The transformation of work values in Israel. *Monthly Labor Review, 122 (5)*, 46–50.

Hattiangadi, A. U. (2000). *Paid family leave: At what cost?* Washington, DC: The Employment Policy Foundation.

Hofstede, G. H. (1980). *Culture's consequences: International differences in work-related values.* Beverly Hills, CA: Sage Publications.

Hofstede, G. H. (1994). *Cultures and organizations: Software of the mind intercultural cooperation and its importance for survival.* London: Harper Collins.

Hofstede, G. H. (1998). A case of comparing apples with oranges: International differences in values. *International Journal of Comparative Sociology, 39(1)*, 16–25.

Imamura, I. (1998). Leisure & Nature. Retrieved from: www.miyazaki-mic.ac.jp/students/clas98/IMAMURA/practice/paid.htm.

Japanese Ministry of Health, Labor and Welfare. (2012). Working conditions and the labour market (January, 2001). The 1999 General Survey on Wages and Working Hours system reveals Increasing difficulty in taking paid leave. *Japan Labor Bulletin, 40.*

Kelly, I. & Dobbin, F. (1999). Civil rights law at work: Sex discrimination and the rise of maternity leave policies. *American Journal of Sociology, 105(2)*, 455–492.

Makita, T. & Ida, M. (2001). Highlights of value change in Japan. *International Journal of Public Opinion Research, 13(4)*. Retrieved from: www3.oup.co.uk/intpor/hdb/Volume_13/Issue_04/.

Mattes, K. (1992). In pursuit of leisure: Employee want more time off. *HR Focus, 69 (5)*, 1, 7.

Ministry of Labor, Social Affairs, and Social Services. (2018). *Advancing equal pay for women and men in the workplace, employer's manual.* Retrieved from: https://employment.molsa.gov.il/publications/publications/doclib/gpgemployerguide_en.pdf.

OECD. (2012). Various statistics downloaded in 2012, OECD iLibrary: https://www.oecd-ilibrary.org/statistics.

Radian, R. (1998). The moderating effect of workaholism and Type A on the impact of vacation on job stress and burnout. M.Sc., Tel Aviv University, Israel. [Hebrew].

Ray, R., Sanes, M., & Schmitt, J. (2013). No-vacation nation revisited. Report by: CEPR Report. Retrieved from: http://cepr.net/publications/reports/no-vacation-nation-2013.

Richter, J. (1992). Balancing work and family in Israel. In S. Zedeck (Ed.), *Work, families and organizations* (pp. 362–394). San Francisco, CA: Jossey-Bass.

Robinson, G. (2003, Sept. 7). Tragedy's quiet side: Understated travelling exhibit leaves room for overpowering emotion, *Star-Telegram*. Retrieved from: www.dfw.com/mld/dfw/entertainment/visual_arts/6698530.htm.

Rosenberg, J. Z. (2001, May). Gone fishin': Workers want more time off. Vault.com. Retrieved from: http://vault.com/blog/workplace-issues/gone-fishin-workers-want-more-time-off/.

Sato, H. (2000). The current situation of "family-friendly" policies in Japan. *Japan Labor Bulletin, 39*(February), 5–10.

Schor, J. B. (1992). *The overworked American: The unexpected decline of leisure*. New York: Basic Books.

Schwartz, S. H. (1992). The universal content and structure of values: Theoretical advances and empirical tests in 20 countries. In M. Zanna (Ed.), *Advances in experimental social psychology* (Vol.*25*, pp. 1–65). New York: Academic.

Schwartz, S. H. (1994). Beyond individualism-collectivism: New cultural dimensions of values. In U. Kim, H. C. Triandis, C. Kagitcibasi, S. C. Choi, & G. Yoon (Eds.), *Individualism and collectivism: Theory, method and applications* (pp. 85–119). London: Sage Publications.

Schwartz, S. H. (1995). Values in the west: A theoretical and empirical challenge to the individualism-collectivism cultural dimension. *World Psychology, 1(2)*, 99–122.

Schwartz, S. H. (1999). Cultural value differences: Some implications for work. *Applied Psychology: An international Review, 48(1)*, 23–47.

Shaviv, T. (1998). The impact of an organized tour vacation on work stress and burnout among managerial and non-managerial vacationers. M.Sc., Tel Aviv University, Israel. [Hebrew].

Super, D. E. & Sverko, B. (1995). *Life roles, values, and careers: International findings of the work importance study*. San Francisco, CA: Jossey-Bass.

Triandis, H. C. (1994). Theoretical and methodological approaches to the study of collectivism and individualism. In U. Kim, H. C. Triandis, C. Kagitcibasi, S. C. Choi, & G. Yoon (Eds.), *Individualism and collectivism: Theory, method and applications*. Thousand Oaks, CA: Sage Publications.

United States Code. Family and medical leave Act [On-line]. Retrieved from: http://caselaw.lp.findlaw.com/casecode/uscodes/29/chapters/28/sections/section_2601.html.

US Bureau of Labor Statistics. (2000). Employee benefit survey: Technical note - compensation and working conditions. U.S. Department of Labor. Retrieved from: http://stats.bls.gov/opub/ils/pdf/opbils09.pdf.

US Bureau of Labor Statistics. (2012). National Compensation Survey: Employee Benefits in the United States. Bulletin 2773. Retrieved from: www.bls.gov/ncs/ebs/benefits/2012/ebbl0050.pdf.

Vogel, R. J. (1993). *Financing health care in sub-Saharan Africa*. Westport, CT: Greenwood Press.

Watanabe, S., Takahashi, K., & Minami, T. (1997). The emerging role of diversity and work-family values in a global context. In P. C. Early & M. Erez (Eds.), *New perspectives on international industrial/organizational psychology* (pp. 276–318). San Francisco, CA: The New Lexington Press.

Working Conditions and the Labor Market. (2001a). The 1999 general survey on wages and working hours system reveals increasing difficulty in taking paid leave. *Japan Labor Bulletin, 40*(January). Retrieved from: www.jil.go.jp/bulletin/year/2001/vol40-01/02.htm#2.

Working Conditions and the Labor Market. (2001b). Men and women indifferent to men's child-care leave. *Japan Labor Bulletin, 40*(July). Retrieved from: www.jil.go.jp/bulletin/year/2001/vol40-07/02.htm.

5
OUTCOMES

Do respites live up to their reputation as an opportunity to recover and replenish resources? Respite studies amass to a body of data clearly evidencing post-respite psychological well-being, as well as post-respite physiological well-being. This has been demonstrated for respites of various natures and lengths from an evening at home to a year-long sabbatical. However, beneficial respite outcomes affect more than the respiting individual; similar to their stress, strain, and burnout antecedents, they also crossover to family members at home, as well as to team members at work. The respite effect, it must be acknowledged, is short-lived. The beneficial post-respite influence found on stress and burnout typically lifts within a month or so when levels of these types of strains return to their pre-respite levels.

Post-respite psychological well-being

Respite research is typically conducted as a repeated-measures method in which various indices of workers are measured in different combinations of time points, including during routine work periods, just prior to a respite of any given length, during that respite, immediately after the respite, and some time into the future after the respite, where routine working conditions are assumed to have resumed. On each occasion, workers rate subjective stress to validate the presumed high or low stress of the occasion, and psychological strain as a hypothesized response to stress. This time-series design allows for causal inferences assessing the influence of stress over strain. Results have been consistent in describing the rise and fall of subjective job stress and strain in the presence or absence of objective job stress (Eden, 2001). Indeed, many of our studies have established a strong positive correlation between stress and burnout throughout study measures. Patterns of stress and strain rising during work periods and falling during or immediately after respites have been consistently demonstrated for respites of various nature and length, including vacations (e.g., Etzion, 2003; Lounsbury & Hoopes, 1986), reserve military service (Etzion et al., 1998), business trips (Westman & Etzion, 2002), and even a short respite of evenings at home (e.g., Davidson et al., 2010; Sonnentag & Bayer, 2005). Respites following night shifts, however, seem to create adverse psychological effects with respect to mood, feelings, and cognitive-performance tasks, especially on the first rest day following the night shift (Totterdell, Spelten, Smith, Barton, & Folkard, 1995).

Empirical evidence has been accumulating that different types of respites positively influence psychological well-being and on-the-job behavior (Eden, 2001; Etzion et al., 1998; Fritz & Sonnentag, 2005; Sonnentag, 2001, 2003; Westman & Eden, 1997; Westman & Etzion, 2002). Evidence is primarily amassed regarding the alleviating effect of respite over perceived job stress and burnout (Aviv, 1998; Etzion, 2003; Etzion & Boklis, 2007; Sonnentag, 2003; Westman & Etzion, 2001), especially where those are chronic (e.g., Eden, 1990; Etzion, 2003; Etzion et al., 1998; Etzion & Sapir, 1997; Frankenhaeuser et al., 1989; Westman & Eden, 1997; Westman & Etzion, 2001). It was estimated that most workers perceive respites as having positive outcomes such as escapism, release from tension, personal development and the opportunity to engage in satisfying activities (Rubinstein, 1980). Workers often believe that respites are an absolute necessity, preventing them from burring out while improving the quality of their personal lives and subsequent job performance, and rendering them more amiable and creative (Schlossberg, 1991). In fact, we found evidence that vacation decreases absenteeism while increasing employees' perceived performance (Westman & Etzion, 2001).

Additional alleviating effects include relief from the negative consequences of work and effort expenditure (Fritz & Sonnentag, 2005), a sense of relaxation and a good mood (Frankenhaeuser et al., 1989), as well as higher performance and reduced levels of absenteeism at the workplace (Westman & Aharon-Madar, 1998). Other psychological strains alleviated by respite include depression, anxiety, dissatisfaction, ill mood, and burnout (Westman, Etzion, & Chen, 2009).

A study we conducted (Etzion, Eden, & Lapidot, 1994; Etzion et al., 1998) in a large industrial enterprise operating in the Israeli public sector documented a decline in the feelings of stress and burnout among men who served in military reserve duty in comparison with a matched group of men who carried on in their normal routines during this period. We also found that the more the reserve duty service was perceived as a positive experience, the greater the decline in the levels of stress and burnout upon returning to work. We additionally noted that the greater the level of detachment reported from the routine of civilian life, the greater the decline in the level of pressures reservists experienced upon their return to work. We found support of the effect of detachment from work and of involvement with non-work activities on post-leave burnout among female military employees who returned from maternity leave Ben Yoseff (2004) and similar results were also found among hi-tech employees (Sraya-Wallin, 1999).

In a different study (Etzion & Sapir, 1997) in which the matched group comparison method was used, we found that vacationing employees (compared to employees who stayed at work during the same period) experienced a decline in stress and burnout measures immediately after the vacation. These results were repeated in a later matched group comparison study (Etzion, 2003), where vacation length was positively related to the degree of relief in stress while also tending to moderate the relationship between vacation satisfaction and levels of perceived stress and burnout.

Chen et al. (2009) found significant differences in pre- and post-weekend respite among nurses whose exhaustion and negative affect were reduced immediately following the weekend. A respite's power of influence, however, is dependent on being free of non-work hassles and other negative experiences as in their presence, post-respite burnout and poor general well-being might increase (Fritz & Sonnentag, 2005). Thus, where levels of psychological well-being remain unchanged or turn to the worse, it would be well worth investigating the properties of the respite itself.

Even in particularly long types of respites where the non-work nature of the respite may be lost, such as sabbaticals, an ameliorating effect is recorded on well-being. Davidson et al. (2010) found that among sabbatical-takers resource loss declined and resource gain and well-being rose during the sabbatical where the comparison group showed no change. This study offered evidence for the salutogenic effects of the sabbatical on the full range of subjective well-being, including both the mitigation of negative well-being and the enhancement of positive well-being.

Business trips as a form of working respites were also shown to have an ameliorating effect of stress and burnout, albeit presenting a pattern different to that of non-work-related respites (Westman, Etzion, & Yoffe, 1998). Studying employees of Israeli hi-tech companies accustomed to overseas travel, averaging at four annual trips lasting six days each, we found that both job stress and burnout declined after the business trip upon employees' return to the permanent work site compared to its high levels before and even during the trip (Westman & Etzion, 2002). We also found that for women, in the face of work-family or family-work conflicts, burnout levels were the lowest during their business trip (Westman et al., 2008; Westman, Etzion, & Gortler, 2004).

Even in studies that yielded no direct vacation effect on stress and burnout (e.g., Ron, 2003; Shaviv, 1998), we are able to find other positive psychological effects. For example, we found that vacation satisfaction was positively associated with stress originating in lack of positive characteristics yet negatively associated with self-efficacy and respite in the form of leisure activities taken prior to a vacation was associated with lower burnout levels.

Post-respite physiological well-being

The physiological effect of stress as an instigator of the activation of the sympathetic nervous system was first noted by Cannon (Cannon, 1914; 1932), who coined the *fight-or-flight* term. Selye (1956) described the physiological response as the General Adaptation Syndrome, which includes three stages of alarm, resistance, and exhaustion, leading to cumulative damage to the organism. Measuring epinephrine 17-ketosteroid excretion of single worker daily for four years, Halberg, Engeli, and Hamburger (1965) found a crest at mid-week and a decline during weekends. Physiological systems have been since included in different psychological stress models (Eller et al., 2009; Kahn & Byosiere, 1992; Melamed et al., 2006). The persistence of stress contributes to long-lasting illness; heart disease, stroke, cancer, diabetes, obesity, and arthritis. Chronic stress results in lowered immune function, increased incidence of depression, and the comprehensive reduction in brain function. Stress attacks the main types of immune cells, the two types of white blood cells; lymphocytes and phagocytes that reduces the immune system's ability to fight off disease (Kelley, 2009).

In essence, two physiological stress systems respond to stress in regulatory processes of the body that are designed to mobilize energy stores and cardiovascular tone to facilitate adaptation. The sympathetic-adrenergic medullary (SAM) responds immediately by sympathetic activation such as the release of catecholamines (e.g., epinephrine and norepinephrine), which regulate an increase in blood pressure and heart rate. The hypothalamic-pituitary-adrenal (HPA) axis responds more slowly by releasing of corticosteroids (e.g., cortisol), resulting in immune suppression. Both systems activate and interact during stress episodes (De Vente, Olff, Van Amsterdam, Kamphuis, & Emmelkamp, 2003; Seeman & McEwen, 1996). When dysregulation occurs, leading to hormone secretion, increased cortisol level (hypercortisolism) and finally, hypocortisolism, work stress is physiologically translated into various ailments.

Exposure to chronic stress, including burnout and vital exhaustion, may impair cardiovascular health (Belkic, Landsbergis, Schnall, & Baker, 2004; Hemingway & Marmot, 1999; Kopp & Rethelyi, 2004; Marmot, Theorel, & Siegrist, 2002; Melamed et al., 2006; Strike & Steptoe, 2004) and increase the risk for ischemic heart disease (Eller et al., 2009; Kivimaki et al., 2006; Schuitemaker, Dinant, Van Der Pol, & Appels, 2004), as well as cardiovascular disease and coronary heart diseases (e.g., Hallman, Thomsson, Burrell, Lisspers, & Seetterlind, 2003; Prescott, Holst, Schnohr, & Barefoot, 2003). Chronic stress and repeated acute episodes may additionally interfere with the normal immune inflammatory response, as well as with blood coagulation processes (Hourani, Council, Hubal, & Strange, 2011). Under chronic stress, ongoing secretion of cortisol from the adrenal glands may eventually damage brain cells. Chronic stress additionally leads to a rise in blood sugar gastrointestinal inflammation which then cyclically contributes to physiological stress. Stress also drastically reduces zinc levels, causing individuals to be more irritable and less focused and exacerbate anxiety, hence accentuating the effects of the original stressors (Kelley, 2009). The work of managers is often associated with acute stress and strain, leading managers to experience stress-related health impairments, which might force early retirement among such managers (Robinson & Inkson, 1994; Zalowski, Hall, & Baum, 1992).

When stress responses are not turned off after duty and are overused, a cumulative 'wear and tear' on the body's physiological systems (neuroendocrine system, autonomic nervous system, and immune system) creates allostasis, or allostatic load, ultimately leading to pathological changes (McEwen, 2003). Adverse health symptoms have been found to be stronger on work days compared to non-work days (e.g., Halberg et al., 1965; Schlotz, Hellhammer, Schulz, & Stone, 2004), as well as on particularly stressful work days compared to less straining or post-vacation days (Glaser, Kiecolt-Glaser, Speicher, & Holiday, 1985). Anxiety, depression and heart rate were lower after the summer vacation compared to five months earlier, during the stressful period of high workload (Caplan & Jones, 1975; Eden, 1990). Blood pressure, heart rate, and epinephrine excretions were lower on a day off at home compared to a day on the job (Frankenhaeuser et al., 1989; Goldstein, Shapiro, Chicz-DeMet, & Guthrie, 1999), as well as during work periods characterized by low stress compared to work periods characterized by high stress (Ritvanen, Veikko, Pertti, Sari, & Osmo, 2006). Air-traffic controllers experiencing increases in job stressors (perceived workload, objective traffic volume, distressing social interactions) also experienced during the same day deterioration in physical and psychological well-being in the form of health complaints, more negative moods, and fewer positive moods (Repetti, 1993). Blood pressure and heart rate were also found to be higher during work time than in the evenings or on leisure days (Evans & Steptoe, 2001).

Subjective physical well-being, however, seems to be worse immediately after one's vacation than before it or during work time (Laski, 2013). Physical fatigue does not significantly differ before and after the vacation; if anything, the vacation leads to slightly higher fatigue (Ben Yoseff, 2004). Although check-ins at hospital emergency rooms drop during non-work days, younger employees tend to check-in on non-work days due to work-related considerations. Check-ins increase significantly on work days immediately following non-work days. Men at working ages tend to check-in at hospital emergency rooms more often than women for stress-related symptoms. Asthma and headaches drive patients to the emergency room more often on non-work days (Tuval, 2001).

A cessation of ongoing job demands may slow down, reduce, or even halt the allostatic load. Differences found between personality Types A and B are suggestive of the idea that reduction of stress is associated with better health (Lee et al., 1988; Matthews, 1982; Rosenman et al., 1970). Neuroendocrine measures and work characteristics were predictors for the amount of perceived need for recovery after work, as well as for health status (Sluiter et al., 2001). Hence, it has been proposed that temporary relief between two periods of work and stress enables psycho-physiological systems to rebalance to more optimal levels of activation (de Bloom et al., 2010; Sluiter et al., 1999; Zijlstra & Cropley, 2006).

Glaser et al. (1985) measured changes in students' physical resistance to viruses at three points in time: the day before a final exam; the day of a final exam; and immediately upon returning from a semester break. They found significant differences in the body's resistance at these three points. After a semester break, the resistance levels were highest, indicating a low level of stress.

However, the physiological effects of vacation have been only infrequently studied. Physiological strains known to be reduced following respite include cardiovascular and neuro-endocrine malfunctions, physical pains, and sleep disturbances (e.g., Frankenhaeuser et al., 1989; Strauss-Blasche et al., 2004). Among participants studied before and after a three-week vacation in a health resort in Austria, higher levels of perceived job demands were associated with higher lipid concentrations (LDL/HDL cholesterol and triglycerides). Following their respite, participants experiencing higher demands showed a greater decrease in serum lipids (Strauss-Blasche et al., 2003). In a longitudinal study conducted over nine years with the participation of over 12,000 middle aged men at high risk for coronary heart disease, frequent annual vacations were associated with reduced risk for all-cause mortality, including mortality attributed to coronary heart disease (Gump & Matthews, 2000). In other words, recovery is expected to result in a reduction of physiological strain indicators (Totterdell et al., 1995). In fact, in a study of Austrian industrial workers, results showed improvement in physical complaints, the quality of sleep, and mood after a vacation compared to their pre-vacation state (Strauss-Blasche et al., 2000).

Recovery is not equivalent for all persons. Physiological tendencies measured among leisure travelers show more pronounced increase in the levels of CgA during the trip and of cortisol after the trip for those individuals who reported a more health-related lifestyle. These findings suggest that the effects of travel were more beneficial for persons with positive characteristics related to health-related lifestyle (Toda, Makino, Kobayashi, & Morimoto, 2007). In contrast, our own study revealed that individuals experiencing pre-hypertension blood pressure prior to their vacation experienced a significant decline in blood pressure during their vacation, whereas individuals medically defined as having normal blood pressure prior to their vacation experienced no significant change. This result suggests that the vacation may contribute to those who are in a greater need of physiological relief (Laski, Etzion, & Westman, 2015). Gender also seems to play a role in recovery. For example, men showed a smaller decrease in HDL and a greater decrease in LDL/HDL compared to women on post-respite measurements (Strauss-Blasche et al., 2003). Similarly, female managers seemed to unwind more slowly in after-work hours (Frankenhaeuser et al., 1989).

Crossover

Crossover is a phenomenon where psychological experiences of one person affects the psychological experiences of another. Traditionally, the term crossover relates to the contagion of stress originating in one spouse at their workplace to the other spouse in the family domain (Westman, 1997), whereas effects carrying over from one domain to the other are considered to spillover (Piotrkowski, 1979; Westman, 1997). Most studies have investigated the crossover of psychological strains, such as burnout (e.g., Bakker & Schaufeli, 2000), distress (Barnett, Raudenbush, Brennan, Pleck, & Marshall, 1995), maladjustment (Takeuchi, Yun, & Teslu, 2002), work–family conflict (Hammer, Allen, & Grigsby, 1997), marital dissatisfaction (Westman, Vinokur, Hamilton, & Roziner, 2004). In our context, we studied the crossover of both negative and positive psychological experiences between spouses. That is, the contagious influence of an experience originating in the workplace by one spouse as it leads to a similar experience by the other spouse who does not share the same workplace (Westman, 1997; Westman & Etzion, 1995). This process emphasizes the fact that boundaries between the work and family domains are permeable.

Different mechanisms have been suggested in explanation of the crossover experience. In the case of negative experiences such as strain, it is possible that demands experienced by one spouse affects demands experienced by the other spouse. For example, specific occupational demands of administrators were found to positively relate the occupational demands of their spouses (Burke et al., 1980) while stress experienced by individuals at their workplace led to stress being experienced by their spouses (Bolger, DeLongis, Kessler, & Wethington, 1989). Other suggested mechanisms involve an indirect transfer, such as via the transfer of moods between spouses. Moods may be transferred between spouses, the mood of one spouse may influence the experience of demands of the other, or demands experienced by one spouse may influence the mood of the other. For example, a positive relationship was found between police officers' perceived job stress and their wives' marital dissatisfaction as reported by both husbands and wives, that is, the police officers' job stress exacted an emotional price on their family (Jackson & Maslach, 1982). Similarly, empathy may open a path from crossover (Riley & Eckenrode, 1986). In a different study, the distress experienced by one spouse was identified as a source of ongoing stress for the other spouse (Cronkite & Moos, 1984). Other studies have found a crossover effect in both stress, and burnout (Bakker & Schaufeli, 2000).

We ourselves found crossover of depression among spouses (Vinokur et al., 1996; Westman & Vinokur, 1998), as well as anxiety (Westman, Etzion, & Horovitz, 2004), marital dissatisfaction (Westman et al., 2004), and work–family conflict (Westman et al., 2004). Where negative experiences such as stress or burnout are concerned, one's own negative experience may be a drive for negative behaviors, such as negative affect or social undermining, toward the spouse, which are experienced by the recipient as an additional source of stress. We and others found evidence for such crossover in several studies (Conger et al., 1993; Vinokur et al., 1996; Westman & Vinokur, 1998). Notwithstanding, it is possible that symptoms may cross over between spouses through a process of modelling or that both spouses are exposed to a common cause of psychological experiences (Westman & Etzion, 1995). For example, we found a unidirectional crossover of strain from husbands to wives working in the same organization during a period of layoffs (Westman, Etzion, & Danon, 2001). Research has exposed crossover both unidirectionally (e.g., Hammer et al., 1997; Jones & Flecher, 1993) and bidirectionally (e.g., Barnett et al., 1995; Westman, Etzion, & Segev, 2002; Westman & Vinokur, 1998).

Crossover may be hazardous to receiving spouses' health. Thus, wives of husbands who experienced high occupational demands were found more likely to consume more cigarettes, coffee, and alcohol (Burke et al., 1980). In a different study, men married to women who were holding white-collar jobs were found to be three times more likely to develop heart disease compared to men whose wives held blue-collar jobs or who stayed at home (Haynes, Eaker, & Feinleib, 1983). Similarly, an association was found between mortality of married women and their husbands' occupations (Fletcher, 1983; 1988). The same study also revealed that spouses tend to die of the same disease. Additional health conditions crossing over between spouses include anxiety (Westman, Etzion, & Horovitz, 2004), depression (Howe, Levy, & Caplan, 2004), and physical and mental health complaints (Gorgievski-Duijvesteijn, Giessen, & Bakker, 2000).

Our findings show that crossover transpires not only from the realm of the workplace to that of the family, but also from that of the family to that of the workplace. We may thus say that negative occupational-related experiences are not merely a work concern but rather a problem that spills over and becomes a family issue (Westman & Etzion, 1995). In our studies, we have indeed confirmed the crossover of burnout (Westman & Etzion, 1995), occupational stress (Westman & Etzion, 1999), and anxiety (Horovitz, 2002) from one spouse to the other. The stronger the stress, the more intensive the crossover effect (Etzion & Westman, 2001; Westman, Etzion, & Horovitz, 2004). The crossover of burnout was established after controlling for the primary (study participant) spouses' own job stress and resistance resources. There seems to be a self-perpetuating cycle of contagion where the secondary spouse's burnout serves as an additional source of stress to the primary spouse. In dual-career families, where both spouses are working, non-work resources may perhaps be dwindled due to their sharing between spouses. For example, in our study (Westman & Etzion, 1995) of male military officers' whose wives were educated professionals comparable to their husbands, many of the wives chose to assume jobs that would have been considered traditionally feminine. It is possible that their choice of jobs was strongly influenced by the fact that their husbands had highly demanding and inflexible jobs that prevented them from participating equally in family tasks. Additionally, the great demands placed upon military officers and their spouses (such as social disruption, relocation, periods of separation, and the ever-present threats presented by training and combat) may explain the high burnout experienced by the wives in our study. We further supported these findings in a later study of married couples where the wife was the one who assumed a military job (Westman et al., 2002). We were also able to establish that crossover occurred bidirectionally not only between spouses but also between team members at the workplace (Westman & Etzion, 1999).

By similar mechanisms, and especially that of empathy, it would be logical to assume that positive experiences and not merely negative ones would also crossover between spouses (Bakker & Demerouti, 2009). Positive experiences that may be surmised to crossover between spouses include job satisfaction, and good moods (Westman et al., 2009). Positive psychological experiences that have indeed been found to crossover between spouses include work engagement (Bakker & Demerouti, 2009; Bakker, Le Blanc, & Schaufeli, 2005), life satisfaction (Demerouti, Bakker, & Schaufeli, 2005; Powsthavee, 2007; Prince, Manolis, & Minetor, 2007), and vigor (Westman et al., 2009). Vigorous individuals are characterized by high levels of energy and mental resilience while working, willingness to invest effort in work, and persistence; they often take initiative, generate their own positive feedback, and tend to invest energies in non-work life domains (Schaufeli & Bakker, 2004). We also found the positive influence of the crossover of one's sense of control (Vinocour-Hai, 2002). It appears that one's sense of control mitigates not only their own experiences of stress and burnout but also helps to strengthen the spouse's sense of control, which in turn helps the spouse in alleviating their own experiences of stress and burnout (Westman & Etzion, 1995; Westman et al., 2001). Our studies also revealed a crossover effect of job induced tension within team members: from principals to teachers of elementary schools and *vice versa* (Westman & Etzion, 1999).

The presence of crossover was found in strenuous working environments but not in laxer environments (Jones & Flecher, 1993). Now the question arises if vacation might have beneficial effects not only on the experiences of the vacationing individual but also on any negative crossover between that individual and the people with whom they have relationships, whether they are one's spouse, children, or team members. We have conducted several studies that looked into this issue. Westman and Etzion (1999) analyzed the relationships between stress and burnout among dual-career couples on two occasions, one stressful and the other a less stressful respite. Evidence was found for the crossover of strain between partners on the stressful occasion but not during the respite. Vinocour-Hai (2002) established the same effect for stress attributable to lack of positive work characteristics but not for burnout. In contrast, both Shaviv (1998) and Goshen-Albocher (1997) found, in different settings, that vacation had an ameliorating influence on the crossover of burnout between spouses. Even where strain existed, it did not cross over in more relaxed situations. That is to say that the extent of contagiousness of stress (in the form of burnout) was a function of stress characterizing one's environment. The more stressful the environment, the more contagious the crossover effect, whereas a tranquil climate helped stop the vicious cycle of crossover of strain, back and forth repetitively, from one spouse to another (Westman & Etzion, 2001). Prior to the vacation, burnout crossed over between spouses bidirectionally, whereas the crossover of post-vacation burnout was reduced to a mere tendency and only from husbands to wives but not *vice versa* (Shaviv, 1998). Possibly gender-based differences in the magnitude of crossover are related to gender-based variance in perceptions regarding work–non-work life balance and the importance of either realm in one's life, especially *vis-à-vis* traditional perceptions regarding gender roles in work and non-work (particularly family) realms. It is also possible that wives are more sensitive and empathetic to their husbands' emotional state thereby becoming more susceptible to their husbands' emotional states (Westman et al., 2001; Westman & Vinokur, 1997).

In a different study, Etzion, Westman, and Vinocour-Hai (2002) demonstrated the ameliorating effect of vacation on job stress. Here, too, crossover was significantly reduced (by half on average) after an institutional vacation of the primary study participant compared to its level prior to the vacation. Perceiving one's stress as a family issue rather than an issue of the individual alone, it appears, then, that respites from work offer an opportunity of recovery for all family members where the quality of the entire family's life may be elevated. We may additionally suggest that the process of changes during a respite where spouses spend nearly all of their time together in comparison to daily life where they spend most waking hours separated by work. Interestingly, while the intensity of stress contagiousness is intensified in stressful working conditions, the opposite seems to be true of stress arising among the unemployed. Two months into the state of unemployment, anxiety crosses over more from spouses to the unemployed than from the unemployed to their spouses (Horovitz, 2002). This is perhaps reflective of some adjustment to unemployment as an ongoing condition. That is, this finding may be akin to the effect we found in another study where during post-vacation routine conditions spouse-to-employee crossover continued to decline, whereas employee-to-spouse crossover rose (Etzion et al., 2002).

Fadeout of the respite effect

While we have evidenced the beneficial effect of vacation of stress, burnout, and other conditions, both physical and physiological, we have also witnessed the limited lifespan of such influences. We are not alone in this observation. Meta-analyzing seven vacation studies, de Bloom et al. (2009), for example, found evidence for fadeout periods of two to four weeks after vacation and joined other researchers in calling for further examination of the fadeout process, its length and nature.

In our own studies, Magen-Neuhaus (1993) and Westman and Eden (1997) recorded a post-vacation gradual fadeout effect as burnout levels among employees returning to work climbed three days after the vacation part way toward its pre-vacation level and all the way by three weeks after the vacation. Some of our studies found that job stress and burnout, as well as behavioral facets such as reduced performance were alleviated by a vacation but returned to their pre-vacation levels within approximately four weeks (e.g., Etzion & Sapir, 1997; Westman & Etzion, 2001; Westman, Etzion, & Aharon-Madar, 1999). Breaking down burnout to its components, we found exhaustion returned to its pre-vacation level three weeks after the vacation but cynicism was only relieved during the vacation, bouncing back to its pre-vacation level immediately after the vacation (Laski et al., 2015). Other psychological states also benefited from the vacation, such as job satisfaction, but returned to their pre-vacation level during the subsequent routine work period (Laski, 2002). We also noted a physiological health promotion aspect of post-vacation relief measured by both systolic and diastolic blood pressure attesting to an adaptive autonomic response towards physiological balance (Laski et al., 2015).

Absenteeism presented an interesting pattern as the rate of health-related absenteeism remained rather constant before the vacation and immediately after it yet dropped significantly four weeks after vacation. Compared to its pre-vacation level, absenteeism for other reasons was lower two weeks after vacation, as well as four weeks after it albeit slowly rising during this period (Westman & Aharon-Madar, 1998; Westman et al., 1999). In a different study, absenteeism presented a pattern similar to that of burnout level, returning to its pre-vacation levels within four weeks upon employees' return to work (Westman & Etzion, 2001). Perhaps these results are related to the respite aspect of absenteeism akin to the findings of Wright and Cropanzano (1998), who found respite could be taken in the form of costly turnover.

Etzion (2003) recorded a reinstatement of pre-vacation stress levels within one month after employees' return to work, whereas burnout remained low at that same point in time. In other words, the ameliorative effect of a vacation on burnout persisted beyond is effect on stressors, and it was independent of the length of vacation. Similar findings emerged in an earlier study where it was also found that the employees taking longer vacations experienced less stress than those taking short vacations before the vacation, immediately after it, and upon the reinstatement of routine conditions later on (Sapir, 1995). The length of vacation influenced the levels of post-vacation stress, which were lower following a longer vacation. While we have not investigated this further, these results may indicate that individuals experiencing less stress are more at ease about taking longer respites. This idea may also be related to our findings that some groups enjoyed greater relief through the vacation than others. For example, Westman and Eden (1997) noted that women and those who were satisfied with their vacations experienced the greatest relief. They were nevertheless also the quickest to experience fadeout of this relief. Among men on active reserve duty, Etzion et al. (1998) found that the greater the detachment experienced by these men, the stronger the relief of stress and burnout they reported. Among teachers, one of our studies revealed a longer fadeout process for job demands, of at least five weeks (Laski et al., 2015). Vacations may have an extended effect on teachers due to their integral facet of the job and frequently scheduled along the school year. It is possible that the longer fadeout could be associated with the situational characteristics of the study participants as being women and mothers. An alternative explanation, however, for the extended period of fadeout found among this group may be the perception of the vacation as part of the benefits of one's vocation and expecting this benefit to recur for as long as the vocation is maintained.

The onset of job demands after vacation may make a vacation's positive effects temporary. The conditions pertaining to the workplace, the vacation, and other situational factors, as well as personal ones seem to influence the rate of fadeout, but no case seems to be completely devoid of this effect. Perhaps the most significant factor in determining fadeout is one's baseline levels of stress and burnout. We found different patterns of relief among those whose baseline exhaustion level was low compared to those for whom it was high. In the first group, exhaustion declined during vacation and rose back to its initial level upon employees' return to work, whereas in the second group, exhaustion significantly dropped during vacation and stayed at that lower level for five weeks afterward, showing a long fadeout process. In other words, individuals who were more exhausted experienced a longer fadeout process compared to those who were initially less exhausted. Cynicism, too, revealed different patterns based on baseline levels: individuals experiencing low baseline levels of cynicism documented a constant level of cynicism over the five weeks that followed the vacation, whereas those who experienced high baseline levels of cynicism demonstrated a gradual decline during vacation that rose partway a week after the vacation and returned close to its initial level three weeks after vacation (Laski et al., 2015).

Collectively, then, our results trace a resource drain that leads to psychological and physiological impairment while removal of demands lead to gains. The incorporation of the pre-vacation (baseline) individual state in respite research enables the reinforcement of the Conservation of Resources theory (Hobfoll, 1989; 2001) corollaries, indicating that those who possess greater resources are less vulnerable to resource loss and are more capable of resource gain and *vice versa*. In other words, vacation may contribute most to those who are in a greater need of physiological relief (Laski et al., 2015). It might also be that the higher the need for recovery, the longer it takes to recover as indicated by teachers who were highly strained and took longer to unwind compared to less strained colleagues (Sonnentag & Zijlstra, 2006).

These processes seem to be fully compatible with Conservation of Resources theory whereby resource loss arouses a stronger motivational power than resource gain. Loss experiences are likely to evoke loss prevention strategies that may be employed during such an opportunity as an organizational vacation, especially by those who have greater resource deficiencies, such as those initially more burned out. Individuals who perceived themselves to be high in job resources experienced a decline in job resources during vacation that lasted, at a decreasing rate, for five weeks after the vacation, whereas individuals who perceived themselves to be low in job resources experienced no change in their perception during or after the vacation. We might presume that during an organizational shut down due to an institutional vacation, a natural segmentation process takes place so that investments in resources during vacation are targeted mainly on non-work domains. This separation and distance may result in a reduction in job-oriented resources so that those who are in a greater need of recovery manage to gain more resources while away from the job setting. The primacy of loss stipulated by the Conservation of Resources theory implies that loss experiences are likely to evoke loss prevention strategies rather than an active quest for opportunities of resource gains and new investments. Actual or anticipated resource loss may produce an inhibitory impact on the process of engagement, which is critical for future resource gain (Gorgievski & Hobfoll, 2008). While going on vacation, then, employees may first focus on the reduction of stress, while resource gain may show a lagged response as they are secondary to loss prevention.

References

Aviv, N. (1998). The meaning of vacation among women working shifts: The effect of vacation on job stress, burnout and work-home conflict. M.Sc., Tel Aviv University, Israel. [Hebrew].

Bakker, A. B. & Demerouti, E. (2009). The crossover of work engagement between working couples: A closer look at the role of empathy. *Journal of Managerial Psychology, 24(3)*, 220–236.

Bakker, A. B., Le Blanc, P. M., & Schaufeli, W. B. (2005). Burnout contagion among nurses who work at intensive care units. *Journal of Advanced Nursing, 51(3)*, 276–287.

Bakker, A. B. & Schaufeli, W. B. (2000). Burnout contagion processes among teachers. *Journal of Applied Social Psychology, 30(11)*, 2289–2308.

Barnett, R. C., Raudenbush, S. W., Brennan, R. T., Pleck, J. H., & Marshall, N. L. (1995). Change in job and marital experiences and change in psychological distress: A longitudinal study of dual-earner couples. *Journal of Personality and Social Psychology, 69 (5)*, 839–850.

Belkic, K. L., Landsbergis, P. A., Schnall, P. L., & Baker, D. (2004). Is job strain a major source of cardiovascular disease risk? *Scandinavian Journal of Work and Environmental Health, 30(2)*, 85–128.

Ben Yoseff, N. (2004). The effect of maternity leave on burnout among women serving in the I.D.F. M.Sc., Tel Aviv University, Israel. [Hebrew].

Bolger, N., DeLongis, A., Kessler, R., & Wethington, E. (1989). The contagion of stress across multiple roles. *Journal of Marriage and Family, 51(1)*, 175–183.

Burke, R. J., Weir, T., & Dunwors, R. E. (1980). Work demands on administrators and spouse well-being. *Human Relations, 33*, 253–278.

Cannon, W. B. (1914). The emergency function of the adrenal medulla in pain and the major emotions. *American Journal of Physiology, 33(2)*, 356–372.

Cannon, W. B. (1932). *The wisdom of the body*. New York: Norton.

Caplan, R. D. & Jones, K. W. (1975). Effects of work load, role ambiguity, and Type A personality on anxiety, depression and heart rate. *Journal of Applied Psychology, 60(6)*, 713–719.

Chen, S., Westman, M., Etzion, D., & Dach-Zahavi, A. (2009). The impact of a short respite on well-being and performance (Working Paper). Tel Aviv University, Henry Crown Institute of Business Research, Faculty of Management, Tel Aviv, Israel.

Conger, R. D., Lorenz, F. O., Elder, G. H., & Simon, R. L. (1993). Husband and wife differences in response to undesirable life events. *Journal of Health and Social Behavior, 34 (1)*, 71–88.

Cronkite, R. C. & Moos, R. H. (1984). The role of predisposing and moderating factors in the stress illness relationship. *Journal of Health and Social Behavior, 25(4)*, 372–393.

Davidson, O. B., Eden, D., Westman, M., Cohen-Charash, Y., Hammer, L. B., Kluger, A. N., et al. (2010). Sabbatical leave: Who gains and how much? *Journal of Applied Psychology, 95(5)*, 953–964.

de Bloom, J., Compier, M., Guerts, S., de Weerth, C., Toon, T., & Sonnentag, S. (2009). Do we recover from vacation? Meta-analysis of vacation effects on health and well-being. *Journal of Occupational Health Psychology, 51(1)*, 13–25.

de Bloom, J., Geurts, S. A. E., Taris, T. W., Sonnentag, S., de Weerth, C., & Kompier, M. A. J. (2010). Effects of vacation from work on health and well-being: Lots of fun, quickly gone. *Work & Stress, 24(2)*, 196–216.

de Vente, W., Olff, M., Van Amsterdam, J. G., Kamphuis, J. H., & Emmelkamp, P. M. G. (2003). Physiological differences between burnout patients and healthy controls: Blood pressure, heart rate, and cortisol responses. *Occupational and Environmental Medicine, 60 (1)*, i54-i61.

Demerouti, E., Bakker, A. B., & Schaufeli, W. B. (2005). Spillover and crossover of exhaustion and life satisfaction among dual-earner parents. *Journal of Vocational Behavior, 67(2)*, 266–289.

Eden, D. (1990). Acute and chronic job stress, strain and vacation relief. *Organizational Behavior and Human Decision Processes, 45(2)*, 175–193.

Eden, D. (2001). Vacations and other respites: Studying stress on and off the job. In C. L. Cooper & I. T. Robertson (Eds.), *International review of industrial and organizational psychology* (pp. 121–146). Chichester, UK: Wiley.

Eller, N. H., Netterstrom, B., Gyntelberg, F., Kristensen, T. S., Nielsen, F., & Steptoe, A. (2009). Work-related psychosocial factors and the development of ischemic heart disease: A systematic review. *Cardiology in Review, 17(2)*, 83–97.

Etzion, D. (2003). Annual vacation: Duration and relief from job stress and burnout. *Anxiety, Stress, and Coping, 16(2)*, 213–226.

Etzion, D. & Boklis, V. (2007). Recovery from job-stress and burnout – The effects of respite from work: A meta-analytic study. Paper presented at the XIIIth congress for the European Congress of Work and Organizational Psychology, Stockholm, Sweden.

Etzion, D., Eden, D., & Lapidot, Y. (1994). Active reserve duty as an ameliorative respite from civilian job-stress and burnout. Paper presented at The Fifth International Conference on Social Stress Research, Honolulu, HI.

Etzion, D., Eden, D., & Lapidot, Y. (1998). Relief from job stressors and burnout: Reserve service as respite. *Journal of Applied Psychology, 83(4)*, 577–585.

Etzion, D. & Sapir, O. (1997). Annual vacation: Duration of relief from job stress and burnout. Paper presented at the 12th Annual Conference for The Society for Industrial–Organizational Psychology, St. Louis, MO.

Etzion, D. & Westman, M. (2001). Job stress, vacation and the crossover of strain between spouses – Stopping the vicious cycle. *Men and Work, 11(1–2)*,51–62 (Hebrew version), 106–118 (English version).

Etzion, D., Westman, M., & Vinocour-Hai, V. (2002). The impact of high & low stress periods on the intensity of burnout crossover between spouses: A longitudinal study. Paper presented at The XXV International Congress of Applied Psychology (ICAP), Singapore, Asia.

Evans, O. & Steptoe, A. (2001). Social support at work, heart rate and cortisol: A self-monitoring study. *Journal of Occupational Health Psychology, 6(4)*, 361–370.

Fletcher, B. (1983). Marital relationships as a cause of death: An analysis of occupational mortality and the hidden consequences of marriage—Some U.K. data. *Human Relations, 36(2)*, 123–134.

Fletcher, B. (1988). Occupation, marriage and disease-specific mortality concordance. *Social Science & Medicine, 27(6)*, 615–622.

Frankenhaeuser, M., Lundberg, U., Fredrikson, M., Melin, B., Tuomisto, M., Myrsten, A.-L., et al. (1989). Stress on and off the job as related to sex and occupational status in white-collar workers. *Journal of Organizational Behavior, 10(4)*, 321–346.

Fritz, C. & Sonnentag, S. (2005). Recovery, health, and job performance: Effects of weekend experiences. *Journal of Occupational Health Psychology, 10(3)*, 187–199.

Glaser, R., Kiecolt-Glaser, J. K., Speicher, C. E., & Holiday, J. E. (1985). Stress, loneliness, and changes in herpesvirus latency. *Journal of Behavioral Medicine, 8(2)*, 25–49.

Goldstein, I. B., Shapiro, D., Chicz-DeMet, A., & Guthrie, D. (1999). Ambulatory blood pressure, heart rate and neuroendocrine responses in women nurses during work and off work days. *Psychosomatic Medicine, 61(3)*, 387–396.

Gorgievski, M. J. & Hobfoll, S. E. (2008). Work can burn us out or fire us up: Conservation of Resources in Burnout and Engagement. In J. R. B. Halbesleben (Ed.), *Handbook of stress and burnout in health care* (pp. 7–22). New York: Nova Publishers.

Gorgievski-Duijvesteijn, M. J., Giessen, C. W., & Bakker, A. B. (2000). Financial problems and health complaints among farm couples: Results of a 10-year follow-up study. *Journal of Occupational Health Psychology, 5(3)*, 359–373.

Goshen-Albocher, M. (1997). Crossover of stress and burnout between executive managers and their spouses. M.Sc., Tel Aviv University, Israel.

Gump, B. B. & Matthews, K. A. (2000). Are vacations good for your health? The 9-year mortality experience after the multiple risk factor intervention trial. *Psychosomatic Medicine, 62(5)*, 608–612.

Halberg, F., Engeli, M., & Hamburger, C. (1965). The 17-ketosteroid excretion of healthy men on weekdays and weekends. *Experimental Medicine and Surgery, 23(1)*, 61–64.

Hallman, T., Thomsson, H., Burrell, G., Lisspers, J., & Seetterlind, S. (2003). Stress, burnout and coping: Differences between women with coronary heart disease and healthy matched women. *Journal of Health Psychology, 8(4)*, 433–445.

Hammer, L. B., Allen, E., & Grigsby, T. D. (1997). Work-family conflict in dual-earner couples: Within-individual and crossover effects of work and family. *Journal of Vocational Behavior, 50(2)*, 185–203.

Haynes, S. G., Eaker, E. D., & Feinleib, M. (1983). Spouse behavior and coronary heart disease in men: Prospective results from the Framingham Heart Study: I. concordance of risk factors and the relationship of psychosocial status to coronary incidence. *American Journal of Epidemiology, 118(1)*, 1–22.

Hemingway, H. & Marmot, M. (1999). Evidence based cardiology: Psychosocial factors in the etiology and prognosis of coronary heart disease; systematic review of prospective cohort studies. *British Medical Journal, 318(7196)*, 1460–1467.

Hobfoll, S. E. (1989). Conservation of resource – A new attempt at conceptualizing stress. *American Psychologist, 44(3)*, 513–524.

Hobfoll, S. E. (2001). The influence of culture, community, and the nested self in the stress process: Advancing Conservation of Resources Theory. *Applied Psychology: An International Review, 50(3)*, 337–421.

Horovitz, S. (2002). The relationship of economic stress and social support and crossover anxiety between couples in times of unemployment. M.Sc., Tel Aviv University, Israel. [Hebrew].

Hourani, L. L., Council, C. L., Hubal, R. C., & Strange, L. B. (2011). Approaches to the primary prevention of posttraumatic stress disorder in the military: A review of the stress control literature. [Retrieved from the Walden Library using the ProQuest Central database]. *Military Medicine, 176(7)*, 721–730.

Howe, G., Levy, M., & Caplan, R. (2004). Job loss and depressive symptoms in couples: Common stressors, stress transmission, or relationship disruption? *Journal of Family Psychology, 18(4)*, 639–650.

Jackson, S. E. & Maslach, C. (1982). After effects of job related stress: Families as victims. *Journal of Occupational Behavior, 3(1)*, 63–77.

Jones, F. & Flecher, B. C. (1993). An empirical study of occupational stress transmission in working couple. *Human Relations, 46(7)*, 881–903.

Kahn, R. L. & Byosiere, P. (1992). Stress in organizations. In M. D. T. Dunnette & L. M. Hough (Eds.), *Handbook of industrial and organizational psychology* (pp. 571–651). Palo Alto, CA: Consulting Psychology Press.

Kelley, D. (2009). The effects of exercise and diet on stress. *Nutritional Perspectives: Journal of the Council on Nutrition, 32(1)*, 37–39.

Kivimaki, M., Leino-Arjas, P., Kaila-Kangas, L., Luukkonen, R., Vahtera, J., & Elovainio, M. (2006). Is incomplete recovery from work a risk marker of cardiovascular death? Prospective evidence from industrial employees. *Psychosomatic Medicine, 68(3)*, 402–407.

Kopp, M. S. & Rethelyi, J. (2004). Where psychology meets physiology: Chronic stress and premature mortality – The Central-Eastern European health paradox. *Brain Research Bulletin, 62(5)*, 351–367.

Laski, S. (2002). The effect of summer break activities and personality variables on stress and burnout among high school teachers in Israel. M.Sc., Tel Aviv University, Israel. [Hebrew].

Laski, S. (2013). Recovery during vacation and its fadeout – A longitudinal study. Ph.D., Tel Aviv University, Israel.

Laski, S., Etzion, D., & Westman, M. (2015). Who gains more from vacation – Recovery during vacation and its fadeout. Paper presented at the 36th Annual Conference for the Stress and Anxiety Research Society (STAR), Tel Aviv, Israel.

Lee, C., Early, P. C., & Hanson, L. A. (1988). Are Type A better performers? *Journal of Organizational Behavior, 9(3)*, 263–269.

Lounsbury, J. W. & Hoopes, L. L. (1986). A vacation from work: Changes in work and non-work outcomes. *Journal of Applied Psychology, 71(3)*, 392–401.

Magen-Neuhaus, T. (1993). The effect of vacation and feelings of stress at work on emotional burnout. M.Sc., Tel Aviv University, Israel. [Hebrew].

Marmot, M., Theorel, T., & Siegrist, J. (2002). Work and coronary heart disease. In S. A. Stanfeld & M. Marmot (Eds.), *Stress and the heart* (pp. 51–71). London: BMJ Books.

Matthews, K. A. (1982). Psychological perspectives on the Type A behavior pattern. *Psychological Bulletin, 91(2)*, 293–323.

McEwen, B. S. (2003). Interacting mediators of allostasis and allostatic load: Towards an understanding of resilience and aging. *Metabolism, 52(10)*, 10–16.

Melamed, S., Shirom, A., Toker, S., Berliner, S., & Shapira, I. (2006). Burnout and risk of cardiovascular disease: Evidence, possible causal paths, and promising research directions. *Psychological Bulletin, 132(3)*, 327–353.

Piotrkowski, C. (1979). *Work and the family system: A naturalistic study of working-class and lower-middle-class families.* New York: The Free Press.

Powsthavee, N. (2007). Do we transmit our long-term happiness? Longitudinal evidence of life-satisfaction crossover within marriage (Research Paper). Institute of Education, University of London, London. Retrieved from: https://iser.essex.ac.uk/bhps/publications?a=Nattavudh+Powdthavee&action=bhps_search&controller=publications&r=Research+Paper&s=Well+Being.

Prescott, E., Holst, C. G., Schnohr, P. J. G., & Barefoot, J. (2003). Vital exhaustion as a risk factor for ischaemic heart disease and all-cause mortality in a community sample. A prospective study of 4084 men and 5479 women in the Copenhagen City Heart Study. *International Journal of Epidemiology, 32(6)*, 990–997.

Prince, M., Manolis, C., & Minetor, R. (2007). Life satisfaction crossover among couples. In R. J. Estes (Ed.), *Advancing quality of life in a turbulent world* (pp. 191–208). Dordrecht, Netherlands: Springer.

Repetti, R. L. (1993). Short-term effects of occupational stressors on daily mood and health complaints. *Health Psychology, 12(2)*, 125–131.

Riley, D. & Eckenrode, J. (1986). Social ties: Costs and benefits with different subgroups. *Journal of Personality and Social Psychology, 51(4)*, 770–778.

Ritvanen, T., Veikko, L., Pertti, H., Sari, V., & Osmo, H. (2006). Response of the autonomic nervous system during periods of perceived high and low work stress in younger and older female teachers. *Applied Ergonomics, 37(3)*, 311–318.

Robinson, P. & Inkson, K. (1994). Stress effects on the health of chief executive of business organizations. *Stress Medicine, 10(1)*, 27–34.

Ron, A. (2003). Effects of organizational vacation on occupational stress and burnout among high-tech employees. M.Sc., Tel Aviv University, Israel. [Hebrew].

Rosenman, R. H., Friedman, M., Straus, R., Jenkins, C. D., Zyzanski, S. J., & Wurm, M. (1970). Coronary heart disease in the Western Collaborative Group Study: A follow-up experience of 4 and one-half years. *Journal of Chronic Diseases, 23(3)*, 173–190.

Rubinstein, C. (1980). Vacations, expectations, satisfactions, frustrations, fantasies. *Psychology Today, 14(5)*, 62–66, 61–76.

Sapir, O. (1995). The effect of vacation on stress and burnout at work. M.Sc., Tel Aviv University, Israel. [Hebrew].

Schaufeli, W. B. & Bakker, A. B. (2004). Job demands, job resources and their relationship with burnout and engagement: A multi-sample study. *Journal of Organizational Behavior, 25(5)*, 293–315.

Schlossberg, H. (1991). Hyatt Survey: If you take a vacation, do it the right way. *Marketing News, 25*, 21–22.

Schlotz, W., Hellhammer, J., Schulz, P., & Stone, A. A. (2004). Perceived work overload and chronic worrying predict weekend-weekday differences in the cortisol awakening response. *Psychosomatic Medicine, 66(2)*, 207–214.

Schuitemaker, G. E., Dinant, G. J., van der Pol, G. A., & Appels, A. (2004). Assessment of vital exhaustion and identification of participants at increased risk of myocardial infraction in general practice. *Psychosomatics, 45(5)*, 414–418.

Seeman, T. E. & McEwen, B. S. (1996). Social environment characteristics and neuroendocrine function: The impact of social ties and support on neuroendocrine regulation. *Psychosomatic Medicine, 58*, 459–471.

Selye, H. (1956). *The stress of life.* New York: McGraw-Hill.

Shaviv, T. (1998). The impact of an organized tour vacation on work stress and burnout among managerial and non-managerial vacationers. M.Sc., Tel Aviv University, Israel. [Hebrew].

Sluiter, J. K., Frings-Dresen, M. H. W., van der Beek, A. J., & Meijman, T. F. (2001). The relation between work-induced neuroendocrine reactivity and recovery, subjective need for recovery and health status. *Journal of Psychosomatic Research, 50(1)*, 29–37.

Sluiter, J. K., van der Beek, A. J., & Frings-Dresen, M. H. W. (1999). The influence of work characteristics on the need for recovery and experienced health: A study on coach drivers. *Ergonomics, 42(4)*, 573–583.

Sonnentag, S. (2001). Work, recovery activities and individual well-being: A diary study. *Journal of Occupational Health Psychology, 6(3)*, 196–210.

Sonnentag, S. (2003). A new look at the interface between non-work and work recovery, work engagement, and proactive behavior recovery, work engagement, and proactive behavior. *JAP, 88(3)*, 518–528.

Sonnentag, S. & Bayer, U. V. (2005). Switching off mentally: Predictors and consequences of psychological detachment from work during off-job time. *Journal of Occupational Health Psychology, 10(4)*, 393–414.

Sonnentag, S. & Zijlstra, F. (2006). Job characteristics and off-job activities as predictors of need for recovery, well-being and fatigue. *Journal of Applied Psychology, 91(2)*, 330–350.

Sraya-Wallin, O. (1999). The effect of vacation characteristics and the fit between job and career-anchors on stress and burnout among employees in a hi-tech environment. M. Sc., Tel Aviv University, Israel. [Hebrew].

Strauss-Blasche, G., Ekmekcioglu, C., & Marktl, W. (2000). Does vacation enable recuperation? Changes in well-being associated with time away from work. *Occupational Medicine, 50*, 167–172.

Strauss-Blasche, G., Ekmekcioglu, C., & Marktl, W. (2003). Serum lipids responses to a respite from occupational and domestic demands in participants with varying levels of stress. *Journal of Psychosomatic Research, 55(6)*, 521–524.

Strauss-Blasche, G., Muhry, F., Lehofer, M., Moser, M., & Marktl, W. (2004). Time course of wellbeing after a three-week resort-based respite from occupational and domestic demands: Carry-over, contrast and situation effects. *Journal of Leisure Research, 36*, 293–309.

Strike, P. C. & Steptoe, A. (2004). Psychosocial factors in the development of coronary artery disease. *Progress in Cardiovascular Disease, 46(4)*, 337–347.

Takeuchi, R., Yun, S., & Teslu, P. T. (2002). An examination of crossover and spillover effects of spouse and expatriate cross-cultural adjustment on expatriate outcomes. *Journal of Applied Psychology, 87(4)*, 655–666.

Toda, M., Makino, H., Kobayashi, H., & Morimoto, K. (2007). Health-related lifestyle and patterns of behavior related to health effects of leisure travel. *Social Behavior and Personality, 35(3)*, 287–294.

Totterdell, P., Spelten, E., Smith, L., Barton, J., & Folkard, S. (1995). Recovery from work shifts: How long does it take? *Journal of Applied Psychology, 80(1)*, 43–57.

Tuval, A. (2001). Patterns of stress-related ER hospitalization at Sheba Medical Center between January 1997 and September 2000. M.Sc., Tel Aviv University, Israel. [Hebrew].

Vincour-Hai, V. (2002). Stress and burnout before and after vacation and their crossover between spouses. M.Sc., Tel Aviv University, Israel. [Hebrew].

Vinokur, A. D., Price, R. H., & Caplan, R. D. (1996). Hard times and hurtful partners: How financial strain affects depression and relationship satisfaction of unemployed persons and their spouses. *Journal of Personality and Social Psychology, 71(1)*, 166–179.

Westman, M. (1997). A model of stress crossover (Working Paper). Tel Aviv University, Henry Crown Institute of Business Research, Faculty of Management, Tel Aviv, Israel.

Westman, M. & Aharon-Madar, M. (1998). The impact of a vacation on stress, burnout, and performance. Paper presented at The Seventh International Conference on Social Stress Research, Budapest, Hungary.

Westman, M. & Etzion, D. (1995). Crossover of stress, strain and resources from on spouse to another. *Journal of Organizational Behavior, 16(2)*, 169–181.

Westman, M. & Eden, D. (1997). Effects of a respite from work on burnout: Vacation relief and fade-out. *Journal of Applied Psychology, 82(4)*, 516–527.

Westman, M. & Etzion, D. (1999). The crossover of strain from school principals to teachers and vice versa. *Journal of Occupation and Health Psychology, 4(3)*, 269–278.

Westman, M. & Etzion, D. (2001). The impact of vacation and job stress on burnout and absenteeism. (Special issue on burnout). *Psychology & Health, 16(5)*, 595–606.

Westman, M. & Etzion, D. (2002). The impact of short overseas business trips on job stress and burnout. *Applied Psychology: An International Review, 51(4)*, 582–592.

Westman, M., Etzion, D., & Aharon-Madar, M. (1999). Vacation, burnout and absenteeism (Working Paper No. 0027/99). Tel Aviv University, Henry Crown Institute of Business Research, Faculty of Management, Tel Aviv, Israel.

Westman, M., Etzion, D., & Chen, S. (2009). The crossover of positive experiences from business travelers to their spouses. *Journal of Managerial Psychology, 24(3)*, 269–284.

Westman, M., Etzion, D., & Danon, E. (2001). Job insecurity and crossover of burnout in married couples. *Journal of Organizational Behavior, 122(5)*, 467–481.

Westman, M., Etzion, D., & Gattenio, E. (2008). International business travels and the work-family interface – A longitudinal study. *Journal of Occupational and Organizational Psychology, 81(3)*, 459–480.

Westman, M., Etzion, D., & Gortler, E. (2004). The work-family interface and burnout. *International Journal of Stress Management, 11(4)*, 413–428.

Westman, M., Etzion, D., & Horovitz, S. (2004). The toll of unemployment does not stop with the unemployed. *Human Relations, 57(7)*, 823–824.

Westman, M., Etzion, D., & Segev, K. (2002). Leave and leisure worldwide – A cross-cultural comparison among the USA, the EU, Japan and Israel (Working Paper). Tel Aviv University, Henry Crown Institute of Business Research, Faculty of Management, Tel Aviv, Israel.

Westman, M., Etzion, D., & Yoffe, Y. (1998). The effect of overseas business trips on job stress and burnout (Working Paper). Tel Aviv University, Henry Crown Institute of Business Research, Faculty of Management, Tel Aviv, Israel.

Westman, M. & Vinokur, A. (1998). Unravelling the relationship of distress levels within couples: Common stressors, empathic reaction, or crossover via social interaction? *Human Relations, 51(2)*, 137–156.

Westman, M., Vinokur, A., Hamilton, M., & Roziner, I. (2004). Crossover of marital dissatisfaction during military downsizing among Russian army officers and their spouses. *Journal of Applied Psychology, 89(5)*, 769–790.

Wright, T. A. & Cropanzano, R. (1998). Emotional exhaustion as a predictor of job performance and voluntary turnover. *Journal of Applied Psychology, 83(3)*, 486–493.

Zalowski, S., Hall, M. H., & Baum, A. (1992). Stress, stress management, and the immune system. *Applied and Preventive Psychology, 1(1)*, 1–13.

Zijlstra, F. & Cropley, M. (2006). Recovery after work. In F. Jones, R. J. Burke, & M. Westman (Eds.), *Work-life balance: Psychological perspective* (pp. 390–420). New York: Psychology Press.

6

DISCUSSION AND CONCLUSIONS

Respites have emerged as a complex, multifaceted phenomenon. Over four decades, respite research has attempted to delineate the influence of different types of respite upon stress and burnout, including computer shutdowns (e.g., Eden, 1990), reserve military duty (Etzion et al., 1994; 1998), breaks as short as coffee breaks (e.g., Doerr et al., 1996), weekends, or a day off (e.g., Frankenhaeuser et al., 1989), as well as longer respites (e.g., Etzion, 2003; Etzion & Sapir, 1997; Westman & Eden, 1997) and even leaves lasting for months such as maternity leave (e.g., Ben Yoseff, 2004; Silverberg, 2007) or sabbaticals (e.g., Davidson et al., 2010). Despite the fact that not all of our studies are in agreement with each other or with studies of other researchers, in nearly all studies, the studied respite was found to be beneficial to varying degrees to the employees in question. Factors and moderators ranging from personal to vocational and situational variables, as well as vacation-specific variables revealed a plethora of future avenues for research in order to optimize the respite's beneficial influences of employees and their families or co-workers.

Our studies are both quantitative and qualitative, as well as theoretical and conceptual. While focusing on the independent relationship between respites on the one hand and stress and burnout on the others, respite research has introduced a plethora of variables that may be conceptually or empirically associated with respites and their influence on either of these major outcomes. Demonstrating the general progress and developments that have transpired in this field of inquiry, it may be useful to overview the chronological unfolding of our respite studies, which collectively represent a significant portion of respite research.

Respite research

Two studies can really be said to have kickstarted the field of respite research. One is that of Lounsbury and Hoopes (1986) that empirically shows vacation reduced work engagement while increasing overall satisfaction, which was also found related to job satisfaction and vacation satisfaction. In the other, just prior, Glaser et al. (1985) showed among students that the body's immune system was at its peak immediately after the term vacation in correlation with lower stress experienced by these students. Our early studies have also set the grounds for respite research as we slowly shed light on the relationships between stress, burnout, social support, work and non-work conflicts, as well as the influences of culture, gender, managerial positions, and other conditions on these relationships (Etzion, 1984; 1987; 1988a; 1988b; Etzion & Chagi-Niv, 1985; Etzion et al., 1982; Etzion & Pines, 1986).

In a four-measure study, Eden (1990) found that during a forced respite due to a computer shutdown among university faculty staff, stress was lowest during the respite although strain remained unchanged. Employees' perceptions regarding critical events had also altered during the respite as the acuteness of these events lowered during the respite, suggesting that the degree of experienced stress determines the degree of perceived stress. That is, that a respite has the power not only to change the degree of experienced stress but also the degree of stress attached to possible events that had not transpired. In a five-measure study, Magen-Neuhaus (1993) first recorded evidence toward the fadeout process. Stress was lowest during the respite, whereas burnout was lowest just prior to it. Workaholism, vacation satisfaction, and gender were also introduced in our studies (Magen-Neuhaus, 1993; Westman et al., 1994).

Lapidot (1992) found that military reserve duty was effective as a stress-relieving respite, explained through the detachment of employees from their routine, daily work, and work life (Etzion et al., 1998). Schmilovitch (1992) examined how leisure fit and vocational fit influenced the experience of burnout. In tandem. In a three-measurement study, Sapir (1995) discerned between intrinsic and extrinsic stressors both of which were alleviated during respite. Active leisure during vacation was found to reduce general and intrinsic stress compared to a passive leisure, whereas such differences were insignificant regarding the influence of vacation over extrinsic stress. A replacement at work, however, impeded the beneficial influence of the vacation over the vacationers' level of stress. The post-vacation fadeout effect was found to take place in a shorter period for stress compared to burnout (Etzion & Sapir, 1996; 1997). Etzion and Westman (1994) began exploring how social support and sense of control moderated the relationship between stress and burnout, whereas we also looked into the crossover of stress, strain and resourcefulness between spouses (Westman, 1997; Westman & Etzion, 1995; 1999; Westman, Etzion, & Danon, 1998; Westman & Vinokur, 1997; 1998).

Yoffe (1997) turned his attention to the influence of business travel as a form of respite, asserting that business travel indeed reduced stress and burnout emerging from negative work stressors (Westman, Etzion, & Yoffe, 1998). Ofek (1997) discerned between respites based on employee's control over their timing.

She found that an organizational shutdown had a beneficial effect over stress originating in the work–non-work conflict, as well as over burnout (Etzion & Ofek, 1998). In a three-measurement study, Aharon-Madar (1997) noticed the beneficial influence of the vacation not only over stress and burnout but also over absenteeism and subjective performance appraisal – all of which faded out within a month after the vacation and return to work (Westman & Etzion, 2001; Westman et al., 1999). In a five-measure study, Westman and Eden (Westman, 1999; Westman & Eden, 1997) were the first to tie the influence of vacation to predictions posited by Hobfoll (1989; Hobfoll & Freedy, 1993; Hobfoll & Shirom, 1993) and the Conservation of Resources theory regarding developments in the absence of stress. They, too, found evidence of the fadeout effect occurring within three weeks for post-vacation relief found in both stress and burnout.

In a three-measure study, Radian (1998) found similar vacation effects over the stress and burnout experienced by municipality employees where fadeout was noted only in stress but not burnout. Aviv (1998) examined the influence of vacation over the stress, burnout, and work–non-work conflict experienced by women working in shifts. Only burnout seemed to be sensitive to the vacation effect, both before and after the vacation. Shaviv (1998) found no vacation effect among vacating managers who vacated individually each in an organized trip, suggesting that managers and non-managers were subject to different vacation effects. In a three-measurement study, Kremer-Fuks (1999) found vacation effect over burnout but not over stress. However, she found group differences between indigenous and immigrant Israeli citizens determined overall life and work satisfaction, as well as vacation satisfaction. Whereas Sraya-Wallin (1999) found no vacation effect over neither stress nor burnout (perhaps due to the small sample size), she was able to detect a moderating effect of role-career anchor fit so that those with better fit experienced a post-vacation rise in burnout while those with looser fit experienced a post-vacation decrease in burnout. She also found that vacation satisfaction alleviated all measured types of stress. Etzion, Zedeck, Maslach, and Owen (2001) furthered research interest into the influence of planned compared to actual respite activities. In a two-measurement study, Zeevi-Ohana (2000) discerned between stress originating from the presence of negative work characteristics and that which originated from the absence of positive work characteristics, noting the former decreased the experience of stress while the latter increased it. This finding was replicated in a different three-measurement study by Laski (2002), which also recorded the fadeout effect. Burnout was additionally found to alleviate following a vacation.

The study of crossover grew wings during this time. We extended our studies into the crossover effect of stress and burnout among co-workers Westman and Etzion (1999) and examined the influence of over the crossover between spouses of both the vacation effect (Vinocour-Hai, 2002; Westman & Etzion, 1999) and job insecurity as in the case of unemployment (Westman et al., 2001). We also started turning our attention to the physiological effects of stress and burnout (Tuval, 2001). We continued our studies into the influence of short overseas business trips on job stress and burnout further noting the difference between the effect of the presence of negative and the absence of positive working characteristics (Koniak, 2003; Westman, 2004; Westman & Etzion, 2002) while also looking into the influence of the longer maternity leave (Ben Yoseff, 2004) and particularly long respites, namely, sabbaticals (Davidson et al., 2010). We further studied the impact of different leisure activities during one's respite over stress and burnout (Laski, 2002; Moshkowitz, 2003; Raviv, 2002; Zvi, 2004), especially as a coping method (Zack, 2005), the role of personal characteristics in the formulation of the vacation effect (Levi, 2002; Zvi, 2004), as well as the relationship between economic stress, social support, and crossover between spouses (Horovitz, 2002; Westman, Etzion, & Horovitz, 2004; Westman et al., 2002). Finally, during this time, we touched upon the cultural differences found in nations worldwide with respect to the formal and informal approaches to time off work (Dar, 2002). We also looked at the effect of an organizational (Ron, 2003) compared to personally-determined (or self-imposed) vacation (Etzion, 2003) reestablishing the positive influence of employees' ability to determine when and for how long they will vacate while further supporting the growing evidence of the fadeout effect.

We revisited these themes in the following years, looking further at organizational stress through the lens of conservation of resources (Westman et al., 2005) and ways by which burnout may be alleviated through respite (Etzion, 1987; 2005). A climax in our research was achieved with a meta-analysis of our previous work (Boklis, 2006). Some of our research themes evolved more slowly, such as our study of the work–non-work conflict and burnout among working mothers after maternity leave examining situational and personal characteristics as moderators (Silverberg, 2007). Other studies found new angles to old themes such as the impact of organizational vacation on the amount of regular working hours, overtime, sick days, and absenteeism (Avni, 2007) or the study on work features, serious leisure, and burnout among flight attendants (Ben-Moshe, 2006). And yet others looked at new factors which had not previously studied such as the group of vacation objectors (Braun (Cohen), 2007) or the effect of a weekend respite on burnout and stress (Morad, 2009). In our last studies we further developed the influence of business travel as a unique type of respite (Westman, Chen, & Etzion, 2011; Westman et al., 2009a) and their interface with the work–non-work conflict (Westman et al., 2008). Further investigating the interface of the work–non-work conflict and burnout (Westman, Etzion, & Gortler, 2004), we also turned to assess the crossover of positive psychological experiences thereby bringing together our crossover research trajectory with our business travel as respite studies (Westman et al., 2009b). Finally, in a six-measurement study, in addition to studying the psychological well-being of vacationers in relation to their leisure activities, we also explored physiological aspects of well-being and paid particular attention to the fadeout process (Laski, 2013).

The power of respites has long been recognized as restorative and stress preventive. A respite from work comprises any absence from the work setting when everyday job-related pressures are lessened or completely absent. Empirically introduced by Lounsbury and Hoopes (1986), the influence of vacation has since been incorporated into respite research and conceptualized as a prominent and powerful form of respite. Respites and vacations were initially used as control occasions in the study of stress-strain relationship. Over time, however, their study has emerged as a topic in its own right, its research focusing on psychological and physiological outcomes of time away from work. Results have consistently indicated a decrease of perceived job stress and burnout (Etzion, 2005), as well as a plethora of additional positive outcomes emerging from resource gain (e.g., Davidson et al., 2010). The effect of vacation over some variables, such as burnout, lasts longer than its effect over other variables, such as stress.

Our respite studies began under the auspices of the Henry Crown Institute of Business Research in Israel. Many of our studies have been published or presented at professional conferences while some remained documented in working papers. Overall, we examined multiple facets of both the respite and respiting individuals. We looked into the nature of the respite itself including its duration and location, respite activities, the degree of pleasure and satisfaction derived from the respite, the degree of physical and emotional detachment from the demands of daily routine, differences in psychological and physiological outcomes between the respite period and the work routine. We explored how personal characteristics such as hardiness, personality type, workaholism, socioeconomic background, gender, and situational employment conditions promote or demote the potential relief that respite can bring to employees. We assessed the influence of organizational flexibility, legitimacy, and general family-friendly policies on the benefit of respites. Finally, we explored some of the outcomes of respite such as crossover and fadeout effects. We learned much about the ways by which respites can support and enhance the psychological and physiological well-being of employees. Have we learned enough to formulate resting guidelines for employees and their organizations?

Recovery and coping

A bipolar, unidimensional approach to the integration of positive and negative psychology presumes that positive and negative affects are inversely related and that one cannot simultaneously feel the positive and the negative end-poles of an affect. Being high on one side of the dimension, the approach goes, necessarily implies being low on the other side (Feldman Barrett & Russell, 1998). The bivariate approach, in contrast, posits that negative and positive affects form two independent factors, representing components of two different facilitation systems. The question of recovery would be treated differently by each approach.

Recovery was classified on the bases of time that has elapsed since work: micro recovery refers to recovery that occurs ten minutes to one hour after task performance; meso recovery refers to recovery that occurs between one hour and up to two days after the work was performed; and macro recovery refers to recovery that takes place more than two days after the work has been completed

(Sluiter, Frings-Dresen, Meijman, & van der Beek, 2000). Within this framework, vacation may be regarded as a preeminent opportunity for macro recovery. Two underlying mechanisms may explain the recovery potential presented by vacation: a passive mechanism comprises relief from daily job demands while an active mechanism comprises active engagement in activities that bear recovery potential. These mechanisms may be theoretically anchored in Conservation of Resources theory and the Job Demands-Resources model.

Conservation of Resources theory (Hobfoll, 1989; 2001) posits resources are essential units for understanding human behavior in general and stress conditions, in particular. Resources include personal characteristics (e.g., self-efficacy), objects (e.g., car, house, computer), conditions (e.g., position, rank, tenure), and energies (e.g., money, time, knowledge). Loss or gain of resources results in stress or well-being, respectively. Stress occurs when individuals are either threatened with resource loss, actually lose resources, or fail to regain resources following an investment of resources. Individuals are inherently motivated to actively seek to obtain, retain, foster, and protect resources that are valued in their own right or as a means to obtain centrally valued ends (Hobfoll, 2002). The goal is to accomplish gains or to offset resource loss and enhance resource reserves. Where this goal cannot be accomplished, the inevitable escalation of resource loss ensues, accompanied by impaired psychological and physical well-being including one's physical and mental health (Gorgievsky et al., 2010; Hobfoll, 1989; 1998; 2001; Hobfoll & Shirom, 2000; Westman et al., 2005). Within the framework of this theory, resource loss represented by burnout is an ultimate state of impoverishment (Hobfoll & Shirom, 2000). Similarly, the Job Demands-Resource model (Demerouti et al., 2001) also postulates the coexistence of two underlying psychological processes. One is represented by job demands and the other by job resources, both of which are physical, social, or organizational aspects of the job. Whereas job resources are functional in achieving work-related goals and reducing demands, job demands require sustained effort such as workload, role overload, or emotional demands. Individuals sway back and forth between these two components or their working environment. When chronic job demands exhaust employees' mental and physical resources, they lead to the depletion of energy and health. The presence of job resources, in contrast, leads to engagement. Within the framework of this model, resource loss represented by burnout is a final stage of the energetic process.

Resources have motivational potential as they make work meaningful, draw on the employee's responsibility for work processes and outcomes, and provide feedback on actual work results while stimulating personal growth and development (Bakker, Demerouti, Taris, Schaufeli, & Schreurs, 2003; Bakker et al., 2004; Bakker et al., 2007; Demerouti et al., 2001; Hackman & Oldham, 1976; 1980; Xanthopoulou et al., 2007a; 2007b). Job resources may be found within the organization (e.g., salary, career), social relations (e.g., social support, team climate), the organization of work (e.g., role clarity, participation in decision making), and the task level (e.g., performance feedback, skill variety). They may buffer the impact of job demands on negative job-related outcomes such as strain, disengagement, and burnout. The primacy of loss posited by the Conservation of Resources theory renders resource loss to be disproportionately more salient than resource gain. Strategies employed, however, to revert such loss may lead to further losses. Individuals who possess greater resources are less vulnerable to resource loss and more capable of resource gain, so that loss may drive further loss while gain promotes further gain. However, since loss is more potent than gain, loss cycles are more impactful and more accelerated than gain cycles.

Faced with the threat of resource loss, individuals tend to focus on their weaknesses and losses and search for loss prevention strategies rather than new opportunities for gains. When individuals embark on a loss spiral, for example under chronic stress, they naturally endeavor to counterbalance the perceived energy depletion. Burnout results from a slow, stressful process of resource loss that is not counterbalanced by resource gains, thus accumulating significant losses. Accordingly, if personal efforts to counteract loss and regain resources fail, a state of ultimate resource impoverishment, namely burnout, occurs (Freedy & Hobfoll, 1994; Hobfoll & Freedy, 1993; Hobfoll & Shirom, 2000; Westman et al., 2005). In fact, both actual and anticipated loss of significant resources will have an inhibitory impact on the process of engagement so that prevention of loss is critical for fostering engagement. Individuals who are rich in both quantity and variety of resources are more capable of sustaining work engagement. Such individuals may experience gain cycles as resource gain, leads to positive well-being rendering individuals to be more capable of further investment of resources. Stronger resource pools may sooth the perception of risk, allowing individuals to engage in risk – an approach that is crucial in many highly demanding work environments (Gorgievsky et al., 2010).

The degree to which coping is successful determines one's subjective experience of stress and hence their physical and psychological well-being (Laski, 2013; Westman, 2004). Coping is the process through which individuals handle challenging conditions and events comprising the expenditure of cognitive and behavioral efforts aimed at managing both external and internal demands that are perceived as taxing or exceeding one's personal resources (Lazarus, 1993). Individuals draw upon psychological resources that dim the degree of stress experienced thereby lowering levels of strain. They may do this both reactively and proactively, as a preventive measure. Proactive coping is considered to derive from one's perception of conditions as challenging and stimulating, whereas reactive coping stems from risk appraisal (Greenglass et al., 1999). Proactively coping individuals acquire and maintain resource reservoirs regardless of threats, align their conditions and resources, reduce or modify impending stressful events, and act upon early signs of threat to resources (Aspinwall & Taylor, 1997; Hobfoll, 2001). Superimposing coping style on one's approach to resources (acquiring them or preventing their loss), four general coping types of behaviors may be detected: proactive or reactive coping achieved by either the acquisition or loss prevention of resources (Westman, 2004). Proactive and reactive coping may occur concurrently (Hobfoll, 2001) and they may change throughout the lifetime of an event, such as business trip (Westman, 2004).

Restoring lost or threatened resources prompts the need for recovery, namely a process of unwinding and replenishment that allows for psychobiological recuperation (Sluiter et al., 2001; Sonnentag & Zijlstra, 2006). Recovery occurs by either ending the demands on resources or further investing in new resources. A respite from work offers a double opportunity addressing both sides of recovery. While removed from work one may pause the cycle of the resource loss, as well as acquire new resources that enhance one's resource reservoir. That is, employees returning to work following a respite not only experience enhanced well-being by having been removed from the source of their unease but their arsenal of coping resources is greater than what was available to them prior to the respite. Recovery includes experiences such as psychological detachment, relaxation, mastery, control, physical activity, social affiliation, hope and optimism, fun and humor, and self-reward. Recovery facilitates good mood and restores individual well-being while also contributing to work engagement and proactivity; its gains are particularly beneficial when following a period of resource depletion where the perception of importance of resource procurement grows (Hobfoll, 2001; Sonnentag, 2003; Sonnentag & Bayer, 2005). Respite from work, and particularly a vacation, may thus be fertile grounds for recovery gains as the respite typically takes place after a period of chronic resource loss in the job. This is true even for part-time and shift employees (Chen et al., 2009).

Coping is distinct from recovery experiences. It is a means to achieve recovery experience. We indeed found a negative relationship between the proactive coping style and both burnout and exhaustion before and after one's vacation. Proactive coping was additionally found to be positively related to vacation satisfaction. Proactive coping may be a derivative of personality characteristics, as we found it be positively related to conscientiousness and age but negatively related to neuroticism (Silverberg, 2007; Zvi, 2004). In line with previous research (e.g., Norfolk, 1989), we found that various forms of proactive coping were found to be related to managerial positions, tenure, and level of education. It seems that individuals who have greater proactive coping inventories, preventive or reflective coping styles, and tendencies toward strategic planning cope more easily with work-related stress factors, rendering them superior candidates for managerial positions (Silverberg, 2007). Alternatively, individuals ranking higher in the organizational hierarchy may enjoy greater coping resources (Westman, 1992). It also seems that these skills may be developed over time, as suggested by their positive relationship with tenure and education. Notwithstanding, women seeking instrumental support experience greater work–family conflict suggesting that traditional gender role perceptions may inhibit their potential to seek support in their social rings or enjoy such support when offered (Silverberg, 2007). Feedback, which is received instantaneously when coping with strenuous conditions, allows employees to continuously enhance and improve the efficacy of their coping methods, triggering a cycle of proactive coping choices (Abdel-Halim, 1978).

Sense of control has been found to have such an effect (Karasek, 1979; Westman, 1992) leading to a negative relationship with stress and to burnout (Westman & Etzion, 1995). The more in control one perceives themselves to be, the less likely they are to experience burnout; the less in control one perceives themselves to be, the more likely they are to experience burnout. This relationship may be bidirectional: individuals experiencing burnout may perceive themselves to have less control over the environment. Less obvious, we found an intriguing effect of sense of control among spouses. The sense of control of one spouse contributes to the prediction of the other spouse's experience of burnout. This finding indicates that individuals may be able to empower their spouses through their own sense of control, rendering sense of control as an external resource available to others, or perhaps only to life partners (Westman & Etzion, 1995). Enhancing the sense of control in both spouses, then, might serve as a coping mechanism for the attenuation of burnout (Leiter, 1991), especially in two-income families.

The effect of a respite from work on stress and strain is not unequivocal. In the equation between leisure and job-related stress and burnout, personal characteristics may play a role either side of the equation. Questions posed by Freudenberger (1983) are illuminative in our understanding of the role of personal characteristics in the experience of stress and burnout, one's (in)vulnerability to stress and burnout, and the choice one may make to enhance the respite effect over stress and burnout. Hindering the respite effect are traits, such as workaholism, that limit one's ability to detach from their work and full enjoy the respite (Eden, 1990; Machlowits, 1980; Rubinstein, 1980). Faced with respite from work, some individuals might even develop leisure sickness and weekend maladies (Vingerhoets et al., 2002). The inescapable power of gender has also been found to influence the respite effect as women were found more likely to feel less relaxed on vacation than men because of their more acute experience of the work–non-work (family) conflict (Etzion, 2005). Also inescapable are technological advances that obscure the line between work and non-work realm, making it ever thinner and thinner as employees become accessible to supervisors, subordinates, colleagues, and customers even when not at work (Etzion et al., 1998).

There may be a hidden causality between personal traits of individuals who take vacation from work and their ability to draw gains from the respite experience. For example, holiday-taking individuals displayed higher global life satisfaction and experienced less negative and unpleasant feelings so that they were happier with specific life domains (e.g., family, economic situation, health), compared to non-holiday-taking individuals (Gilbert & Abdullah, 2002). In other words, personal traits determine how individuals interpret demands, opportunities, and the meaning of a vacation (Westman & Eden, 1997).

Social and organizational support

One's ability to recover is integrated in the availability of recovery and coping resources. Coping resources are related as the availability of one resource leads to better use of another resource (Hobfoll, 1989; Lefcourt, Martin, & Saleh, 1984; Westman & Etzion, 2002). Some resources have been studied more often than others with respect to their influence on employees' response to job demands. A particular coping resource that alleviates the perception of stress and prevents the onset of psychological strain is perceived social support (Etzion, 1984; House, 1981). That is, individuals may be able to draw upon not only psychological but also social resources that enable them to appraise stress as being less distressing and thereby experience lower levels of strain.

Social support is a descriptor of a supportive social environment that facilitates individual coping with stress, strain, and other challenges. The concept of social support is multidimensional spanning both mental and behavioral types of support, such as information, emotional assistance, or the bestowment of a sense of belonging up individuals offered both formally and informally (Horovitz, 2002). Informal support originates in one's circle of personal relationship such as friends, family members, or other members of the organization or community, whereas formal support originates in community agencies and authorities (House, 1981). Social support may be offered in its potential form rather than its actual form, which is drawn upon in need; the mere potential of drawing upon it renders it a resource (Wallston et al., 1983). Regardless of its form or nature, social support becomes a resource when the social relationships involved are perceived to be beneficial (Cohen, 1988; Wethington & Kessler, 1986). We found social support to be negatively related to perceived job stress and positively related to perceived health (Koniak, 2003).

Emotional support is the support received by others in the form of affection, sympathy, and love. Individuals typically gain such support while spending time with others who offer emotional support during different leisure activities, including entertainment, rest, or relaxation. Emotional support reduces stress by addressing the need of relatedness. Instrumental support is the support received through material assistance or through services rendered. This type of support reduces stress by offering a direct resolution extinguishing the source of stress or by supplying required resources, such as time, that may be used in the procurement of relevant resources. Informative support is offered through information concerning the environment that help frame the magnitude and acuteness of events. This information facilitates better understanding and coping possibilities. Appraisal support is received through a communication of information allowing the individual to enhance self-evaluation thereby conferring greater sense of control (House, 1981; Taylor & Arnow, 1988). These are the foundations upon which support groups are built. The individual's social support and sense of control have been singled out as the most common coping resources that alleviate the perception of stress and prevent the onset of psychological strain. The two are reciprocally related as individuals characterized by a high sense of control are likely to be perceived by others as valuable, whereas their social attraction may promote their confidence in their ability to influence their environment (Hobfoll, 1989).

The need for social support may be biological (Bowlby, 1969), a fundamental innate need for the purpose of promoting one's physiological and psychological well-being. Social ties are essential for one's physical security, as well as emotional satisfaction (Ganster & Bart, 1988). Deprivation of social support leads to negative psychological outcomes such as anxiety or depression (Cohen & Wills, 1985). Indeed, the presence of social support has been found to be negatively related to stress-related anxiety experienced by unemployed individuals (Ullah et al., 1985). The main effect of social support is activated whether or not the individual is feeling stress. By its mere presence in one's life social support instills a positive feeling in the individuals, a sense of stability, and recognition in self-value and self-esteem. The buffering effect of social support is felt more clearly when the individual feels stress. In such cases, social support acts as a buffer, preventing loss of health and functionality that would otherwise emerge from stressful events, especially when the individual feels unequipped to cope with the event. In such cases, social support may alleviate the threatening perception of the event, instill a sense of adequacy in coping thereby leading to the reassessment of the situation as less threatening, or both (Cohen & Wills, 1985).

Social support has been found to alleviate stress mostly where there was some fit between the source of unease, the source of support, and the type of support offered. For example, with respect to job stress, while women tend to prefer social support offered from within the family, men tend to prefer support offered by work-related sources (Etzion, 1984). Among spouses, we found a negative relationship between general social support and anxiety originating in the unemployment of one spouse. Moreover, we found that social support intensified during the period of (economic) stress (Horovitz, 2002). Lack of support can become an acute source of stress (Friedman, 1995). Similarly, negative content in social support (social undermining) also produces detrimental effects (Burke & Greenglass, 1995). The adverse effect of the negative content of support are greater than the beneficial effects of positive support.

Social support within the organization is no less important than non-organizational social support. The less stressed and burned out employees are, the more productive they become, the longer they will stay in the workforce, extending their experience and knowledge (Toomey & Connor, 1988). Studies about organizational culture and climate abound (e.g., Ostroff, Kinicki, & Tamkins, 2004; Schein, 2004). We ourselves consider most workplaces to be plagued by a burnout climate, where cross-organization stress propagates burnout experiences among employees (Westman & Eden, 1997). Moderation of the stress–burnout relationship through social support has been widely demonstrated (Etzion, 2003). Intimately tied to this is the bidirectional crossover effect we recorded between co-workers (Westman & Etzion, 1999). At the extreme of stress outcomes is the Japanese death by overwork, *karoshi* phenomenon (Maruyama & Tokuda, 2017; Rienhold, 1996), for which strong social support within the organization may act as a preventive measure.

Hence support, both formal and informal, within the organization, has substantial potential to alleviate pressure. Different types of supports extended to employees by supervisors and co-workers have been repeatedly found to moderate perceptions of stress and alleviate outcomes of perceived stress such as anxiety and depression (Payne, 1980). Social support within the organization was also found to alleviate burnout (Etzion, 1984). We found commander and peer support, especially of the affective and instrumental types, generated the greatest respite effect for military leave among Israeli combat soldiers performing their compulsory service. In a different context, the decision of women to return to work following their maternity leave was found to be positively influenced by social support extended by supervisors and colleagues during the pregnancy (Glass & Riley, 1998). To instill equanimity among employees and achieve the welfare of families and communities, organizations must consider their employees' private and public life in unity (Bailyn, 1993). For example, our findings that respites mitigate the crossover effect of burnout among spouses suggests that it would in the organization's best interests during stressful times to facilitate cross-partner relief, that is, by reducing the secondary spouse's strain, burnout of the primary spouse who is the organizational member may be alleviated as well (Eden, 2001; Etzion et al., 2002). This may apply to the context of team members and other co-workers as well (Westman & Etzion, 1999).

Organizations may address the conjoined issue of stress and support through the prism of role quality, which has been shown to be more detrimental to psychological well-being of employees than long working hours (Klein et al., 1998). From a broader perspective, then, we would encourage organizations to identify stressful events and occupations that accelerate the crossover so they may draw more accurate conclusions on both the intra- and extra-(familial) organizational consequences of stress and find appropriate preventive measures. We would encourage organizations to consider one specific type of stress: that which originates in the presence of negative work characteristics. We have repeatedly found this type of stress to have more detrimental effects on resource replenishment than the absence of positive work characteristics (e.g., Raviv, 2002; Zeevi-Ohana, 2000), one of which may be considered social support (Laski, 2002; Levi, 2002). Alleviating organizational pressures may be useful in post-respite recovery as well, as research also shows that the workload experienced by employees negatively affects their post-vacation well-being, including quality of sleep, social activities, and mood (Strauss-Blasche et al., 2002).

Organizational support is required in two additional, partially intertwined, realms in order to reap the benefits of time off work, or respites, of any form: one would be the encouragement of legitimacy perceptions toward employees taking time off, which might also be embedded in the other, namely, family-friendly policies. In the face of growing demands on employees' time, leisure time has become a coveted luxury and paid leave has been recorded as the preferred benefit (Radian, 1998; Ron, 2003; Shaviv, 1998). In fact, we found a positive relationship between accumulated, unrealized paid leave days and both burnout and the urge to work. That is, employees who do not realize their paid leave or are more strongly propelled to work experience greater burnout, exhaustion, and cynicism (Braun (Cohen), 2007; Harris, 1984). Respite avoidance may emerge in the face of economic pressure, organizational dismissal waves, ambition, or work addiction or it may develop as a result of increased post-respite workload for themselves or burdening co-workers while away (Dar, 2002).

The need for recovery may be daily. Due to the power of continuous exposure to daily stress, burnout experts consider the occasional time out from work as a means of recuperation (Cherniss, 1981; Freudenberger, 1974; Maslach, 1976; Pines et al., 1981). Employees' daily work engagement was found to vary among others as a function of one's subjective feeling of being recovered in the morning; the availability of personal resources, such as energy, fostered work engagement over and above the benefits of job resources available at any given day (Kühnel, 2009). Personal resources acquired in the evening through sports activity were found to be important antecedents for work engagement on the following working day. These findings suggest that the development of personal resources during daily leisure time might be beneficial for the work environment so that organization may benefit from encouraging employees to practice active leisure (Naegel & Sonnentag, 2011). Our finding that leisure spent on family and personal errands was related to greater stress originating in the absence of positive work characteristics further emphasizes the point (Laski, 2002). One's ability to revitalize and "charge their batteries" (Eden, 2001; Etzion, 1995) may also apply to short breaks during the work day, such as rest periods, coffee breaks, mealtimes, and breaks for exercise in the gym or swimming pool, and to similar recreational activities after work and at the weekend (Dar, 2002). Similarly, understanding the impact of long weekends on employees' stress may contribute to the welfare of both the organization and its employees. Promoting respites of any length, then, is a measure that organizations could adopt in order to promote employees' well-being and health (Chen et al., 2009).

The feeling of respite legitimacy, that is socially acceptable in the eyes of supervisors and fellow employees to be absent on leave, appears to be key in employees' ability to truly enjoy their leave and relax. Through its formal and informal policies, the organization may also contribute to various vacation variables that do not dictate the content and character of the vacation rather the frame within which those are applied, such as timing, duration, and the availability of accumulating time off (Utechet & Aldag, 1989). The legitimacy of one's choice to take time off (as opposed to time off taken due to organizational shutdowns) may be positively related to constructive post-respite perceptions regarding work (Ben Yoseff, 2004). Our interviews with men on military reserve duty (Etzion et al., 1998) demonstrated that the positive respite effect emanated mostly from the legitimacy afforded to the absence of the men from work as their reserve duty was compulsory. In fact, the relieving effect of the military reserve duty was found greater than the reliving effect of vacation (Etzion, 2005). While organizations may not be able to send their personnel off to reserve duties, the lesson of legitimacy may extend to one's right for leave, as well as the length and frequency of the respite.

Encouraging respites may be beneficial from an additional standpoint. We found that within the month upon return from their vacation, employees had worked fewer hours than in the time prior to their vacation or at the end of the fadeout process. However, during that same vacation effect period we found absenteeism had also declined not only among vacationers but also among non-vacationers (Avni, 2007). While encouraging respites, however, organizations will also benefit from allowing employees to fully detach from their work, as we repeatedly found that the greater the level of detachment from the routine work life, the greater the decrease in stress levels experienced upon one's return to work (Etzion, 2005).

Vacation entitlement and consumption may be nationally ingrained and tied to welfare regimes. Evidence is found that democratic welfare regimes lead to high levels of vacation entitlement and consumption, such as in Europe, whereas more liberal welfare regimes are associated with lower vacation entitlement and greater inequalities in the distribution of leisure time (Richards, 1999). Other national influences are affected by cultural attitudes toward gender roles. A vast body of research on the conflict between careers or work and the family points to the effect that women often experience this conflict more acutely than men due to traditional perceptions of their respective roles (Etzion & Bailyn, 1994; Etzion & Rechnitz-Elpeleg, 2002). While the potential of family-friendly developments in legislation may be culturally dependent, organizations may nevertheless promote such programs, thereby facilitating greater gender equality. Family friendly policies that assist employees in managing their work–non-work balance promote successful participation in the labor market and engagement in the workplace while enhancing one's quality of life, allowing employees to be involved in caregiving for children, elders, or other family members to pursue non-work enriching activities such as education, volunteering, leisure, and self-care (Ernst Kossek & Ollier-Malaterre, 2012).

Organizations certainly do not have to wait for staggering statistics to change legislation. This was the case of the *karoshi* phenomenon that led the Japanese Ministry of Health, Labor and Welfare in 2002 to modify its standards and approve more applications for workers' compensation for cardio-vascular diseases and mental disorders (International Labour Organization, 2013). Statistics on compensation paid for stress-related diseases appear to delineate a similar path for "the overworked American," as termed by Juliette Schor (1992) in her book by the same title. Organizational initiatives emerging from the understanding of the organization benefits produced by a healthy, stress-free workforce might still change the ending of the story. Considering that week-long respites had the same ameliorating effects as longer respites, it may be constructive to encourage employees to take several short respites throughout the year rather than accumulating their time into one longer respite. In this way organizations will be able to achieve a maximal positive long-term effect (Etzion, 2003; Etzion & Sapir, 1997). In fact, a longitudinal (nine-year follow-up) study conducted in a sample of over 12,000 individuals established an association between frequent annual vacations and reduced risk of all-cause mortality, as well as mortality attributed to coronary heart disease (Gump & Matthews, 2000). Based on our meta-analysis (Appendix 1), we might further state that respites from work do not present a proper treatment for the problem of work-related stress and burnout, and that this complex problem requires a comprehensive, systemic treatment such as improvements in the work environment, training in stress coping strategies, or the creation of support groups within the workplace (Pines, 1993; Shirom, 2003). Respites may nevertheless serve as an acute means of treatment of one's stress and burnout. Our meta-analysis shows that a respite of six days or more is as effective as longer respite with respect to stress and even more effective than the longer respite with respect to burnout. Hence, when employees or their supervisors sense greater burnout levels than usual, a respite may very well serve as a suitable, immediate alleviator of such experiences. Supervisors may also benefit from the knowledge gained in our research pointing at elevated levels of stress just prior to the respite and immediately after it (approximately three days before and after the respite). To better capitalize on the respite effect, supervisors may aim to purposefully reduce organizational stress during these periods.

Business travel is a special case of respite that organizations will do well to capitalize upon. While business trips comprise a positive experience to the traveler due to the change in atmosphere and the breaking of one's routine, such getaways also entail the negative influence of jet lag or time away from one's family (Striker et al., 2000). While business trips do not have the magnitude of influence that non-work-related respites have, they do nevertheless alleviate stress and burnout. Stress and burnout were recorded as highest just prior to the business trip. Possibly, anticipation geared toward the overseas trip and work at a less familiar site have a negative impact on the employee. The period spent outside the regular workplace has similarly been characterized by ambiguity and overload that negatively affect the traveling employee.

Our findings accordingly show that during the trip burnout decreases while stress does not. Our findings further reveal that detachment from the regular work environment and trip satisfaction enhance the respite effect of business trips (Yoffe, 1997). Business travelers perceive the trip to comprise four phases: pre-trip, journey, stay, and post-trip. Coping strategies of travelers vary between these phases, using proactive individual and organizational coping strategies in the pre-trip phase and a combination of proactive and reactive coping strategies in the other phases (Westman, 2004). Organizations might help orchestrate these pre-respite and during-respite factors to help employees optimize recuperation during such business trips. Organizations might additionally mitigate negative respite-effects of the business travel by presenting travelers with a support system throughout the cycles of the trip – before, during, and after it. We found emotional social support and organizational support to be the single variables to uniquely contribute to both perceived stress of traveling and employees' perceived health (Koniak, 2003). In addition to any operational assistance organizations offer to the traveler (arranging flights, accommodation, etc.), it might help to support employees' families during employees' absence, offer psychological support to the employee and their family, facilitate easily available communication between employees, and enhance the employees' relief by upgrading travel terms and offering a day off for recuperation upon return (DeFrank et al., 2000).

How to rest

Complaints about pressure or fatigue often trigger recommendations dispensed by various advice givers to take a vacation. Psychologists and physicians prescribe occasional time off work as a means of recharging, rejuvenating, and renewing one's resources (Etzion, 2003). The intuitive basis for these responses is clear: vacationers do often feel refreshed at the end of their vacation. The scientific basis for these responses, however, shows that the conditions surrounding the vacation, or any respite, influence one's potential to enjoy the relaxing respite effect. That is, the salutogenic impact of vacation on well-being is neither universal nor uniform (Westman & Eden, 1997). Noting the broad range of factors that appear to influence the beneficial effect of a respite, it might sometimes seem that time away from work does not in itself contribute to psychological and physiological well-being of employees (Sraya-Wallin, 1999). What best defines a respite with respect to the objective of alleviating occupational stress and psychological burnout may vary according to personal, situational, vocational, environmental, and respite-related circumstances. It is thus difficult to prescribe a one-size-fits all respite program, given the plethora of respites types, personal and situation factors, organizational considerations, and environmental and cultural conditions. We are nevertheless able to draw some generalities that may help individuals and their organizations replenish employee resources, enhance their overall well-being, and increase productivity.

Respites are a time where individuals proactively invest resources (e.g., time, money, planning) often with the intention, conscious or not, to gain further resources and increase one's pool of resources. Although experiencing different degrees of stress and burnout, employees are most likely unaware of the resource drain they are experiencing, which leads to psychological and physiological impairment; they may not fully understand the vicious cycle of loss and the potential of the respite. They may thus not fully understand the importance of their time away from work in facilitating psychobiological recuperation and a cycle of resource gain. Mental processes, whether energetic or motivational, are the result of the interaction between job demands and job resources. We suggest that a recess from work interferes with the energetic process, thus thwarting potential health impairment outcomes, whereas resource gains accelerate the motivational process (Laski et al., 2013).

Any respite requires adaptation (Eden, 2001), but some respites require more coping effort than others (Davidson et al., 2010). When discussing the respite effect, we must consider three phases: before, during, and after the respite. However, we must make a distinction in our discussion on how to best reap a positive respite effect between short respites that have some form of regularity – from minutes on the job to post-work hours, to the weekend – as opposed to less regular and longer respites. This distinction rests on the way by which these different types of respites have traditionally been studied. Because of their time span and repeated regularity, the study of short respites is typically conducted from a narrower prism than that involving the study of longer respites. With longer respites, we know that stress and burnout build up toward the respite and then drop down. We have no parallel data for shorter, regularly repeated respites. Similarly, whereas with longer respites, we know the respite effect lasts for approximately a month upon return to work, shorter respites have not traditionally been studied for similar long-term effects. These observations are important because they bear on the ways by which employees and organizations may capitalize on the different types of respites.

What to do before the respite

The first thing to do when planning a respite, is to actually plan it. We found that merely planning to rest and relax during one's vacation was associated with reduced stress both before and after the vacation, as well as with vacation satisfaction (Zvi, 2004). Based on our data, we cannot discern whether planning reduces stress or less stressful individuals are those who take time to plan. Research nevertheless shows this association to be true whether or not action during one's respite matched planning (see Zvi, 2004). Furthermore, poor planning was identified as a key component for individuals preventing them from capitalizing on the respite effect. Planning should relate not only to activities, but also to sleep, which has been identified as another key factor in post-respite feelings of exhaustion and fatigue. Where sleep is concerned, it should be considered several days before a vacation as individuals often lose sleep on the day of travel or the night before it to pack (Gallup, 2002). Additionally, increased job pressure keeps employees longer at their desks prior to their vacations where some planning may be able to relieve such tendencies.

Planning must also extend into finding a fit between one's personal characteristics and features of the respite in order to capitalize on the beneficial respite effect (Etzion, 2006). Employees must be aware of the kind of respite that would best serve them, given personal, vocational, situational, and environmental factors (Sapir, 1995). We additionally encourage employees to seek greater control over characteristics of the respite. We found that greater control over such characteristics was associated with greater satisfaction, which in itself promotes the beneficial outcomes of the respite effect. Employees' sense of control may be related to self-efficacy, that is, employees' belief in their own ability to mobilize the mental, motivational, and emotional resources required in the execution of action that would meet situational demands (Bandura, 1997). In our case, that would be their ability to exercise greater control over their respite plan and execution.

Findings regarding successful coping strategies of business travelers (Westman, 2004) may perhaps be extended to travelers in general and even to non-travelers. Social support, sense of control, personal resources including both psychological and material, can all be used proactively to acquire resources prior to the respite while all of these strategies, as well as time management and anxiety management strategies can be used proactively to prevent the loss of resources prior to the respite. In other words, psychological and behavioral preparations have the power to alleviate stress that would otherwise accumulate prior to the respite. Such strategies employed prior to the respite may also aid employees who refrain from taking time off due to perceived post-respite or consequential co-worker workload to change their perceptions regarding the beneficial value of a respite.

We found that the subjective feeling that one deserves to take a break from work and that co-workers, too, perceive one's break to be legitimate are among most important variables determining the success of the respite in bringing about relief (Etzion, 2006). Whereas it may be the organization's onus to create a climate of legitimacy surrounding employees' respites, not all organizations might step up to the role. Since the feeling that one deserves to take a break from work is subjective, we suggest individuals may promote such feelings through their own sense of control or through their social support networks. Social support from both within and without the workplace have the power to aid individuals in gaining and procuring resources that can then help in their coping with the situation. However, different individuals may gravitate toward specific types of social support or different sources of social support. For example, women seem to gravitate more toward sources of social support external to their workplace while men seem to gravitate more toward social support from within their workplace. We found that a certain degree of verbal communication is necessary for the potential of social support sources in buffering stress to be realized (Etzion, 1984). It may be that some individuals are more talkative than others. Our studies, for example, show that women more than men tend to talk with others as a mode of coping with stress (Etzion & Pines, 1986). The process of cultivating sources of social support may thus represent a process of social transaction demanding an investment of energy that would then be limited to one's area of central life interest. Employees embedded in different cultures have also demonstrated different degrees in their tendencies toward coping strategies that require talking about the source of distress (Etzion et al., 1982;

Etzion, Kafry, & Pines, 1983; Etzion & Pines, 1986). Matching one's source of social support with one's need is an imperative as developing and maintaining social relationships beyond one's need might prove a stressful demand in itself.

Finally, expectations regarding the respite may influence the respite experience and hence the respite effect. As noted by Barbara Bailey Reinhold (1997, p. 104), "expectation edits reality." Expectations have been shown to work as self-fulfilling prophecies, influencing whether or not an event occurs, the extent to which it occurs, and how individuals react to its occurrence (Eden, 2001). The expectancy hypothesis suggests that those who expected more relief will gain more relief from their respite. This may be the basis for findings showing that women who had been initially more depressive gained more than their less depressive counterparts (Hobfoll, Johnson, Ennis, & Jackson, 2003) or for conjectures concerning the greatest contribution of a respite made toward those who are in a greater need of physiological relief (Laski et al., 2015). If employees can be convinced, based on their own data or information provided by a credible source, that they may expect significant and prolonged relief from their respite, they might then reap greater benefit from the respite effect and experience a slower fadeout of the beneficial respite effect.

To summarize

Plan your respite, plan your time, sleep, and workload leading up to the respite, plan for post-respite recuperation, use your social support network and find evidence of the legitimacy of your leave, take as much control as you possibly can over the features of your respite, match your respite to fit your personal and situational needs and preferences, and expect the very best!

What to do during the respite

Working individuals arrive at their vacation with different resource pools and experience their vacation differently. The respite body of research points clearly to the fact that respite satisfaction is key, possibly *the* key, to gaining relief from job stress and burnout and generally enhancing well-being. This appears to be true for all types of respites: a vacation (Ron, 2003; Shaviv, 1998; Westman & Eden, 1997), a sabbatical (Davidson et al., 2010), a business trip (Westman, Etzion, & Yoffe, 1998), teachers' summer vacation (Zeevi-Ohana, 2000), students or teachers facing a mid-term break (Etzion & Zvi, 2004 and Laski et al., 2013 respectively), or daily leisure (Raviv, 2002), to name a few examples. The goal, then, is to enhance satisfaction through a positive respite experience.

Once planned, satisfaction is enhanced when the respite is executed as expected. We often found that employees who experienced active types of leisure reported reduced levels of both occupational stress and burnout relative to workers who went on a passive type of vacation (e.g., Ben-Moshe, 2006; Etzion & Sapir, 1996; Sraya-Wallin, 1999). Sports, sightseeing, and excursions in nature were among the most satisfying and recovery-enhancing activities reported in our studies, even when they comprised only a small portion (4–5%) of the respite activities. Time spent voluntarily with family and friends had a similar effect, which was perhaps also related to the impact of one's social support on one's well-being. Different forms of entertainment were among the least exercised activities that had a positive effect on well-being. In contrast, we also found that engaging in family and personal chores hindered the respite effect (Laski, 2002). For students, it was studying and working for a living that hindered respite satisfaction and consequentially also the respite effect (Zvi, 2004).

We do not, however, encourage that employees follow these guidelines blindly. It is vital that employees consider activities that are best suited to their personality and preferences. We found that different types of activities correlated with different levels of neuroticism and conscientiousness and that personality types moderated the influence of certain activities over stress, burnout, and respite satisfaction (Zvi, 2004). It is thus imperative that employees consider the type of respite that would serve to renew their resources compared to those that might further dwindle resources (Westman, 1999). Moreover, it appears that the personal fit between respite characteristics and personality are crucial in achieving a beneficial respite effect (Eden, 2001), whether personality determines leisure content or the ability to be satisfied (Laski, 2002). For example, we found that vacation satisfaction correlates strongly and positively with extraversion, and also with conscientiousness and openness (Zeevi-Ohana, 2000). Vocation might also be a consideration in the choice of the type of respite one has with respect to both its predetermined length and activities. For example, although we found that teachers engaging during their respite in preparation activities for post-respite school time enjoyed greater relief, not all teachers capitalized on this potential. For example, teachers characterized by greater neuroticism engaged less in school preparation activities during their respite, whereas teachers characterized by greater conscientiousness participated more in such activities. Teachers characterized by greater agreeableness dedicated less time to family and personal chores and more time to leisure activities and entertainment (Laski, 2002; Zeevi-Ohana, 2000). Conditions that cannot be altered for the sake of a respite, such as marital state and age of children, should also be considered when planning one's respite. For example, we found that women might feel less relaxed on vacation than men when vacationing with their family. Similarly, parents to young children experienced greater levels of stress after the respite compared to its level before the respite (Laski, 2002). Degrees of self-efficacy, as well as positive and negative affectivity also have the potential of influencing the respite effect (Davidson et al., 2010; Eden, 2001).

Sleep is an important factor in recovery. Travelers were found to stay up later than normal and wake up earlier than usual during their vacations. Sleep problems occurred in the face of unfamiliar or noisy surroundings, uncomfortable accommodations, and worrying about work, finance, or family-related concerns. Additionally, women were approximately two times more likely than men to experience sleep problems during vacation (Gallup, 2002). It is thus important to keep in mind arrangements that would promote regular sleep. A stay at a health resort may enhance one's mood, quality of sleep, and reduce general feelings ill-being (Strauss-Blasche et al., 2004).

The degree of detachment from work is an additional key factor for achieving a beneficial respite effect (Etzion, 2003; Etzion et al., 1998; Sapir, 1995). The effect of easily available communication with work should be considered not only by the organization, as noted earlier, but also by the employees in managing their respite time, whether daily or on a longer respite. Last, we note the degree of change from one's routine. Some individuals may enjoy a relaxing vacation while others need to stay active in order to enjoy their respite. For example, workaholics do not enjoy as great a relief as non-workaholics as an effect of a respite (Westman et al., 1994). This may be due to their preference to remain psychologically attached to their work and realize fewer vacation days (Braun (Cohen), 2007). It may also be that they experience job stress in a manner that produces less burnout compared to non-workaholics (Eden, 2001). Notwithstanding, consider the practicalities of detachment as a function of respite length and theme, whether an annual vacation, a holiday leave, a maternity leave, military reserve duty, a business trip, or organizational training, as well as a function of organizational culture, climate, and attitude.

To summarize

Enhance satisfaction through a positive experience, execute leisure as planned to better fulfill expectations, detach from work, as well as the daily routine as much as possible, and most importantly – match the respite with personal preferences and needs.

What to do after the respite

An important characteristic often neglected by employees, and hence typically unplanned for, is that of post-respite adjustment. We found that stress develops and contracts differently during a respite according to various respite characteristics, as well as in response to different types of stressors (e.g., the presence of negative work characteristics or the absence of positive work characteristics). It is thus important to allocate resources for post-respite adjustment. Sleep content before and during a respite might also affect the need for post-respite adjustment.

A positive respite experience is one that contributes to recovery. It is additionally one upon which individuals may capitalize after the fact (Davidson et al., 2010). The rosy-view effect implies that individuals exercise not only pre-respite rosy prospection in expecting a positive respite but also a post-respite rosy retrospection, remembering the respite as a better experience than it actually was (Mitchell et al., 1997). The post-respite retrospective rosy-view effect may perhaps be harnessed to enhance the respite effect and prolong the fadeout process of the respite effect (Eden, 2001).

To summarize

Leave time to adjust and rest as needed, enhance positive respite memories, and seek other ways to prolong the respite effect.

Call for the future

Respite is the generally accepted way of combating work-related stress: it allows employees to distance themselves from such stressors for a while. Respites facilitate recovery in two ways: on the one hand, they relieve, fully or partially, the pressure experienced from work-related stressors, while on the other, they allow individuals to regain resources through activities that are energizing, revitalizing, and refreshing to the individual who aims to recharge their batteries (Eden, 2001; Etzion, 1995). This refreshing effect applies to respites of any length. Respites may be taken during the work day in the form of rest periods, coffee breaks, mealtimes, and workout breaks, they may be taken in after-work hours in the form of serious or non-serious leisure activities, they may be applied organizationally, such as weekend or holiday shutdown, or they may be applied individually, on terms dictated by employees.

Respite research is a fairly young doctrine, with fewer than four decades to boast. Empirical evidence of the beneficial, salutogenic effects of respites, leave, and leisure are still accumulating. Do respites really achieve recovery-related objectives? What enhances or hinders such effects? How does taking a respite affect the level of stress and psychological burnout experienced upon the return to work? Do respites have a uniform effect on all employees? Does the way by which individuals spend their respite influence post-respite attitudes or feelings? These are but some questions that have risen over the years and which this book attempts to answer.

We may proudly look back on the knowledge we gained over the years and the broad perspective achieved on the types of variables that can influence the respite effect. Respite research results nevertheless leave many questions unanswered. We suggest that future research relate to both work and non-work demands and resources, so as to gain a broader understanding of their dynamics during work and non-work periods. Lengthy respites, where some new sense of routine is created, merit a deeper breakdown of stress, burnout, and satisfaction dynamics. In other words, there is much room from greater integration of leisure research and recovery and respite research. This is particularly important in light of findings emerging from our meta-analysis showing that stress and burnout levels rise immediately (three days) prior to the respite and following it. Ways to benefit and recover from work stress and burnout under unusual conditions leading to unexpected respites such unemployment or lengthy sick leave have yet to be explored. Post-respite debriefing and adjustments has rarely been studied in our field. Additionally, the inevitable fadeout process requires further research, especially with respect of finding ways to prolong it. Given that those who gain most of respite often experience the quickest fadeout, it would be beneficial to further understand the dynamic of resource gain and loss, particularly after the respite. The accumulative effect of different types of respites over time has yet to be studied. In the same way that regular annual vacations prevent heart disease and death (Gump & Matthews, 2000), it may be that regular annual respites prevent long-term rises in employees' stress and burnout levels. That is, restoration of energy resources may aid in maintaining stress and burnout levels that allow normal functioning.

Cases and conditions in which the respite effect cannot be taken for granted may be further explored, such and job insecurity, economic hardship, negative expectations, or negative family relations. The focus of research should be upon ways to enhance the salutogenic effect of a respite even in the face of adversity and resource-consuming variables. Interventions aimed at purposefully enhancing the respite effect and prolonging its fadeout process have yet to be developed and researched systematically.

The value of such research projects lies not only in the improvement of human well-being and work potential but also in its contribution to the quality of life outside the workplace. Additional respite studies are needed in order to better understand the impact of variables, such as duration and timing of the respite, preferred activities and their matching to personal, vocational, and other situational conditions, ways by which respite pleasure and satisfaction may be enhanced and translated into recovery, the influence of physical and emotional detachment from the demands of the daily routine, and ways by which it would be possible to capitalize on differences in well-being and resource potential between the respite period and everyday work routine.

Our research suggests that the question holding most value as key to the beneficial respite effect is that of the interaction or fit between respite characteristics and persona characteristics. Shedding further light on this question is likely to be extremely constructive in terms of implementation both organizationally and individually. We believe that the research projects we have described above together with the pioneering work we have already done in the area, have the potential to begin answering these questions and point the way to interactive solutions. We encourage others to follow suit.

References

Abdel-Halim, A. A. (1978). Employee effective responses to organizational stress: Moderating effects of characteristics. *Personal Psychology, 31(3)*, 561–579.

Aharon-Madar, M. (1997). The impact of vacation on job stress, burnout, absenteeism, and performance. M.Sc., Tel Aviv University, Israel. [Hebrew].

Aspinwall, L. G. & Taylor, S. E. (1997). A stitch in time: Self-regulation and proactive coping. *Psychological Bulletin, 121(3)*, 417–436.

Aviv, N. (1998). The meaning of vacation among women working shifts: The effect of vacation on job stress, burnout and work-home conflict. M.Sc., Tel Aviv University, Israel. [Hebrew].

Avni, T. (2007). The impact of organizational vacation on the amount of regular working hours, overtime, sick days and absenteeism among the organization's employees. M.Sc., Tel Aviv University, Israel. [Hebrew].

Bailey Reinhold, B. (1997). *Toxic work.* New York: Plume.

Bailyn, L. (1993). *Breaking the mold: Women, men and time in the new corporate world.* New York: Free Press.

Bakker, A. B., Demerouti, E., Taris, T. W., Schaufeli, W. B., & Schreurs, P. J. G. (2003). A multi-group analysis of the job Demands-Resources model in four home-care organizations. *International Journal of Stress Management, 10(1)*, 16–38.

Bakker, A. B., Demerouti, E., & Verbeke, W. (2004). Using the Job Demands-Resources model to predict burnout and performance. *Human Resource Management, 43(1)*, 83–104.

Bakker, A. B., Hakanen, J. J., Demerouti, E., & Xanthopoulou, D. (2007). Job resources boost work engagement, particularly when job demands are high. *Journal of Educational Psychology, 99(2)*, 274–284.

Bandura, A. (1997). *Self-efficacy: The exercise of control.* New York: Freeman.

Ben Yoseff, N. (2004). The effect of maternity leave on burnout among women serving in the I.D.F. M.Sc., Tel Aviv University, Israel. [Hebrew].

Ben-Moshe, L. (2006). Work features, "serious leisure" and burnout among flight attendants. M.Sc., Tel Aviv University, Israel. [Hebrew].

Boklis, V. (2006). The impact of vacation on job-stress and burnout – Meta-analysis. M. Sc., Tel Aviv University, Israel.

Bowlby, J. (1969). *Attachment and loss.* New York: Basic Books.

Braun (Cohen), A. (2007). "Vacation objectors" – The factors behind vacation avoidance in a public organization. M.Sc., Tel Aviv University, Israel. [Hebrew].

Burke, R. J. & Greenglass, E. A. (1995). Longitudinal study of psychological burnout in teachers. *Human Relations, 48(2)*, 187–202.

Chen, S., Westman, M., Etzion, D., & Dach-Zahavi, A. (2009). The impact of a short respite on well-being and performance (Working Paper). Tel Aviv University, Henry Crown Institute of Business Research, Faculty of Management, Tel Aviv, Israel.

Cherniss, C. (1981). Preventing burnout: From theory to practice. In J. W. Jones (Ed.), *The burnout syndrome* (pp. 172–176). London: London House Management Press.

Cohen, S. (1988). Psychosocial models of the role of social support in the etiology of physical disease. *Health Psychology, 7(3)*, 269–297.

Cohen, S. & Wills, T. A. (1985). Stress, social support and the buffering hypothesis. *Psychological Bulletin, 98(2)*, 310–357.

Dar, A. (2002). Law and custom regarding leave and working hours in Israel, Japan and the EU – A cross-cultural comparison. M.Sc., Tel Aviv University, Israel. [Hebrew].

Davidson, O. B., Eden, D., Westman, M., Cohen-Charash, Y., Hammer, L. B., Kluger, A. N., et al. (2010). Sabbatical leave: Who gains and how much? *Journal of Applied Psychology, 95(5)*, 953–964.

DeFrank, R. S., Konopaske, R., & Ivancevich, J. M. (2000). Executive travel stress: Perils of the road warrior. *Academy of Management Executive, 14(2)*, 58–71.

Demerouti, E., Bakker, A. B., de Jonge, J., Janssen, P. P. M., & Schaufeli, W. B. (2001). Burnout and engagement at work as a function of demands and control. *Scandinavian Journal of Work and Environment and Health, 27(2)*, 279–286.

Doerr, K. H., Mitchell, T. R., Klastorin, T. D., & Brown, K. A. (1996). Impact of material flow policies and goals on job outcomes. *Journal of Applied Psychology, 81(2)*, 142–152.

Eden, D. (1990). Acute and chronic job stress, strain and vacation relief. *Organizational Behavior and Human Decision Processes, 45(2)*, 175–193.

Eden, D. (2001). Vacations and other respites: Studying stress on and off the job. In C. L. Cooper & I. T. Robertson (Eds.), *International review of industrial and organizational psychology* (pp. 121–146). Chichester, UK: Wiley.

Etzion, D. (1984). The moderating effect of social support on the relationship of stress and burnout. *Journal of Applied Psychology, 69(4)*, 615–622.

Etzion, D. (1987). Burnout: The hidden agenda of human distress (Working Paper No. 930/87). Tel Aviv, Israel: Henry Crown Institute of Business Research, Faculty of Management, Tel Aviv University.

Etzion, D. (1988a). The experience of burnout and work/non-work success in man and female engineers: A match-pairs comparison. *Human Resource Management, 27(2)*, 163–179.

Etzion, D. (1988b). Understanding the dynamics of burnout through cross-cultural research: Reflection on the comparison between Israelis and Americans (Working Paper No. 965/88). Tel Aviv, Israel: Henry Crown Institute of Business Research, Faculty of Management, Tel Aviv University.

Etzion, D. (1995). Is a break from work an effective means for combating stress and preventing burnout in organizations? *Innovations in Management* – A Publication of the Faculty of Management, Tel Aviv University. [Hebrew].

Etzion, D. (2003). Annual vacation: Duration and relief from job stress and burnout. *Anxiety, Stress, and Coping, 16(2)*, 213–226.

Etzion, D. (2005). Respite from work as a means of recovery from occupational stress and burnout. Paper presented at the 25th annual convention of Scientists Talk About Research (STAR), Halle, Germany.

Etzion, D. (2006). Respite as a means of recovery from occupational stress and burnout. In P. Buchwald (Ed.), *Stress and anxiety – Application to health, community, work place and education* (pp. 186–195). Cambridge: Cambridge Scholar Press.

Etzion, D. & Bailyn, L. (1994). Patterns of adjustment to the career/family conflict of technologically trained women in the U.S. and Israel. *Journal of Applied Social Psychology, 24(17)*, 1520–1549.

Etzion, D. & Chagi-Niv, I. (1985). Life-work overlap as a factor in burnout of kibbutz members holding central positions. Paper presented at the The XXth Scientific Convention of the Israel Psychological Association, Ramat-Gan, Israel.

Etzion, D., Eden, D., & Lapidot, Y. (1994). Active reserve duty as an ameliorative respite from civilian job-stress and burnout. Paper presented at The Fifth International Conference on Social Stress Research, Honolulu, HI.

Etzion, D., Eden, D., & Lapidot, Y. (1998). Relief from job stressors and burnout: Reserve service as respite. *Journal of Applied Psychology, 83(4)*, 577–585.

Etzion, D., Kafry, D., & Pines, A. (1982). Tedium among managers: A cross-cultural American – Israeli comparison. *Journal of Psychology and Judaism, 7(1)*, 30–41.

Etzion, D. & Ofek, A. (1998). The influence of two types of vacation on perceived job stressors and burnout among social workers. Paper presented at The 7th International Conference on Stress Research, Budapest, Hungary.

Etzion, D. & Pines, A. (1986). Sex and culture in burnout and coping among human service professionals. *Journal of Cross – Cultural Psychology, 17(2)*, 191–209.

Etzion, D. & Rechnitz-Elpeleg, O. (2002). Formal and informal support systems for women managers: Israel versus Sweden. In L. Nicolaou-Smokoviti (Ed.), *Scientific yearbook: Essays in honor of Professor Litsa Nicolaou-Smokoviti* (pp. 429–443). Piraeus, Greece: University of Piraeus.

Etzion, D. & Sapir, O. (1996). The effect of an annual vacation in relieving stress and burnout at work (a controlled field study at an industrial enterprise in Israel). Report by: Tel Aviv University, Henry Crown Institute of Business Research, Faculty of Management, Tel Aviv, Israel.

Etzion, D. & Sapir, O. (1997). Annual vacation: Duration of relief from job stress and burnout. Paper presented at the 12th Annual Conference for The Society for Industrial–Organizational Psychology, St. Louis, MO.

Etzion, D. & Westman, M. (1994). Social support and sense of control as moderators of the stress-burnout. relationship in military careers. *Journal of Social Behavior and Personality, 9 (4)*, 639–656.

Etzion, D., Westman, M., & Vinocour-Hai, V. (2002). The impact of high & low stress periods on the intensity of burnout crossover between spouses: A longitudinal study. Paper presented at The XXV International Congress of Applied Psychology (ICAP), Singapore, Asia.

Etzion, D., Zedeck, S., Maslach, C., & Owens, P. (2001). *Students' experience of stress and burnout before and after school breaks* (Working Paper). University of California, Department of Psychology, CA, Berkeley.

Etzion, D. & Zvi, E. (2004). Stress and burnout among students before and after the Passover break: Vacation activities and personality variables as moderators. Report by: Tel Aviv University, Henry Crown Institute of Business Research, Faculty of Management, Tel Aviv, Israel.

Kossek, E. E. & Ollier-Malaterre, A. (2012). Work-family policies: Linking national contexts, organizational practice and people for multi-level change. In S. Poelmans, J. Greenhaus & M. las Heras (Eds.), *New frontiers in work-family research: A vision for the future in a global world* (pp. 1–53). London: Palgrave.

Feldman Barrett, L. & Russell, J. A. (1998). Independence and bipolarity in the structure of current affect. *Journal of Personality and Social Psychology, 74(4)*, 967–984.

Frankenhaeuser, M., Lundberg, U., Fredrikson, M., Melin, B., Tuomisto, M., Myrsten, A.-L., et al. (1989). Stress on and off the job as related to sex and occupational status in white-collar workers. *Journal of Organizational Behavior, 10(4)*, 321–346.

Freedy, J. R. & Hobfoll, S. E. (1994). Stress inoculation for reduction of burnout: A conservation of resources approach. *Anxiety, Stress and Coping, 6(4)*, 311–325.

Freudenberger, H. J. (1974). Staff burnout. *Journal of Social Issues, 30(1)*, 159–165.

Freudenberger, H. J. (1983). Burnout: Contemporary issues, trend, and concern. In B. A. Farber (Ed.), *Stress and burnout in human service profession*. New York: Pergamon Press.

Friedman, I. A. (1995). Student behaviour patterns contributing to teacher burnout. *Journal of Educational Research, 88(5)*, 281–289.

Ganster, D. C. & Bart, V. (1988). The impact of social support on mental and physical health. *British Journal of Medical Psychology, 61(1)*, 17–36.

Gilbert, D. & Abdullah, J. (2002). A study of the impact of the expectations of a holiday on an individual's sense of well-being. *Journal of Vacation Marketing, 8(4)*, 352–361.

Glaser, R., Kiecolt-Glaser, J. K., Speicher, C. E., & Holiday, J. E. (1985). Stress, loneliness, and changes in herpesvirus latency. *Journal of Behavioral Medicine, 8(2)*, 25–49.

Glass, J. L. & Riley, L. (1998). Family responsive policies and employee retention following childbirth. *Social Forces, 76(4)*, 1401–1435.

GALLUP. (2002). How to rest on vacation. Web news and product reviews. https://news.gallup.com/poll/6400/Need-Vacation-From-Your-Vacation.aspx.

Gorgievsky, M. J., Bakker, A. B., & Schaufeli, W. B. (2010). Work engagement and workaholism: Comparing the self-employed and salaried employees. *Journal of Positive Psychology, 5(1)*, 83–96.

Greenglass, E., Schwarzer, R., Jakubiec, D., Fiksenbaum, L., & Taubert, S. (1999). The Proactive Coping Inventory (PCI): A multidimensional research instrument. Paper presented at the 20th International Conference of the Stress and Anxiety Research Society (STAR), Cracow, Poland.

Gump, B. B. & Matthews, K. A. (2000). Are vacations good for your health? The 9-year mortality experience after the multiple risk factor intervention trial. *Psychosomatic Medicine, 62(5)*, 608–612.

Hackman, J. R. & Oldham, G. R. (1976). Motivation through the design of work: Test of a theory. *Organizational Behavior and Human Performance, 16(2)*, 250–279.

Hackman, J. R. & Oldham, G. R. (1980). *Work redesign*. Reading, MA: Addison Wesley.

Harris, P. L. (1984). Assessing burnout: The organizational and individual perspective. *Family and community Health, 6(4)*, 32–43.

Hobfoll, S. E. (1989). Conservation of resource – A new attempt at conceptualizing stress. *American Psychologist, 44(3)*, 513–524.

Hobfoll, S. E. (1998). *Stress, culture and community: The psychology and philosophy of stress*. New York: Plenum.

Hobfoll, S. E. (2001). The influence of culture, community, and the nested self in the stress process: Advancing Conservation of Resources Theory. *Applied Psychology: An International Review, 50(3)*, 337–421.

Hobfoll, S. E. (2002). Social and psychological resources and adaptation. *Review of General Psychology, 6(4)*, 307–324.

Hobfoll, S. E. & Freedy, J. (1993). Conservation of resources: A general stress theory applied to burnout. In W. B. Schaufeli, C. Maslach, & T. Marek (Eds.), *Professional burnout: Recent developments in theory and research* (pp. 115–129). Washington, DC: Taylor & Francis.

Hobfoll, S. E., Johnson, R. J., Ennis, N., & Jackson, A. P. (2003). Resource loss, resource gain, and emotional outcomes among inner city women. *Journal of Personality and Social Psychology, 84(3)*, 632–643.

Hobfoll, S. E. & Shirom, A. (1993). Stress and burnout in the workplace: Conservation of resources. In R. T. Golembiewski (Ed.), *Handbook of organizational behavior* (pp. 41–60). New York: Marcel Dekker.

Hobfoll, S. E. & Shirom, A. (2000). Conservation of resources theory: Applications to stress and management in the workplace. In R. T. Golembiewski (Ed.), *Handbook of organization behavior* (pp. 57–81). New York: Marcel Dekker.

Horovitz, S. (2002). The relationship of economic stress and social support and crossover anxiety between couples in times of unemployment. M.Sc., Tel Aviv University, Israel. [Hebrew].

House, J. S. (1981). *Work stress and social support*. Reading, M.A: Addison Wesley.

International Labour Organization. (2013, April 23). World Day for Safety and Health at Work 2013 – Case Study: Karoshi: Death from overwork. Retrieved from: www.ilo.org/safework/info/publications/WCMS_211571/lang–en/index.htm.

Karasek, R. (1979). Job demands, job decision latitude, and mental strain: Implication for job redesign. *Administrative Science Quarterly, 24(2)*, 285–309.

Klein, H. M., Hyde, J. S., Essex, M. J., & Clark, R. (1998). Maternity leave, role quality, work involvement, and mental health one year after delivery. *Psychology of Women Quarterly, 22(2)*, 239–266.

Koniak, M. (2003). Overseas business trips: The impact of business trip purpose, frequency of business trip and warning time before business trip on stress and health. M.Sc., Tel Aviv University, Israel. [Hebrew].

Kremer-Fuks, S. (1999). The impact of vacation and slack work periods on job stress and burnout among newcomers and veteran Israelis. M.Sc., Tel Aviv University, Israel. [Hebrew].

Kühnel, J. (2009). Benefits of recovery for work engagement. D.Sc., University of Konstanz, Germany.

Lapidot, Y. (1992). The effects of a military-reserve service on work stress and burnout. M. Sc., Tel Aviv University, Israel. [Hebrew].

Laski, S. (2002). The effect of summer break activities and personality variables on stress and burnout among high school teachers in Israel. M.Sc., Tel Aviv University, Israel. [Hebrew].

Laski, S. (2013). Recovery during vacation and its fadeout – A longitudinal study. Ph.D., Tel Aviv University, Israel.

Laski, S., Etzion, D., & Westman, M. (2013). Psychological and physiological recovery from work stressors during vacation and its fadeout: A longitudinal study. Paper presented at the 121st Annual Convention for the American Psychological Association (APA), Honolulu, Hawaii.

Laski, S., Etzion, D., & Westman, M. (2015). Who gains more from vacation – Recovery during vacation and its fadeout. Paper presented at the 36th Annual Conference for the Stress and Anxiety Research Society (STAR), Tel Aviv, Israel.

Lazarus, R. S. (1993). Coping theory and research: Past, present and future. *Psychosomatic Medicine, 55(3)*, 234–247.

Lefcourt, H. M., Martin, R. A., & Saleh, W. E. (1984). Locus of control and social support: Interactive moderators of stress. *Journal of Personal Social Psychology, 47(2)*, 378–389.

Leiter, M. P. (1991). Coping patterns as predictors of burnout: The function of control and escapist coping patterns. *Journal of Organizational Behavior, 12(2)*, 123–144.

Levi, M. (2002). The impact of summer break and personal variables on job stress and burnout among teachers in a Kibbutz regional school. M.Sc., Tel Aviv University, Israel. [Hebrew].

Lounsbury, J. W. & Hoopes, L. L. (1986). A vacation from work: Changes in work and non-work outcomes. *Journal of Applied Psychology, 71(3)*, 392–401.

Machlowits, M. (1980). *Workaholics, living with them, working with them.* Massachusetts: Adison Wesley Publishing.

Magen-Neuhaus, T. (1993). The effect of vacation and feelings of stress at work on emotional burnout. M.Sc., Tel Aviv University, Israel. [Hebrew].

Maruyama, I. & Tokuda, Y. (2017). Changing work styles among Japanese healthcare professionals. *Journal of General and Family Medicine, 18*(*4*). Retrieved from: http://onlinelibrary.wiley.com/doi/10.1002/jgf2.117/full.

Maslach, C. (1976). Burnout. *Human Behaviour, 5*(*9*), 16–22.

Mitchell, T. R., Thompson, L., Peterson, E., & Cronk, R. (1997). Temporal adjustments in the evaluation of events: The "rosy view". *Journal of Experimental Social Psychology, 33 (4)*, 421–448.

Morad, N. (2009). The effect of a weekend respite on burnout and stress among hospital nurses. M.Sc., Tel Aviv University, Israel. [Hebrew].

Moshkowitz, S. (2003). The relationship between work-overload, leisure activities and burnout among high-tech employees. M.Sc., Tel Aviv University, Israel. [Hebrew].

Naegel, I. J. & Sonnentag, S. (2011). The relationship between leisure time sports activity and personal resources. Paper presented at the Annual APA Convention (August 4–7), Washington, DC.

Norfolk, D. (1989). *Executive stress.* New York: Warner Books.

Ofek, A. (1997). The effect of organizational vacation and personal vacation on the connection between stress and burnout. M.Sc., Tel Aviv University, Israel. [Hebrew].

Ostroff, C., Kinicki, A. J., & Tamkins, M. M. (2004). Organizational culture and climate. In W. C. Borman, D. R. Ilgen, & R. J. Klimoski (Eds.), *Handbook of psychology* (Vol. 12, Industrial and organizational psychology, 11, pp. 566–593). Hoboken, NJ: Wiley.

Payne, R. (1980). Organizational stress and social support. In C. L. Cooper & R. Payne (Eds.), *Current concerns in occupational stress* (pp. 269–298). New York: Wiley.

Pines, A. (1993). Burnout: An existential perspective. In W. B. Schaufeli, C. Maslach, & T. Marek (Eds.), *Professional burnout: Recent development in theory and research* (pp. 33–51). Washington, DC: Taylor & Francis.

Pines, A., Aronson, E., & Kafry, D. (1981). *Burnout: From tedium to personal growth.* New York: Free Press.

Radian, R. (1998). The moderating effect of workaholism and Type A on the impact of vacation on job stress and burnout. M.Sc., Tel Aviv University, Israel. [Hebrew].

Raviv, E. (2002). Effects of leisure activity on occupational stress and burnout among high-tech employees. M.Sc., Tel Aviv University, Israel. [Hebrew].

Richards, G. (1999). Vacations and the quality-of-life patterns and structures. *Journal of Business Research, 44(3)*, 189–198.

Rienhold, B. B. (1996). *Toxic work.* New York: Plum/Penguin.

Ron, A. (2003). Effects of organizational vacation on occupational stress and burnout among high-tech employees. M.Sc., Tel Aviv University, Israel. [Hebrew].

Rubinstein, C. (1980). Vacations, expectations, satisfactions, frustrations, fantasies. *Psychology Today, 14(5)*, 62–66, 61–76.

Sapir, O. (1995). The effect of vacation on stress and burnout at work. M.Sc., Tel Aviv University, Israel. [Hebrew].

Schein, E. H. (2004). *Organizational culture and leadership* (3rd ed.). San Francisco, CA: Jossey-Bass.

Schmilovitch, H. (1992). The burnout phenomenon and its relationships to vocational fit and to leisure fit. M.Sc., Tel Aviv University, Israel.

Schor, J. B. (1992). *The overworked American: The unexpected decline of leisure.* New York: Basic Books.

Shaviv, T. (1998). The impact of an organized tour vacation on work stress and burnout among managerial and non-managerial vacationers. M.Sc., Tel Aviv University, Israel. [Hebrew].

Shirom, A. (2003). Job-related burnout. In J. C. Quick & L. E. Tetrick (Eds.), *Handbook of occupational health psychology.* Washington, DC: American Psychological Association.

Silverberg, S. (2007). Home/work conflict and burnout among working mothers after maternity leave – Situational and personal characteristics as moderators. M.Sc., Tel Aviv University, Israel. [Hebrew].

Sluiter, J. K., Frings-Dresen, M. H., Meijman, T. F., & van der Beek, A. J. (2000). Reactivity and recovery from different types of work measured by catecholamines and cortisol: A systematic literature overview. *Occupational Environmental Medicine, 57(5)*, 289–315.

Sonnentag, S. (2003). A new look at the interface between non-work and work recovery, work engagement, and proactive behavior recovery, work engagement, and proactive behavior. *JAP, 88(3)*, 518–528.

Sonnentag, S. & Bayer, U. V. (2005). Switching off mentally: Predictors and consequences of psychological detachment from work during off-job time. *Journal of Occupational Health Psychology, 10(4)*, 393–414.

Sonnentag, S. & Zijlstra, F. (2006). Job characteristics and off-job activities as predictors of need for recovery, well-being and fatigue. *Journal of Applied Psychology, 91(2)*, 330–350.

Sraya-Wallin, O. (1999). The effect of vacation characteristics and the fit between job and career-anchors on stress and burnout among employees in a hi-tech environment. M. Sc., Tel Aviv University, Israel. [Hebrew].

Strauss-Blasche, G., Ekmekcioglu, C., & Marktl, W. (2002). Moderating effects of vacation on reactions to work and domestic stress. *Leisure Sciences, 24(2)*, 237–249.

Strauss-Blasche, G., Ekmekcioglu, C., Marktl, W., Riedmann, B., Riedmann, G., Humpeler, E., et al. (2004). Vacation at moderate and low altitude improves perceived health in individuals with metabolic syndrome. *Journal of Travel Medicine, 11(5)*, 300–307.

Striker, J., Dimberg, L., & Liese, B. H. (2000). Stress and business travel: Individual, managerial, and corporate concerns. *National Productivity Review, 20(1)*, 3–9.

Taylor, B. C. & Arnow, B. A. (1988). *The nature and treatment of anxiety disorder.* New York: The Free Press.

Toomey, E. L. & Connor, J. M. (1988). Employee sabbaticals: Who benefits and why. *Personnel, 65(4)*, 81–84.

Tuval, A. (2001). Patterns of stress-related ER hospitalization at Sheba Medical Center between January 1997 and September 2000. M.Sc., Tel Aviv University, Israel. [Hebrew].

Ullah, P., Banks, M., & Warr, P. (1985). Social support, social pressures and psychological distress during unemployment. *Psychological Medicine, 15(2)*, 283–295.

Utechet, K. M. & Aldag, R. J. (1989). Vacation discrepancy: Correlates of individual differences and outcomes. *Psychological Reports, 65(3)*, 867–882.

Vingerhoets, A. J. J. M., van Huijgevoort, M., & van Heck, G. L. (2002). Leisure sickness: A pilot study on its prevalence, phenomenology, and background. *Psychotherapy and Psychosomatics, 71(6)*, 311–317.

Vincour-Hai, V. (2002). Stress and burnout before and after vacation and their crossover between spouses. M.Sc., Tel Aviv University, Israel. [Hebrew].

Wallston, B. S., Alagna, S. W., De-Villis, B. M., & De-Villis, R. F. (1983). Social support and physical health. *Health Psychology, 2(1)*, 367–391.

Westman, M. (1992). Moderating effects of decision latitude on stress-strain relationship: Does organizational level matter? *Journal of Organizational Behavior, 13(7)*, 713–722.

Westman, M. (1997). A model of stress crossover (Working Paper). Tel Aviv University, Henry Crown Institute of Business Research, Faculty of Management, Tel Aviv, Israel.

Westman, M. (1999). Applying Hobfoll's conservation of resources (COR) theory to respite research. Paper presented at The Academy of Management Symposium, Chicago, IL.

Westman, M. (2004). Strategies for coping with business trips: A qualitative exploratory study. *International Journal of Stress Management, 11(2)*, 167–176.

Westman, M., Chen, S., & Etzion, D. (2011). Recovery during business trips. Paper presented at the European Association of Work and Organizational Psychology (EAWOP), Maastricht, The Netherlands.

Westman, M. & Eden, D. (1997). Effects of a respite from work on burnout: Vacation relief and fade-out. *Journal of Applied Psychology, 82(4)*, 516–527.

Westman, M., Eden, D., & Magen-Neuhaus, T. (1994). Effects of vacation on job stress and burnout: Relief and fade-out. Paper presented at The 15th International Conference on Social Stress Research, Honolulu, HI.

Westman, M. & Etzion, D. (1995). Crossover of stress, strain and resources from on spouse to another. *Journal of Organizational Behavior, 16(2)*, 169–181.
Westman, M., Etzion, D., & Aharon-Madar, M. (1999). Vacation, burnout and absenteeism (Working Paper No. 0027/99). Tel Aviv University, Henry Crown Institute of Business Research, Faculty of Management, Tel Aviv, Israel.
Westman, M., Etzion, D., & Chen, S. (2009a). Are business trips a unique kind of respite? In P. Perrewé, D. Ganster, & S. Sonnentag (Eds.), *Research in occupational stress and well-being* (pp. 167–204). Oxford: JAI Press/Elsevier.
Westman, M., Etzion, D., & Chen, S. (2009b). The crossover of positive experiences from business travelers to their spouses. *Journal of Managerial Psychology, 24(3)*, 269–284.
Westman, M., Etzion, D., & Danon, E. (1998). Downsizing and the crossover of burnout in married couples employed by the same organization (Working Paper). Tel Aviv University, Henry Crown Institute of Business Research, Faculty of Management, Tel Aviv, Israel.
Westman, M., Etzion, D., & Danon, E. (2001). Job insecurity and crossover of burnout in married couples. *Journal of Organizational Behavior, 122(5)*, 467–481.
Westman, M., Etzion, D., & Gattenio, E. (2008). International business travels and the work-family interface – A longitudinal study. *Journal of Occupational and Organizational Psychology, 81(3)*, 459–480.
Westman, M., Etzion, D., & Gortler, E. (2004). The work-family interface and burnout. *International Journal of Stress Management, 11(4)*, 413–428.
Westman, M., Etzion, D., & Horovitz, S. (2004). The toll of unemployment does not stop with the unemployed. *Human Relations, 57(7)*, 823–824.
Westman, M., Etzion, D., & Segev, K. (2002). Leave and leisure worldwide – A cross-cultural comparison among the USA, the EU, Japan and Israel (Working Paper). Tel Aviv University, Henry Crown Institute of Business Research, Faculty of Management, Tel Aviv, Israel.
Westman, M., Etzion, D., & Yoffe, Y. (1998). The effect of overseas business trips on job stress and burnout (Working Paper). Tel Aviv University, Henry Crown Institute of Business Research, Faculty of Management, Tel Aviv, Israel.
Westman, M., Hobfoll, S. E., Chen, S., Davidson, B. O., & Laski, S. (2005). Organizational stress through the lens of conservation of resources (COR) theory. In P. Perrewé & D. Ganster (Eds.), *Research in occupational stress and well-being* (pp. 167–220). Oxford: JAI Press/Elsevier Science.
Wethington, E. & Kessler, R. C. (1986). Perceived support, received support, and adjustment to stressful life events. *Journal of Health and Social Behavior, 27(1)*, 78–89.
Xanthopoulou, D., Bakker, A., Dollard, M. F., Demerouti, E., Schaufeli, W., & Taris, T. W. (2007a). When do job demands particularly predict burnout? *Journal of Managerial Psychology, 22(8)*, 766–786.
Xanthopoulou, D., Bakker, A. B., Demerouti, E., & Schaufeli, W. (2007b). The role of personal resources in the Job Demands-Resources model. *International Journal of Stress Management, 14(2)*, 121–141.
Yoffe, Y. (1997). The effects of overseas business trips on job stress and burnout. M.Sc., Tel Aviv University, Israel. [Hebrew].
Zack, I. (2005). Proactive coping and involvement in "serious leisure" among flight attendants. M.Sc., Tel Aviv University, Israel.
Zeevi-Ohana, A. (2000). The impact of summer break activities and personality variables on job stress and burnout among high school teachers. M.A., The Hebrew University, Jerusalem, Israel.
Zvi, E. (2004). Stress and burnout among students before and after the Passover break – Vacation activities and personality variables as moderators. M.Sc., Tel Aviv University, Israel.

APPENDIX 1

Meta-analytic study of respite and fade-out effects

The purpose of this study (Etzion & Boklis, 2006) was to systematically investigate the conditions under which a respite from work produced beneficial effects on employees' well-being. We also examined the conditions in which such respites did not necessarily produce the desired beneficial effects, and considered what might be done to enhance the recovery effect of the respite. We incorporated in this study data collected in Israel and other countries over ten years from 23 studies – whether published, unpublished, or in process – that had been designed to capture the respite effect by using a longitudinal, repeated measure procedure, involving two or three measurements for each respondent, before the respite, after it, and approximately one month after the respite, assuming respondents had by then returned to the routine of their work or studies. Results document the respite effect (decline in the stress and burnout measures immediately after the vacation), as well as the fade-out effect (post-respite rise of stress and burnout usually to pre-respite levels).

Method

Our sample was based on a broad search of library catalogues (PsycInfo, ProQuest, and Medline) and using the following search words: stress, burnout, respite, annual vacation, time off, leisure, and leave. We included in our analyses results reported in various sources: professional literature, graduate and post-graduate dissertations, and other unpublished academic working papers. Partially successful, we aimed at finding sources that not only measured the levels of stress and burnout but which also referred also to the correlation between these variables per measurements. The search results elicited only a few studies conducted outside of our own respite project. Only one request to such authors to obtain their full datasets was answered favorably (Fritz & Sonnentag, 2006), so other potential sources had to be omitted. Thus, of 30 studies found, our sample includes only 23, and the number of studies included in each meta-analytic analysis ranges from 5 to 22, whereas the number of samples (or effects) included in each analysis ranges from 10 to 66 while the number of cases in each analysis varies from 1004 to 5815. Studies produced more than one effect because they tested multiple stress and burnout measures over multiple samples, occasionally with respect to multiple types of respites, and more often than not had used multiple types of measurements for their stress and burnout measures.

Data coding and elementary calculations required prior to our meta-analyses were conducted using Microsoft Excel. Coding was performed according to the type of effect we aimed to discover and the type of moderating variables that we included. In all cases we coded author names, year in which the paper was written, and sample size. According to the main variable tested, stress or burnout, we additionally coded each of the two variable measurements and the correlation between variable scores in the two measurements. Some data were recovered not through publication but by contacting authors. Where data were not available for two measurements, the average of correlations established in the other studies served as substitute for the missing data. For example, in a study where the correlation between the burnout scores based on the Burnout Measure (BM) (Pines et al., 1981) in the first and third measurement was missing, the average correlations of studies that compared a first and third measurement of burnout and used the same burnout measure served as a substitute. In cases where we examined the correlation between stress and burnout (at either one of the three measurements), additional data coded for each sample included the reliability of stress scores at the relevant measurement, reliability of burnout scores at the relevant measurement, and the correlation between stress and burnout scores at the same measurement. We used reliability specified for each measure where possible, or overall reliability where reliability was not specified per measure.

Given the nature of available data, moderating variables of three types were examined: demographic variables, respite variables, and methodological variables. These moderation analyses were restricted only to the respite effect as analyses involving the fade-out effect would have been based on too few samples. Among demographic variables, all but one categorical variable (gender) were continuous variables that were converted to categorical variables as the small number of samples in each analysis prevented us from examining moderation by continuous variables. Division of the variables into levels or categories was based on the range of values for each variable and the availability of its subcategories in the sample. We also considered the number of samples that fall into each category in order to reach a similar number of samples in each category of each variable. Our final classification was as follows: *gender* (men/women), *average age in sample* (young/medium/old), *years in office* (many/few), *percentage of married people in the sample* (small to medium/large), and *percentage of managers in the sample* (low/medium to high). The vacation variables tested were: *type of vacation* (organizational/personal) and *vacation length* (weekend/short vacation/long vacation). Finally, the methodological variables tested were: *publishing status* (published/not published), *timing of first measurement* (long/short/very short time before the vacation), *timing of second measurement* (very short/short/long time after the vacation), and *the combination of the timing of the first measurement and the timing of the second measurement.* The variable that combines the timing of the first and second measurements was examined in order to obtain a clearer picture of the effect of the timing of measurements on changes in stress and burnout. As the analysis examining the correlation between stress and burnout at the third measurement was based on too small a number of samples, the moderation effect was tested only for hypotheses regarding the first and second measurements.

Analyses

The meta-analyses were conducted using the Comprehensive Meta-Analysis program (Borenstein & Rothstein, 1999). For the purpose of performing analyses testing the effect of vacation on stress and burnout at work, we calculated ***d*** (standardized mean difference) for each sample. In our single-group pre/post-test design, we focused on the difference between the individual's scores and the change scores, where change scores were based on the variances of scores in the pre- and post-measurements, as well as on the correlation between pre- and post-scores:

$$\mathrm{d} = (\mathrm{Mpost} - \mathrm{Mpre})/\mathrm{SDcs}$$

$$\mathrm{SD^2cs} = \mathrm{SD^2pre} + \mathrm{SD^2post} - (2)(\mathrm{r})(\mathrm{SDpre})(\mathrm{SDpost})$$

where
d – Standardized mean difference
SD^2cs – Variance of the change scores
Mpre – Average of the scores in the before measurement
Mpost – Average of the scores in the after measurement
SD^2pre – Variance of the scores in the before measurement
SD^2post – Variance of the scores in the after measurement
r – Correlation between the before and after scores

(Morris & DeShon, 2002)

To minimize the sample error, we calculated for each sample the bias function value $c(df)$:

$$c(df) = 1 - 3/(4df - 1)$$

In the case of a single-group pre/post-test design, the number of degrees of freedom equals the number of paired observations (sample size, ***n***) minus one, therefore:

$$c(n-1) = 1 - 3/(4n-5)$$

For each sample, *d* was multiplied by $c(n-1)$, which created the corrected standardized mean difference (*dc*) (see also Hunter & Schmidt, 1990; Morris & DeShon, 2002):

$$dc = d^{*}c(n-1)$$

When examining the respite effect on work-related stress and burnout, we imported into the program the coded data relevant to all analyses and additionally: corrected standardized mean difference values, sizes of the samples. The analyses were performed in the framework of the random effects model with the underlying assumption that since samples were taken from different populations, overall variance would not reflect merely error variance but rather real differences between the groups. Whereas the program allows to correct standardized mean differences, in the current study *d* was already corrected before it was imported into the program. We thus chose Cohen's unadjusted *d* to be used in our analyses. Employing Forrest plots for data presentation in each analysis, we found no samples with extreme effects (outliers).

When examining the correlation between stress and burnout, we imported into the program the coded data relevant to all analyses and additionally: the correlation coefficients between stress and burnout, reliabilities of the stress scores, reliabilities of the burnout scores, and sample sizes. Under the assumption that the variance between the groups exceeded 0, the analyses examining the correlation between stress and burnout were conducted under the framework of the random effects model, similarly to the analyses examining the effect of vacation on stress and on burnout. Employing Forrest plots for data presentation in each analysis, we found no samples with extreme effects (outliers).

The meta-correlations between stress and burnout were examined by Hunter and Schmidt's (1990) statistical procedure, which includes calculations of weighted mean correlations (r), their variance, and the percentages of variance explained by the artifacts. Weighted mean correlations (r) consider the sampling errors, but not the measurement errors, in the stress and burnout measure. For this reason, we used the adjustment of Hunter and Schmidt (1990), which allows for the calculation of the estimated true correlations (ρ) based on the reliabilities of the stress and burnout scores. The estimated true correlations (ρ) consider both the sample errors and the measurement errors. In spite of the fact that in the current study we had all the information regarding the reliabilities of the stress, as well as the burnout scores, the version we used of the Comprehensive Meta-Analysis program allowed us to apply the Hunter and Schmidt adjustment only in its general format (for all the samples) and not adjustment of the artifacts sample by sample.

In addition to the weighted mean correlations, the estimated true correlations and the percentage of variance explained by the sampling error and the measurement error (% due to artifact), all of which were reported by the software, we also calculated confidence intervals and credibility intervals, which aid in estimating the true correlation between the variables (Whitener, 1990). Confidence intervals relate to the correlations after adjustment for the sampling error (r) while credibility intervals relate to the correlations after adjustment for both the sampling error and the measurement error (ρ). The lower and upper limits of the 95% confidence intervals were calculated based on r, the standard deviation of r, and the number of samples in each analysis. The lower and upper limits of the 90% credibility intervals were calculated based on ρ, the standard deviation of ρ, and the number of samples in each analysis. As the analysis relating to the correlation between stress and burnout at the third measurement was based on only 21 samples, in the calculation of the confidence intervals and credibility intervals we used the t values rather than the z values.

The effect of possible moderating variables was examined in the framework of the fixed effects model with the underlying assumption that differences would exist between groups, but not within groups, in each moderating variable. All analyses were based on existing data other than that of gender moderating effect, in which corrected standardized mean difference values were calculated based on subsamples of men and women. These were imported into the program along with sample sizes and identifying details of men and women samples. Employing Forrest plots for data presentation, we found two womens' samples (Magen-Neuhaus, 1993) with extreme effects (outliers), both of which were excluded from the analysis. Moderating effects were examined using the Q homogeneity test, which is intended for examining moderation of categorical variables. A significant Q test indicated heterogeneity of the samples and the possibility that moderating variables exist. Thus, the Q test for homogeneity between groups was conducted for each variable, and where found significant, the variable was considered to have a moderating effect. The distribution of the samples, demonstrated by the Forrest plots, was also key in determining of homogeneity. For moderation concerning the correlation between stress and burnout, two additional homogeneity tests were used: wide credibility intervals that included zero (Whitener, 1990) and percentages of explained variance by artifacts that were lower than 75% (Gilboa, 2005). In the case of an inconsistency between the different homogeneity tests, the Q test was treated as the main criterion due to its ability to balance between type I error and statistical significance (Cortina, 2003; Gilboa, 2005).

Results

Respite effect

As hypothesized, the current meta-analytic study showed a slight, though significant drop in employees' stress (d = –0.13) and burnout (d = –0.14) during the respite. These results provided additional support for the existence of the respite effect described in previous studies. The interpretation of the strength of the effects is based on the guidelines of Cohen (1988) guidelines, which were found to be relevant to meta-analytic effects as well (Lipsey & Wilson, 1993).

Fade-out effect

In line with previous research, we examined the fade-out effect in two ways, once by testing whether employees' stress and burnout levels rise after returning from the vacation and once by testing whether the stress and burnout levels return to their original levels within a certain time period after the vacation. We found the rise in stress and burnout after employees' return from their respite (second measurement) was insignificant. We also found no difference between the levels of stress and burnout measured before the respite and after it (third measurement). In other words, by the time of the third measurement, stress and burnout returned to their original levels thereby exhibiting the fade-out effect.

It is nevertheless important to remember that once the respite effect has been established the fade-out effect is inevitable. The real question, then, is not of its presence, but rather of its duration. Previous research has shown that stress returns to its regular level within three to four weeks after vacation, whereas burnout seems to have a longer period of fade-out. To estimate the duration of the respite effect before it completely fades out would require further inspection of those studies that conducted a third measurement. In our small sample, however, this was impossible.

The relationship between stress and burnout

In the current meta-analysis, we found a medium to high positive correlation (ρ = 0.29 to 0.38) between stress and burnout on all measurement occasions (before the respite, immediately after it, and at a lengthy time period after it). These results are in line with the theoretical notion of burnout being an outcome of stress and corroborate similar results of previous studies on occupational stress and burnout that are not included in the current meta-analysis (Lee & Ashforth, 1996; Pines, 1993; Pines et al., 1981; Schaufeli & Enzmann, 1998).

Moderating effects

Demographic variables

None of the five demographic variables examined were found to have a significant moderating effect on the impact of vacation on stress and burnout. Our hypotheses regarding the moderating effects of demographic variables were based on previous studies that found higher levels of stress and burnout in certain groups than in others. However, apart from the studies that were included in our analyses, very little in-depth work has been conducted on the respite effect in different groups of individuals. We might suggest then that lacking a broad theoretical basis, some assumptions underlying the moderating effect of demographic variables may have been misguided. Methodological problems regarding the moderating effect of demographic variables on the respite effect may form an additional source of confusion. For example, the need to transform continuous variables thereby hindering the use of weighted least squares (WLS) regression analyses (Neter, Wasserman, & Kutner, 1989), which would have been the appropriate method by which to treat continuous variables. Reduced value range poses another challenge. For example, the variable *number of years in office* ranged only between five and 17 years. Such shortcomings, which stem mostly from the small number of samples that were available at the time for analysis, impair the validity of the findings and obscure any conclusions regarding the respite effect among different groups of people.

Vacation variables

Type of vacation. The research hypothesis that a personally initiated respite had a greater effect on stress and burnout than an organizationally enforced respite was not confirmed. We consider two issues to have hindered the finding of such an effect: sample size and potentially inappropriate assumptions regarding the true meaning of employees' control over the timing and duration of a respite. With respect to sample size available for the investigation of this potential moderator, the comparison of stress and burnout levels was made over 29 and 33 (respectively) samples in which respites were initiated organizationally, and only 11 and 12 (respectively) samples in which respites were initiated personally. That is, the number of samples in the organizationally enforced respite category was three times greater than the number of samples in the personally initiated respite category. With respect to the true meaning of employees' control over the timing and duration of a respite, the logic behind this hypothesis was that the personally initiated respite has the advantage of enabling employees to arrange the timing of the respite according to their personal needs. However, other factors that influence the timing of employees' respites besides their personal needs were not considered. Often, organizations dictate the periods during which respites may be taken – and these are usually relatively low-pressure periods, while during very loaded periods, in which employees' stress increases and their burnout levels rise, personally initiated respites are not approved. Additional factors such as children's vacation time also influence the convenience of the timing of work respites. Therefore, it often happens that employees, even when allowed to control the timing and duration of the respite, do not take a respite when they most need it, but when it meets the needs of the organization and or the family.

Vacation length. We confirmed that the respite effect was greater for short respites (6–18 days) compared to longer respites (over two months) only for burnout but not for stress. Stress levels decline significantly after the respite disregarding the length of the respite.

Methodological variables

Publishing status. As hypothesized, we found that reductions in stress and burnout levels were more substantial when reported in published studies compared to unpublished studies. Additionally, the positive correlations between stress and burnout on all measurement occasions were found to be stronger in published studies. These findings show that as in other research fields, respite research suffers from the publishing bias: published studies do not necessarily reflect all of the research conducted in the field and studies with insignificant or unexpected results are less likely to be published (Fried & Ager, 1998). The current study, however, did not suffer from the publishing bias, as our sample included a greater number of unpublished compared to published studies.

The timing of the first measurement. We found that the timing of the first (pre-respite) measurement indeed moderates the effect of vacation on stress. The difference between the pre- and post-respite stress measurements was found to be small but significant when the former was performed relatively a long time (over a month) prior to the respite compared to only one or two weeks before it ($d = -0.1$ in both cases). The difference between the two stress measurements was more evident when the first measurement was performed close to the time of the respite, around three days before it ($d = -0.28$). A similar pattern was established for the moderating effect of the timing of the first measurement on burnout as well. The respite effect over burnout was insignificant when the first measurement was performed at a distance from the time of the respite, it was significant but small when the first measurement was performed one to two weeks before the respite ($d = -0.15$), and was more substantial when the first measurement was performed around three days prior to the respite ($d = -0.25$). Respectively, the timing of the first measurement also moderated the correlation between stress and burnout, the effect of which grew stronger as the timing of measurement came closer to the respite ($r = 0.25$, $r = 0.35$, and $r = 0.45$ respectively). This pattern of findings can indicate that the regular stress and burnout levels is maintained until about a week or two before the respite, rising substantially close to the respite and declines to a level slightly lower than usual at the end of the respite. Notwithstanding, the pre-post respite variance used in our analyses were recorded in different samples, thereby hindering inferences of any valid conclusions regarding the change in stress levels between the three pre-respite measurements (long/short/very short time before the respite). This finding is of great importance as it may indicate that measurements taken too close to the respite (as done in about a quarter of our samples) do not reflect the regular levels of stress and burnout, but rather the extent to which the respite itself is a source of stress. We might thus recommend that measurements be taken one to two weeks before the respite, although a more accurate recording of the stress-respite dynamic would be recorded in all three measurements from a month to three days before the respite.

The timing of the second measurement. The decline in stress and burnout levels between the two measurements was significant (*d* = –0.19 and *d* = –0.2 respectively) only when the second measurement was taken a very short time (approximately three days) after the respite (around three days post-vacation). The timing of the second measurement did not moderate the effect of respite on the correlation between stress and burnout. From these important findings we may infer that the greatest potion of the fade-out effect occurs within less than a week after the respite.

The timing of the first and second measurements combined. When the measurements were combined, the differences between them were significant only when the second measurement was performed three days after the respite and the first measurement was performed either three days (*d* = –0.34 for both stress and burnout) or between a week to two weeks before the respite (*d* = –0.14 for stress, *d* = –0.18 for burnout). This pattern of results emphasizes results mentioned above where measurement that is too close to the respite right before it falsely magnifies the difference between the two measurements. These results additionally show that stress and burnout levels declined significantly only when the second measurement was performed around three days after the respite. Not only does this test strengthen the impression that the influence of the respite lasts for only a very short time but they indicate that the fade-out effect is quite similar for both stress and burnout in the period immediately following the respite.

Discussion

Our meta-analytic study yielded a positive, albeit small, respite effect on stress and burnout, lasting mostly for a short period of time only. A moderate to strong positive correlation between stress and burnout was found on all measurements occasions. Demographic variables and type of respite did not moderate the respite effect over stress and burnout. The respite effect was most pronounced after respites that span one to three weeks in length. The timing of both the first and the second measurements was found crucial as the timing of the first measurement was also found to moderate the correlation between stress and burnout. Evidence of the respite effect was strongest when the first measurement of stress and burnout was taken either one to two weeks or three days prior to the respite and the second measurement was taken three days after the respite. Finally, the effect of vacation on burnout and the strength of correlation between stress and burnout were stronger in published studies than in unpublished ones.

Despite growing evidence regarding the salutogenic effect of the respite and consensus among studies regarding questions such as the positive correlation between stress and burnout, answers to many questions have yet to be elucidated. The meta-analytic method we applied can shed light on such questions due to its objective and accurate nature as authors' discretion cannot affect results and differences in sample size, effect size, and research design may be considered (Davies & Crombie, 2003; Egger & Smith, 1997). Yet another remarkable advantage of meta-analysis is that it is allows for replication, facilitating further, critical examination of the data so that readers may reach conclusions on their own (Davies & Crombie, 2003; Egger & Smith, 1997; Lipsey & Wilson, 1993). Meta-analytic research facilitates greater generalization of the acquired knowledge. The current meta-analysis enhances generalizability by integrating a variety of populations and respites in contrast to the limited prism of the individual studies. It is additionally easier to overcome the second-degree error (absence of effect where one exists) and detect between-group variance due to the formation of larger groups. In the current study groups were sufficiently large to facilitate the investigation of numerous moderating effects (Davies & Crombie, 2003; Egger & Smith, 1997; Lipsey & Wilson, 1993).

Meta-analyses help pinpoint limitations in previous research, raising new questions that had not been tackled by the analyzed studies and broadening the basis for next generation research. The most notable limitation found in our research was the distorted variance in measurements resulting from the first measurement of stress and burnout taken too close to the onset of the respite, when these measurements were irregularly high. In this, our meta-analysis may aid in preventing similar bias in future studies. The current research has demonstrated the positive effect of the respite on employees' well-being, reflected in the post-respite decline of employees' levels of stress and burnout levels. Albeit the small and short-lasting effect, even small effects are important in meta-analyses when the problems at hand are difficult to resolve (Lipsey & Wilson, 2001).

The validity of meta-analytic research is very much influenced by the methodological quality of the studies on which it is based (Lipsey & Wilson, 2001). Thus, stress and burnout measurements taken within three days of the respite, both before and after, could have impaired not only the results of the individual studies in our sample but also the results of the current research. Additional bias inherent in the individual studies is the use of self-report questionnaires and self-selection related to low response rates (Schaufeli & Enzmann, 1998). Undoubtedly, most limitations in the current research are a result of the limited sample size in some of our analyses, reflected, for example, in the need to convert continuous variables to categorical ones. Additionally, the shortage of theoretical knowledge regarding moderating effects was felt. The number of samples in which three stress and burnout measurements were taken was also limited. This prevented examination of the moderating effect of the timing of the third measurement, which would have shed light on the fade-out period. Another limitation of the current research is that ultimately it was based mainly on studies conducted under the auspices of our own project thereby excluding representation of the greater body of respite research.

References

Borenstein, M. & Rothstein, H. (1999). *Comprehensive meta-analysis.* Englewood, New Jersey: Biostat.

Cohen, S. (1988). Psychosocial models of the role of social support in the etiology of physical disease. *Health Psychology, 7(3)*, 269–297.

Cortina, J. M. (2003). Apples and oranges (and pears, oh my!): The search for moderators in meta-analysis. *Organizational Research Methods, 6(4)*, 415–439.

Davies, H. T. O. & Crombie, I. K. (2003). What is a systematic review? *"What is …?" Bulletins, 1(5)*, 1–6.

Egger, M. & Smith, G. D. (1997). Meta-analysis. Potentials and promise. *British Medical Journal, 315(7119)*, 1371–1374.

Etzion, D. & Boklis, V. (2006). Meta-analytic study of respite and fade-out effects. Report by: Tel Aviv University, Henry Crown Institute of Business Research, Faculty of Management, Tel Aviv, Israel.

Fried, Y. & Ager, J. (1998). Meta-analysis: Review, integration and recommendations: A guideline for meta-analysts. *International Review of Industrial and Organizational Psychology, 13*, 123–158.

Fritz, C. & Sonnentag, S. (2006). Recovery, well-being, and performance related outcomes: The role of workload and related experiences. *Journal of Applied Psychology, 91 (4)*, 936–945.

Gilboa, S. (2005). A meta-analysis of stress and performance at work: Examining moderators and mediators. Ph.D., Tel Aviv University, Israel.

Hunter, J. E. & Schmidt, F. L. (1990). *Methods of meta-analysis.* Thousand Oaks, CA: Sage Publications.

Lee, R. T. & Ashforth, B. E. (1996). A meta-analytic examination of the correlates of the three dimensions of burnout. *Journal of Applied Psychology, 81(2)*, 123–133.

Lipsey, M. W. & Wilson, D. B. (1993). The efficacy of psychological, educational, and behavioral treatment. *American Psychologist, 48(12)*, 1181–1201.

Lipsey, M. W. & Wilson, D. B. (2001). *Practical meta-analysis.* Thousand Oaks, CA: Sage Publications.

Magen-Neuhaus, T. (1993). The effect of vacation and feelings of stress at work on emotional burnout. M.Sc., Tel Aviv University, Israel. [Hebrew].

Morris, S. B. & DeShon, R. P. (2002). Combining effect size estimates in meta-analysis with repeated measures and independent-groups designs. *Psychological Methods, 7(1)*, 105–125.

Neter, J., Wasserman, W., & Kutner, M. (1989). *Applied linear regression models* (2nd ed.). Homewood, IL: Irwin.

Pines, A., Aronson, E., & Kafry, D. (1981). *Burnout: From tedium to personal growth.* New York: Free Press.

Pines, A. (1993). Burnout: An existential perspective. In W. B. Schaufeli, C. Maslach, & T. Marek (Eds.), *Professional burnout: Recent development in theory and research* (pp. 33–51). Washington, DC: Taylor & Francis.

Schaufeli, W. & Enzmann, D. (1998). *The burnout companion to study and practice.* London: Taylor & Francis.

Whitener, E. M. (1990). Confusion of confidence intervals and credibility intervals in meta-analysis. *Journal of Applied Psychology, 75(3)*, 315–321.

APPENDIX 2

Annotated bibliography

Aron, C. S. (1999). Working at play: A history of vacation at the United States. New York: Oxford University Press.

The book traces the history of vacations and their conceptual and implemental developments from the early 19th through almost the middle of the 20th century in the United States of America. Historical, religious, cultural, sociological, and political evolvements and their influences on people's attitudes toward vacationing and leisure in different historical periods are presented.

Aviv, N. (2000). The meaning of vacation among shift working women: The effect of vacation on job stress, burnout and work-home conflict. M.Sc. Thesis in Management Sciences, Organizational Behavior, Faculty of Management, Tel Aviv University, Israel [Hebrew].

The effect of respite on job stress, burnout, and work–home conflict was examined in a repeated measures design with three occasions (routine work, pre-, and post-vacation). Subjects were 88 female shift workers at hospitals and hotels in Israel (mean age 33). Results indicated that burnout (BM) levels were significantly lower before and after vacation compared to routine work period. No significant differences were found for job stress and work–home conflict across the different occasions.

Ben Yoseff, N. (2004). The *effect of maternity leave on burnout among women serving in the IDF*. M.Sc. Thesis in Management Sciences, Organizational Behavior, Faculty of Management, Tel Aviv University, Israel [Hebrew].

The respite effect of maternity leave on burnout was examined in a repeated measures design for half of the women (two occasions: pre- and post-leave); for the rest of the women, there was only a post-leave report about the two occasions. Subjects were 42 female military workers in the IDF. Results showed an interaction effect of detachment from work and of involvement with non-work activities on post-leave burnout (BM) so that burnout of women who were highly involved with non-work activities and highly detached from work was higher than for women with low involvement in non-work activities and high detachment from work.

Caplan, D. & Jones, K. W. (1975). Effects of work load, role ambiguity and Type A personality on anxiety, depression and heart rate. *Journal of Applied Psychology*, *60*(6), 713–719.

The effects of job stressors (quantitative work load and role ambiguity) on strains (anxiety, depression, resentment, heart rate) and the moderating effect of Type A personality were studied among 73 American male computer system users (graduates, undergraduates, and faculty; mean age 23). The study was longitudinal, with two measurements: three days preceding an announced shutdown (of 23 days) of the system and five months later. Positive associations were found between role ambiguity and anxiety, depression and resentment, between subjective work load and anxiety, which was the greatest for Type A persons, and between anxiety and heart rate. A decline in subjective work load and in all the strain measures was demonstrated on the second occasion as compared to the first.

Cohen, Y. (1997). *The impact of intermissions at work (courses) on work stress, coping and burnout*. M.Sc. Thesis in Management Sciences, Organizational Behavior, Faculty of Management, Tel Aviv University, Israel [Hebrew].

The impact of in-service courses on job stress, burnout, and ways of coping was studied among 140 employees (90% females) of an academic institution, including 62 course participants (mean age 44) and a comparable control group of 78 who did not take part in the course. Measurement occasions ranged from 1–3 pre-course and 1–4 post-course, all in all five measurements per course. Results yielded a significant increase of post-course job stress in both the experimental and the control group. Additionally, general satisfaction was related to job stress and burnout such that the higher the levels of general satisfaction, the lower the levels of stress and burnout and vice versa.

Deem, R. (1996). No time for a rest? An exploration of women's work, engendered leisure and holidays. *Time and Society*, *5*, 5–25.

This review of the concept of work and working time, including respite and leisure, focuses on women's distinctive experiences and time concepts within their social, cultural, and economic contexts. Different forms of holidays are discussed as a way to disrupt intensified life styles or to interrupt the domination of clock time and work routines.

Diddams, M., Surdyk, L. K., & Daniels, D. (2004). Rediscovering models of Sabbath keeping: Implications for psychological wellbeing. *Journal of Psychology and Theology, 32(1)*, 3–11.

A theoretical review yielding three models of Sabbath keeping as respite: Life Segmentation in which people actively segment their lives to create respite; Prescribed Meaning, in which people prescribe positive and religious meaning to life segmentation; Integrated Sabbath, in which Sabbath keeping is an integrated belief system of daily rest, reflection, and relationship development. Implications for well-being, inherent challenges, and research directions are discussed.

Eden, D. (2001). Vacations and other respites: Studying stress on and off the job. In C. L. Cooper & I. T. Robertson (Eds.), *International review of industrial and organizational psychology* (pp. 121–146). Chichester: John Wiley & Sons.

This chapter traces the development of respite research, presenting respite effects on stress, strain, and well-being. The nature, strength, and methodological advantages of repeated measures, longitudinal respite design are presented, as well as theoretical grounds. Lastly, directions for future research are proposed, such as vacation and personal characteristics affecting the quality of the respite experience.

Eden, D. (1990). Acute and chronic job stress, strain, and vacation relief. *Organizational Behavior and Human Decision Processes, 45*, 175–193.

In a longitudinal study conducted among 29 university and non-university workers, two acutely stressful job events, namely computer shutdown and reopening, were compared to routine work and to a holiday vacation (Passover). The acutely stressful job events were perceived as more stressful, and aroused greater psychological (A-state) and physiological (systolic and diastolic blood pressure) strain, than the follow-up period of routine work (chronic job stress). Stress levels were lower during vacation respite than in work time, but strain was as high during vacation as during work.

Eden, D. (1982). Critical job events, acute stress and strain: A multiple interrupted time series. *Organizational Behaviour and Human Performance, 30*, 312–329.

The effects of two Critical Job Events (CJEs) – as acute objective stressors – on subjective stress, were assessed among 39 first year female nursing students in an interrupted time series with five replications including the two CJEs and three low objective stress occasions. A confirmatory pattern of significantly rising and falling strain was found for anxiety (A-state scale), systolic blood pressure, and pulse rate. Similarly, qualitative overload and serum uric acid changed four times out of five.

Etzion, D. (2005, July). *Respite from work as means of recovery from occupational stress and burnout.* Presented at the 26th International Conference of the Stress and Anxiety Research Society (STAR), at Halle (Saale), Germany.

This paper raises questions concerning ways of recovery from occupational stress and psychological burnout of employees in organizations. Because one of the main factors in burnout is continuous exposure to daily stress, many burnout experts recommend taking a vacation every time we complain of pressure or fatigue. They see occasional time out from work as a means of recharging one's batteries and renewing one's strength. Do respites really achieve these objectives? What characterizes them? Do they have the same effect on all kinds of workers? The respite project that is described in this paper, has been systematically exploring these questions during the past ten years. The findings from more than 30 studies are summarized and discussed.

Etzion, D. (2003). Annual vacation: Duration of relief from job stressors and burnout. *Anxiety, Stress and Coping, 16(2)*, 213–226.

The impact of an annual vacation on job stressors and burnout was studied in a longitudinal, repeated measures design with three occasions (one before, two after) among 51 industrial workers (90% men, at all hierarchy levels) who took a summer vacation and 51 matched controls who did not take their vacation during the same period. Results showed that in the vacation group, both stress and burnout (BM) decreased after returning to work. Three weeks later, stress had reverted to its initial level, but burnout remained low. The comparison group showed no change. Duration of vacation had no moderating effect.

Preliminary reports and presentations

Etzion, D. & Sapir, O. (1997, April). *Annual vacation: Duration and relief from job stress and burnout.* Paper presented at the SIOP 12th annual conference, St Louis, MO.

Etzion, D. & Sapir, O. (1996). *The effect of an annual vacation in relieving stress and burnout at work (A controlled field study at an industrial enterprise in Israel).* Working Paper No. 11/96, The Henry Crown Institute of Business Research in Israel, Faculty of Management, The Leon Recanati Graduate School of Business Administration, Tel Aviv University.

Etzion, D. & Sapir, O. (1995). *The effect of an annual vacation in relieving stress and burnout at work.* Working Paper No. 31/95, The Henry Crown Institute of Business Research in Israel, Faculty of Management, The Leon Recanati Graduate School of Business Administration, Tel Aviv University [Hebrew].

Etzion, D. & Sapir, O. (1995, October). *Respite from work as a mechanism for coping with stress and burnout.* Paper presented at the 25th Scientific Convention of the Israel Psychological Association, Beer-Sheva, Israel.

Sapir, A. (1995). *The influence of vacation on work stress and burnout.* M.Sc. Thesis in Management Sciences, Organizational Behavior, Faculty of Management, Tel Aviv University, Israel [Hebrew].

Etzion, D. (1995). Is respite from work an effective way of coping with job stress and buffering burnout in organizations? *Chidushim Benihul* – Newsletter of the Faculty of Management, Tel Aviv University, pp. 4–5 [Hebrew].

Under the heading "research in process," this paper contains an initial description of the respite project and its preliminary research results. It displays the goals and long-term plans for studying different types of respite longitudinally in order to find out their impact on workers' well-being, in terms of recovery from job stress and burnout.

Etzion, D., and Dar, A. (2004). *Leave and leisure worldwide: A cross-cultural comparison among the USA, the EU, Japan and Israel.* Working Paper No. 21/04, The Henry Crown Institute of Business Research in Israel, Faculty of Management, The Leon Recanati Graduate School of Business Administration, Tel Aviv University.

In this paper, the authors examine the association between cultural values, as defined by Hofstede (1980) and Schwartz (1994) and the laws and customs regarding annual leave, working hours and family-friendly policies in the US, Europe, Japan, and Israel. Findings indicated that dimensions of Affective Autonomy, Harmony and Egalitarian Compromise (Schwartz, 1994) were associated with a more generous policy, whereas the values of Hierarchy and Mastery were associated with a more restrictive policy. An association was found between Hofstede's Masculinity–Femininity dimension and family-friendly policies. The conclusion that a change in policy regarding working time requires a primary change in cultural values is discussed.

Preliminary reports and presentations

Etzion, D. & Dar, A. (2004). Patterns of leisure and vacation across cultures. *Panim – Quarterly of Society, Culture and Education, 27*, 23–38 [Hebrew].

Dar, A. (2002). *Vacation laws and practices in Israel, the USA, the EU and Japan – A cross-cultural comparison.* M.Sc. Thesis in Management Sciences, Organizational Behavior, Faculty of Management, Tel Aviv University, Israel [Hebrew].

Etzion, D., Eden, D., & Lapidot, Y. (1998). Relief from job stress and burnout: Reserve service as a respite. *Journal of Applied Psychology, 83(4)*, 577–585.

The respite effect of reserve service on job stress and burnout was studied among technically oriented male employees of a large Israeli industrial enterprise, in a repeated measures design with two occasions (before/after). The subjects were 81 employees (mean age 38) who were called for two to four weeks of active reserve service and were compared to 81 matched controls (mean age 39) who were not called for service during the same period. Results showed a significant decline in job stress and in burnout (BM) among the reservists and no change among the controls. Furthermore, quality of service experience and degree of detachment from work interacted in moderating the respite effects; the greater the detachment, the stronger the effect positive experience had in relieving reservists from stress and burnout.

Preliminary reports and presentations

Etzion, D., Eden, D., & Lapidot, Y. (1997). *Reserve service as a respite from job stressors and burnout.* Working Paper No. 29/97, The Henry Crown Institute of Business Research in Israel, Faculty of Management, The Leon Recanati Graduate School of Business Administration, Tel Aviv University.

Etzion, D., Eden, D., & Lapidot, Y. (1994, June). *Active reserve duty as an ameliorative respite from job stress and burnout.* Paper presented at the 5th International Conference on Social Stress Research, Honolulu, HI.

Lapidot, Y. & Etzion, D. (1993, October). *The effect of military reserve duty on the perception of job stress and burnout among engineers and technicians.* Paper presented at the 24th Scientific Convention of the Israel Psychological Association, Ramat Gan, Israel.

Lapidot, Y., Etzion, D., & Eden, D. (1992). *The effect of military reserve duty on the perception of job stress and burnout among engineers and technicians.* Working Paper No. 45/92, The Henry Crown Institute of Business Research in Israel, Faculty of Management, The Leon Recanati Graduate School of Business Administration, Tel Aviv University [Hebrew].

Lapidot, A. (1992). *Reserve service as a respite from job stressors and burnout.* M.Sc. Thesis in Management Sciences, Organizational Behavior, Faculty of Management, Tel Aviv University, Israel [Hebrew].

Etzion, D. & Ofek, A. (1998, May). ***The influence of two types of vacation on perceived job stressors and burnout among social workers.*** **Paper presented at the Symposium on Respite Research at the 7th International Conference on Social Stress Research.**

The impact of two vacation types (organizational vs. personal) on job stress and burnout was studied among 90 female social workers (mean age 38) in a religious community. A repeated measures design, with three occasions, was employed. Measurements were taken before and after the Passover (organizational) vacation and during the first week after the summer vacation (personal vacation). Results showed a differential effect of the two types on burnout: a decrease after the organizational vacation and an increase after the personal vacation. Job stress decreased after the personal vacation for employees working 80% or more of a full-time job, but increased for those working less than that.

Preliminary reports and presentations

Ofek, A. (1997). *The influence of vacation from work – organized vacation and personal vacation – on stress and burnout.* M.Sc. Thesis in Management Sciences, Organizational Behavior, Faculty of Management, Tel Aviv University, Israel [Hebrew].

Etzion, D. & Westman, M. (2001). Job stress, vacation and the crossover of strain between spouses: Stopping the vicious cycle. *Men and Work, 11,* 106–118.

The impact of acute job stress and vacation on the crossover of burnout between life partners/couples was examined in two longitudinal, repeated measures design studies. One study was conducted among members of the Israel Engineers Association who went on an organized tour abroad (25 couples, mean age 52) and reported before going abroad and a week after return. Another study was carried out among Executive MBA program students and their spouses (32 couples, mean age 40) who reported at the beginning of the academic year and just before the end of term examinations. Results of both studies showed that more burnout crossed over from one partner to the other under stressful conditions than during more relaxed periods. These results indicate that a change from a stressful environment to a tranquil one may stop the crossover of strain from one spouse to another.

Preliminary reports and presentations

Etzion, D. & Westman, M. (1999, August). *Vacation and crossover of strain between spouses: Stopping the vicious cycle.* Presented at the 59th Annual Meeting of the Academy of Management, Chicago, IL.

Shaviv, T. (1998). *The impact of an organized tour vacation on work stress and burnout among managerial and non-managerial vacationers.* M.Sc. Thesis in Management Sciences, Organizational Behavior, Faculty of Management, Tel Aviv University, Israel [Hebrew].

Goshen-Albocher, M. (1997). *Crossover of stress and burnout between executive managers and their spouses.* M.Sc. Thesis in Management Sciences, Organizational Behavior, Faculty of Management, Tel Aviv University, Israel [Hebrew].

Etzion, D., Westman, M., & Kremer, S. (2001, July). *The impact of vacation and slack work periods on job stress and burnout among newcomers and veteran Israelis.* Paper presented at the 22nd Conference of the International Society for Stress and Anxiety Research (STAR), Palma de Mallorca, Spain.

The effect of a routine vacation on job stress and burnout among new immigrants compared to veteran Israelis was studied among 59 employees (31% immigrants, mean age 47, 37% males) of an industrial plant in Israel. A repeated measures design with three occasions was used: after a holiday (a slack period), during a heavy workload period, and following the summer vacation (slack period). Results showed an increase of burnout level between the first and the second occasions and a decrease in its level on the third occasion. No differences in stress levels between the occasions were revealed. In general, job stress and burnout levels were higher, on all occasions, among the new immigrants than among veteran Israelis. Likewise, overall life and work satisfaction and the importance of life outside work were higher among veteran Israelis than their new immigrant counterparts.

Preliminary reports and presentations

Kremer (Fuks), S. (1999). *The impact of vacation and slack work periods on job stress and burnout among newcomers and veteran Israelis.* M.Sc. Thesis in Management Sciences, Organizational Behavior, Faculty of Management, Tel Aviv University, Israel [Hebrew].

Etzion, D., Westman, M., & Vinocour-Hai, V. (2002, July). ***The impact of high and low stress periods on the intensity of burnout crossover between spouses: A longitudinal study.*** **Paper Presented at the 25th ICAP Conference, Singapore.**

The effects of a holiday vacation (nine days) on stress and exhaustion as well as on the crossover of burnout between spouses were studied among 149 couples (mean 21 years of marriage), administrative personnel of an academic Israeli institution (mean age 46, 54% academics) and their spouses (mean age 48, 62% academics). A repeated measures design was applied with three occasions: eight weeks before the vacation (end of the academic year), three days after and three weeks later. Results showed a bi-directional crossover between spouses on all three occasions. Additionally, vacation moderated the crossover effect so that for both spouses, the crossover effects after vacation were reduced. On the third occasion, spouse-to-employee crossover continued to decline whereas employee-to-spouse crossover rose. A general ameliorative effect was found for stress, but not for burnout.

Preliminary reports and presentations

Vinocour-Hai, V. (2001). *Stress and burnout before and after vacation and their crossover between spouses.* M.Sc. Thesis in Management Sciences, Organizational Behavior, Faculty of Management, Tel Aviv University, Israel [Hebrew].

Etzion, D. & Zvi, E. (2004). Stress and burnout among students before and after the Passover break: Vacation activities and personality variables as moderators. Working Paper No. 11/04, The Henry Crown Institute of Business Research in Israel, Faculty of Management, The Leon Recanati Graduate School of Business Administration, Tel Aviv University.

The effect of the Passover break and its content (activities) on job stress and burnout was studied among 228 Israeli undergraduates (mean age 24, 36% males, 70% singles) in a repeated measures design, with two occasions, before and after the vacation. Results showed a decline of stress and burnout levels after the vacation and a moderating effect of vacation satisfaction on burnout such that the higher the vacation satisfaction, the greater the post-vacation burnout. As for vacation activities, "studying" and "working for a living" had positive relations with post-vacation burnout and negative relations with vacation satisfaction; "visiting friends" and "traveling for sightseeing" had negative relations with post-vacation stress and burnout and positive relations with vacation satisfaction.

Preliminary reports and presentations

Etzion, D. & Zvi, E. (2004, July). *Stress and burnout among students before and after the Passover break – Vacation activities and personality variables as moderators*. Paper presented at the STAR 25th International Conference, Amsterdam, The Netherlands.

Zvi, E. (2004). *Stress and burnout among students before and after the Passover break – Vacation activities and personality variables as moderators*. M.Sc. Thesis in Management Sciences, Organizational Behavior, Faculty of Management, Tel Aviv University, Israel [Hebrew].

Frankenhaeuser, M., Lundberg, U., Fredrikson, M., Melin, B., Tuomisto, M., Myrsten, A.-L., Hedman, M., Bergman-Losman, B., & Wallin, L. (1989). Stress on and off the job as related to sex and occupational status in white-collar workers. *Journal of Organizational Behavior, 10*, 321–346.

The impact of work and non-work characteristics on health was examined among two groups of Swedish participants in Volvo: 30 managers and 30 clerical workers (50% women; mean age 41). Data on cardiovascular and neuroendocrine functions, self-reports, and interviews were collected for 12 consecutive hours during a normal day at work (9 am to 5 pm) and after work (6 pm to 9 pm), and 12 hours during a work-free day at home, a week later. Attitudes toward work, workload, and sex identity were examined by questionnaire. All groups showed a moderate increase in cardiovascular and neuroendocrine activity during the work day, whereas after work, slower unwinding in female managers was indicated. Differences related to occupational level and/or sex were found for factors such as autonomy, social support at work, and conflicting demands from work and from outside of work responsibilities.

Fritz, C. & Sonnentag, S. (2005). Recovery, health and job performance: Effects of weekend experiences. *Journal of Occupational Health Psychology, 10(3)*, 187–199.

The effect of weekend experiences (non-work hassles, social activity, and positive work reflection) on health indicators (burnout, general well-being, and job performance) was studied among 87 German emergency service workers (mean age 33, 95.4% males) in a longitudinal study with three measurement occasions: before and after the weekend and at the beginning of the next work week. Results showed that non-work hassles, absence of positive work reflection, and low social activity during the weekend predicted post-weekend burnout (OBI: Exhaustion and Disengagement) and poor general well-being. On the post-weekend occasion, associations were found between lower pursuit of learning and non-work hassles and between positive work reflections and pursuit of learning. Practical implications for individual and organizational optimization of recovery processes are finally related to.

Fritz, C. & Sonnentag, S. (2006). Recovery, wellbeing, and performance related outcomes: The role of workload and related experiences. *Journal of Applied Psychology, 91*(*4*), 936–945.

Vacation effects on well-being (reduced health complaints and burnout) and performance related outcomes (self-reported task performance and reduced effort expenditures) were examined among 221 German, non-academic university employees (mean age 46, 14.5% males, mean job experience 21 years) in a longitudinal, repeated measures design with four occasions: one week before vacation (T1), during vacation (T2), two days (T3), and two weeks (T4) afterward. Specific vacation experiences (positive and negative work reflection, relaxation, mastery experience, and non-work hassles) were assessed during vacation. Results showed changes in well-being and self-reported effort expenditure at T3 and T4 compared to T1, revealing vacation and partial fade-out effects. Additionally, specific vacation experiences and workload (at T2) predicted some of the outcomes.

Gilbert, D. & Abdullah, J. (2002). A study of the impact of the expectation of a holiday on an individual's sense of wellbeing. *Journal of Vacation Marketing, 8*(*4*), 352–361.

The effect of holiday-taking expectation on well-being of UK tourists was studied among 355 holiday-takers compared with 249 non-holiday-takers. Results showed that those who had waited to go on holiday had higher global life satisfaction and experienced less negative/unpleasant feelings (higher positive affect). The holiday-takers were happier with specific life domains, i.e., their family, economic situation, and health, compared to the non-holiday-taking group.

Gump, B. B. & Matthews, K. A. (2000). Are vacations good for your health? The 9-year mortality experience after the multiple risk factor intervention trial. *Psychosomatic Medicine, 62*(*5*), 608–612.

The association between vacation frequency and the risk of post-trial (multiple risk factor intervention) death was studied among 12,338 middle-aged men (mean age 46) at high risk for coronary heart disease (CHD), during a nine-year follow-up period. Participants were patients of 22 clinical centers in 18 US cities, screened for a "Multiple Risk Factor Intervention Trial" (MRFIT). Vacation frequency was reported in the first five years of the trial. Results showed that the frequency of annual vacations was associated with a reduced risk of all-cause mortality and, more specifically, mortality attributed to CHD.

Hendel, D. D. & Solberg, J. (1983, April). *Sabbatical and leave experiences of female and male faculty at a large research university.* Paper presented at the Annual Meeting of the American Educational Research Association, Montreal, Canada.

Sex differences among 1495 American university faculty members' use and evaluation of sabbaticals and leaves (80% males, 78% tenured), were examined. Attention was directed to two faculty programs: a competitive leave (CL) program (one quarter at full pay) and a non-competitive sabbatical (NCS) program (one year at half salary). Results showed that more males than females (CL: 44% vs. 33% respectively; NCS: 39% vs. 31% respectively) had received both sabbatical/leave types. Sex differences were revealed among those who had not taken any sabbatical/leave as to their reasons for doing so; no sex differences were found in sabbatical/leave benefit perceptions among those who did take it.

Haworth, J. T. (1997). *Work, leisure and wellbeing.* London: Routledge.

The book provides a review of research in the areas of work, leisure, and well-being. Central theories and research regarding the value of personal, psychological experience obtained by employment and environmental influences, all related to well-being and health, are presented. In this regard, the important roles of enjoyment and embodiment in forming the quality of life are discussed. Particular reference is made to work–leisure dynamics among young adults, as well as to leisure's multiple merits at the age of retirement. Finally, future scenarios are outlined.

Jarecky, R. K. & Sandifer, M. G. (1986). Faculty members' evaluations of sabbaticals. *Journal of Medical Education, 61*, 803–807.

Faculty members' evaluations of sabbaticals were examined via open interviews with 70 medical school faculty members and 15 others who were eligible but non-takers of sabbaticals. Results showed that at the 19 responding schools, sabbaticals were taken by one-sixth of the faculty members. 80% of sabbatical takers viewed the experience as favorable. The authors evaluated 75% sabbaticals as yielding substantial accomplishments. Respondents related to pre-sabbatical planning as a key to sabbatical satisfaction and indicated that post-sabbatical depression was common.

Kabanoff, B. (1980). Work and nonwork: A review of models, methods and findings. *Psychological Bulletin, 88*, 60–77.

This paper is a review and analysis of theory and research in the field of work/non-work relations, including reference to conceptual and methodological problems. It concludes with the empirical limitations of work/non-work relations paradigms. The importance of objective definitions of attributes common across life spheres is highlighted.

Kabanoff, B. & O'Brien, G. E. (1980). Work and leisure: A task attributes analysis. *Journal of Applied Psychology, 65*, 596–609.

Three models of work–leisure relationships (compensation, generalization, and segmentation) and the moderating effect of locus of control were studied among 1383 participants (63% males, aged 15 and older) of an Australian metropolitan survey. Locus of control was associated with higher leisure influence, variety, skill utilization, and pressure. Work attributes showed positive relationships to leisure attributes for variety and skill utilization. Only partial support for a moderation of control beliefs (I-E scale) on work–leisure patterns was found.

Kinnunen, U. (1988). Teacher stress during an autumn term in Finland: Four types of stress processes. *Work and Stress, 2(4)*, 333–340.

Stress patterns, personality, coping styles, and background variables were examined in a longitudinal study of six repeated measures along a school autumn term in Finland, among 153 teachers. Cluster analysis yielded four different stress-type groups which differed with regard to their personality characteristics and coping strategies.

Laski, S. (2002). *The effect of summer break activities and personality variables on stress and burnout among high school teachers in Israel*. M.Sc. Thesis in Management Sciences, Organizational Behavior, Faculty of Management, Tel Aviv University, Israel [Hebrew].

The respite effect of the two-month summer break of 86 high school teachers on job stress, burnout, and general satisfaction was examined in a longitudinal, repeated measures design with three occasions (before-after-after). Respite effects were found on stress, burnout (BM), and general satisfaction. Moderating effects were found for vacation activities, personal and demographic characteristics, as well as personality traits (Big Five).

Lounsbury, J. W. & Hoopes, L. L. (1986). A vacation from work: Changes in work and nonwork outcomes. *Journal of Applied Psychology, 71*, 392–401.

The effect of vacation on six work and non-work characteristics was studied among 128 workers of 24 industrial organizations, in a longitudinal study of repeated measures with two occasions (before and after vacation). Job involvement–central life interest decreased and both life satisfaction and turnover intention increased after vacation. Vacation satisfaction moderated the relationships between pre- and post-vacation life and job satisfaction such that individuals with higher levels of vacation satisfaction were more likely to have higher post-vacation life/job satisfaction.

Levi, M. (2002). *The impact of the summer break and personal variables on job stress and burnout among teachers in a kibbutz regional school.* M.Sc. Thesis in Management Sciences, Organizational Behavior, Faculty of Management, Tel Aviv University, Israel [Hebrew].

The effect of the two-month summer break on job stress and burnout of 44 Israeli teachers (mean age 42, 23% males) working in a kibbutz school was examined. About half of the teachers were kibbutz members, another 40% were kibbutz residents and the rest were city residents. A longitudinal, repeated measures before/after design was employed. A respite effect was found on job stress (negative work characteristics), which declined on the second occasion. No such effect was found for stress stemming from lack of positive characteristics or for burnout. However, a moderating effect was found for type of residence on burnout: *kibbutz* residents suffered the highest levels of burnout, *kibbutz* members came next, with intermediate levels of burnout, and teachers living outside of the *kibbutz* were characterized by the lowest levels of burnout.

Matthes, K. (1992). In pursuit of leisure: Employees want more time off. *HR Focus, 69(5)*, 1, 7.

This review of worldwide attitudes and concepts of time off work, vacation, and leisure examines vacation benefits for employees and presents a multinational comparison of vacation requirements and recommendations.

Radian, R. (1998). *The moderating effect of workaholism and Type A on the impact of vacation on job stress and burnout.* M.Sc. Thesis in Management Sciences, Organizational Behavior, Faculty of Management, Tel Aviv University, Israel [Hebrew].

The effect of an involuntary, organizational vacation (Passover) among 58 workers (mean age 47, 48% males, 87% managers) in the Israeli public sector was studied in a repeated measures design with three occasions (before vacation, immediately afterward, and six weeks later). Results showed that stress increased after vacation compared to its level before, whereas burnout (BM) decreased and was still the same on the third measurement occasion. Additionally, burnout level of Type As was found to be higher than that of Type Bs on all occasions. It was also found that the greater the detachment from work throughout vacation, the lower the burnout level afterward.

Raviv, E. (2002). ***Effects of leisure activity on occupational stress and burnout among high-tech employees.*** **M.Sc. Thesis in Management Sciences, Organizational Behavior, Faculty of Management, Tel Aviv University, Israel [Hebrew].**

The effect of leisure activity on stress and burnout was studied among 69 employees of five hi-tech companies (mean age 30, seniority 2.2, 57% males). Results showed that leisure satisfaction and perceived leisure freedom were associated with stress (negative correlations). The need for leisure as detachment was positively associated with job stress and burnout (BM). Leisure satisfaction was negatively associated with burnout (BM), stress, and exhaustion (MBI-GS). Among 13 leisure activities, only sports had significant associations with stress (positive) and burnout (BM) (negative).

Repetti, R. L. (1993). Short-term effects of occupational stressors on daily mood and health complaints. ***Health Psychology, 12,*** **125–131.**

The effects of daily changes in two job conditions, namely subjective/objective workload and social interactions with co-workers/supervisors, on daily mood and health complaints were examined among 52 air-traffic controllers at a major international American airport (mean age 34, 77% males, 88% academics). Subjective data were reported and objective indicators were collected on three consecutive days. Results showed that increases in job stressors (perceived workload, objective traffic volume, distressing social interactions) were associated with a same-day deterioration in physical and psychological well-being (health complaints, more negative moods, and less positive moods).

Richards, G. (1999). Vacations and the quality of life: Patterns and structures. ***Journal of Business Research, 44(3),*** **189–198.**

This comparative worldwide review of national and cultural attitudes toward vacations, tourism and leisure, as well as associations with quality of life, shows the substantial differences in the structural position and cultural attitudes toward time off work, relating to different welfare regimes (social democratic vs. liberal) and social rights reflected in them. More specifically, higher levels of vacation entitlement and consumption in Europe can be linked to corporatist or democratic welfare regimes, whereas in the USA and Japan more liberal welfare regimes are associated with lower vacation entitlement and greater inequalities in distribution of income and leisure time. These trends and their resultant work/non-work modes are discussed.

Ron, A. (2003). Effects of organizational vacation on occupational stress and burnout among high-tech employees. M.Sc. Thesis in Management Sciences, Organizational Behavior, Faculty of Management, Tel Aviv University, Israel [Hebrew].

The effect of vacation on stress and burnout was studied among 70 employees of four hi-tech companies (mean age 29, 67% males) in a repeated measures design with two occasions (before/after). No respite effect on stress and burnout was found. Results showed that vacation satisfaction was positively associated with stress (lack of positive characteristics) and negatively associated with self-efficacy (MBI-GS). Additionally, it was found that the more pre-vacation time spent on leisure activities, the lower the burnout level.

Rubinstein, C. (1980). Vacations, expectations, satisfactions, frustrations, fantasies. *Psychology Today, 14*, 62–76.

This comprehensive report of a survey relating to various characteristics of vacations and people's attitudes to non-work and leisure time included more than 2,500 American respondents (mean age 34, 60% females, 90% full-time workers). Different aspects of vacations were examined, relating to types, locations, length, social settings, and the effect of sociodemographic differences. Finally, six distinct vacation profiles based on goals and needs ascribed to them are presented.

Shaviv, T. (1998). *The impact of an organized tour vacation on work stress and burnout among managerial and non-managerial vacationers*. M.Sc. Thesis in Management Sciences, Organizational Behavior, Faculty of Management, Tel Aviv University, Israel [Hebrew].

The respite effect of an organized tour on job stress and burnout was studied among 72 members of the Israeli Engineers Association (mean age 29, 67% men, 90% married, 61% academics), in a repeated measures design with two occasions (seven to ten days before the tour and at its end). No respite effect was found on stress and burnout. Several job and vacation characteristics tended to moderate vacation effect on stress and burnout; post-vacation stress level tended to decrease after vacation for those characterized by higher vacation satisfaction, low job responsibility, high detachment from work, low workload expectations, and for those having no substitutes at work while on vacation. Among the latter group, post-vacation burnout (BM) level decreased and vacation satisfaction was lower. Additionally, a burnout crossover effect between spouses was found only before vacation, not afterward, thus indicating more crossover during stressful periods than relaxed ones.

Sluiter, J. K., Van der Beek, A. J., & Frings-Dresen, M. H. W. (1999). The influence of work characteristics on the need for recovery and experienced health: A study on coach drivers. *Ergonomics, 42,* 573–583.

The effect of work characteristics (job demands and control) on occupationally induced fatigue and health complaints were studied among 363 Dutch coach drivers. Results indicated that need for recovery was a major predictor of psychosomatic complaints (PCS), sleep complaints (SQS), and emotional exhaustion (MBI). The influence of job control and job demands on health problems was moderately confirmed.

Sluiter, J. K., Frings-Dresen, M. H. W., Van der Beek, A. J., & Meijman, T. F. (2001). The relation between work-induced neuroendocrine reactivity and recovery, subjective need for recovery, and health status. ***Journal of Psychosomatic Research, 50(1)*****, 29–37.**

The relationship between subjective need for recovery and perceived health status, on the one hand, and neuroendocrine reactivity during work, neuroendocrine recovery from work and work characteristics, on the other hand, were examined among 59 workers (mean age 39) of four different industrial and municipal Dutch organizations during five consecutive days, three days at work and two days off. Neuroendocrine reactivity and recovery were measured by urinary adrenaline and cortisol six times a day. Results showed that both neuroendocrine measures and work characteristics were predictors for the amount of perceived need for recovery after work, as well as for health status.

Sonnentag, S. (2004). Flight attendants' daily recovery from work: Is there no place like home? ***International Journal of Stress Management, 11(4)*****, 366–391.**

The effect of location and different leisure time activities on well-being at bedtime was examined among 47 flight attendants (mean age 30, 62% females) in a diary study over four consecutive work days. Multilevel analysis showed that staying in a hotel did not impair well-being at bedtime. Spending time on work-related activities during off-job time decreased well-being, whereas spending time on physical activities (sports) and experiencing off-job time activities as recovery improved well-being, also when initial well-being immediately after work was controlled. Time spent on social activities increased depression during the evening.

Sonnentag, S. (2003). Recovery, work engagement, and proactive behavior: A new look at the interface between nonwork and work. ***Journal of Applied Psychology, 88(3)*****, 518–528.**

The effect of recovery during leisure time on work-related outcomes, such as work engagement and proactive behavior was examined in a diary study over five consecutive work days. Subjects were 147 employees (mean age 39, 65% males) of six public service organizations (e.g., administration of justice, of real estate, of public security etc.) performing administrative tasks. Multilevel analyses showed that day-level recovery was positively related to day-level work engagement and day-level proactive behavior (personal initiative, pursuit of learning) during the subsequent work day. The data suggest considerable daily fluctuations in behavior and attitudes at work, with evidence that these are related to prior experience and opportunity for recovery in the non-work domain.

Sonnentag, S. (2001). Work, recovery activities, and individual well-being: A diary study. *Journal of Occupational Health Psychology, 6*, 196–210.

The effects of different types of post-workday activities on well-being at the end of the day were studied among 100 Dutch teachers (mean age 41, 51% males, 38% parents), in a diary study of five consecutive days. Multilevel analyses in which pre-leisure well-being and work situation variables were entered as control variables indicated that work-related activities had a negative effect on individuals' situational well-being before going to sleep, whereas low-effort activities, social activities, and physical activities had a positive effect. Moreover, a lagged effect of high time pressure on poor situational well-being was found. The study showed that leisure time activities and a low-stress work situation contribute independently to an individual's well-being.

Sonnentag, S. & Kruel, U. (2006). Psychological detachment from work during off-job time: The role of job stressors, job involvement, and recovery efficacy. *Journal of Occupational Health Psychology*.

Psychological detachment refers to an individual's experience of being mentally away from work, to make a pause in thinking about work-related issues, thus to "switch off." This study examines job stressors, job involvement, and recovery efficacy as predictors of psychological detachment in a sample of 148 school teachers. Psychological detachment was assessed by self-reports and by ratings provided by family members. Multiple regression analysis showed that workload, job involvement, and recovery efficacy were significant predictors of both self-rated and family-rated detachment. The study findings suggest that with respect to practical implications it is crucial to manage workload and to increase recovery efficacy.

Sorcinelli, M. D. (1986, April). *Sabbaticals and leaves: Critical events in the career of faculty*. Paper presented at the Annual Meeting of the American Educational Research Association, San Francisco, California.

Evaluations of sabbaticals and leaves of 112 faculty members (72% males) were examined via open-ended interviews followed by questionnaires, both focusing on career growth, significant events, and opportunities, and constraints, perceived job characteristics and job/life satisfaction. Results indicated that participants linked critical career events to role and status changes and formal professional opportunities; they valued the essentiality of opportunities to learn new things and take on new challenges and thus associated sabbaticals or leaves with career advancement.

Sraya-Weill, O. (1999). The effect of vacation characteristics and the fit between job and career-anchor on stress and burnout among employees in a high-tech environment. M.Sc. Thesis in Management Sciences, Organizational Behavior, Faculty of Management, Tel Aviv University, Israel [Hebrew].

The effect of vacation on job stress, stress deriving from the need for technological updating and on burnout was studied among 92 Israeli employees (mean age 40, 59% males) in technological fields characterized by career anchors (technical or managerial) congruent or incongruent with their current jobs. A repeated measures design of three occasions (before, after, and routine work time) was conducted. Results showed no significant differences between levels of stress and burnout on the different occasions. Subjects with anchor-job congruence demonstrated a moderate reduction in post-vacation burnout relative to its pre-vacation level. A negative correlation was found between the level of detachment from work and post-vacation job stress. Post-vacation stress and burnout levels decreased for active vacationers, as opposed to passive ones.

Strauss-Blasche, G., Ekmekcioglu, C., & Marktl, W. (2003). Serum lipids responses to a respite from occupational and domestic demands in subjects with varying levels of stress. *Journal of Psychosomatic Research, 55(6)*, 521–524.

The effect of respite at a three-week health resort on serum lipid concentrations was studied among 153 vacationers (mean age 58, 27% males) in Austria, with different levels of perceived job/domestic demands. Different cholesterol types (CHOL, HDL-c, LDL-c), triglycerides (TG), and LDL/HDL, as well as perceived demands were measured at the beginning and end of the stay. Results showed that subjects with a higher level of perceived demands had higher levels of CHOL, LDL/HDL, and TG and lower levels of HDL-c. In response to the respite, subjects experiencing more demands showed a greater decrease of LDL-c and LDL/HDL. Gender moderated these stress-related respite effects such that high demand males showed a smaller decrease in HDL and a greater decrease in LDL/HDL than females on the post-respite occasion.

Strauss-Blasche, G., Ekmekcioglu, C., & Marktl, W. (2002). Moderating effects of vacation on reactions to work and domestic stress. *Leisure Sciences, 24*, 237–249.

The impact of a two-week summer (organizational) vacation on the effect of occupational and domestic stress on well-being was studied in a longitudinal, repeated measures study with two occasions (before/after), among 53 Austrian employees of a hardware company (mean age 34, 30% females). Results showed that workload had a negative effect on post-vacation quality of sleep, social activities, and mood. Domestic stress had a similar negative impact on well-being (complaints, quality of sleep, social activities, and life satisfaction) before and after vacation. Greater vacation recuperation (measured by a direct question) was associated with fewer physical complaints and greater life satisfaction among subjects with high workload. No buffering effect of vacation on the association between occupational/domestic stress and well-being was found.

Strauss-Blasche, G., Ekmekcioglu, C., & Marktl, W. (2000). Does vacation enable recuperation? Changes in wellbeing associated with time away from work. *Occupational Medicine, 50*, 167–172.

Vacation effect on well-being was studied in a repeated measures study with two occasions (before/after) among 53 Austrian industrial workers (70% males) and a small follow-up group five weeks after vacation. Results showed that three days after vacation, physical complaints, the quality of sleep, and mood improved, compared to pre-vacation. Five weeks after vacation subjects still reported less physical complaints than before vacation. The recuperation experienced during vacation explained 56% of the variance in well-being. The amount of personally available time and vacation satisfaction moderated recuperation.

Strauss-Blasche, G., Ekmekcioglu, C., Marktl, W., Riedmann, B., Riedmann, G., Humpeler, E., Schobersberger, W., Waanders, R., Fries, D., & Mittermayr, M. (2004). Vacation at moderate and low altitude improves perceived health in individuals with metabolic syndrome. *Journal of Travel Medicine, 11(5)*, 300–307.

The effect of a three-week vacation at low and moderate altitude vacation sites on aspects of perceived health was examined among 72 Austrian overweight males with increased cardiovascular risk (mean age 57), in a longitudinal design of four repeated measures (two weeks pre-vacation, on day two, on day 18, and seven weeks post-vacation). Results showed that fitness, recreational ability, positive and negative mood, and social activities improved during vacation, independent of altitude, and occupational status. On the fourth occasion, improvements in all reported aspects of health except for social activities were maintained. Active individuals showed a greater long-term improvement in social activities than their retired counterparts.

Strauss-Blasche, G., Muhry, F., Lehofer, M., Moser, M., & Martl, W. (2004). Time course of wellbeing after a three-week resort-based respite from occupational and domestic demands: Carry-over, contrast and situation effects. *Journal of Leisure Research, 36(3)*, 293–309.

The duration of well-being following a three-week stay at a health resort (initiated by personal physicians and covered by insurance) was studied among 298 Austrian subjects (mean age 59, 55% males), by two daily diary reports of mood, quality of sleep, and ill-being evaluations, across 104 days. Results showed that mood and the quality of sleep improved during the stay. The restfulness of sleep showed a gradual decline during the four-week observation period (post-resort time) without reaching pre-respite levels. Positive mood level dropped immediatcly on return without showing any further changes. Ill-being showed a marked increase in the first week at home followed by a decrease to an intermediate level. These response patterns are discussed in terms of carry-over, contrast, and situation effects of well-being.

Strauss-Blasche, G., Reithofer, B., Schobersberger, W., Ekmekcioglu, C., & Marktl, W. F. (2005). Effect of vacation on health: Moderating factors of vacation outcome. *Journal of Travel Medicine, 12(2)*, 94–101.

The effect of vacation content and activities on perceived vacation recuperation was examined a week after vacation, in a cross-sectional study among 191 white-collar Austrian employees (mean age 38, 43% males). Results showed that recuperation could be facilitated by free time, warmer (and sunnier) vacation locations, exercise, good sleep, and making new acquaintances, especially among vacationers reporting higher levels of pre-vacation work strain. Exhaustion was increased by vacation-related health problems and a greater time-zone difference to home, and was reduced by warmer vacation locations.

Tamir, R. (1997). *Effect of vacation on stress and burnout among soldiers in Lebanon*. M.Sc. Thesis in Management Sciences, Organizational Behavior, Faculty of Management, Tel Aviv University, Israel [Hebrew].

The respite effect of military leave was studied among 50 Israeli combat soldiers performing their compulsory service (age range 18–21), in a repeated measures design with three occasions: before setting out on a new mission, during its performance, and before taking leave. No vacation effect on stress and burnout was found. The difference in stress levels between the second and third occasions was significant among soldiers with greater commander and peer support. As for support types, affective and instrumental support had the greatest impact on stress.

Totterdell, P., Spelten, E., Smith, L., Barton, J., & Folkard, S. (1995). Recovery from work shifts: How long does it take? *Journal of Applied Psychology, 80*, 43–57.

The effect of shift work on different aspects of mood (cheerfulness, calmness, and alertness), physical symptoms, social satisfaction, and sleep characteristics was examined among 61 nurses in a longitudinal, daily diary study of 28 consecutive days. The study included self-ratings of mood and feelings, cognitive-performance tasks during work and after-sleep reports. Maintaining a within-subject approach, a pooled time-series analysis was used, including ANCOVA and a multiple regression with dummy variables. Results showed that many measures were worse on rest days that followed a night shift rather than a day shift and tended to be worse on the first rest days compared with subsequent rest days. Alertness was lowest on the first rest day following a night shift. Social satisfaction was better on workdays that were preceded by two rather than one rest day. Reaction time decreased over consecutive night shifts and tended to increase over rest days following night shifts.

Toomey, E. L. & Connor, J. M. (1988). Employee sabbaticals: Who benefits and why. ***Personnel, 65,*** **81–84.**

This theoretical review of organizational benefits of employee sabbaticals includes it being an inducement for high-quality employees, enrichment of knowledge and expertise in the organization, reduction of stress and burnout, thus contributing to a more productive workforce, and reducing the trend toward early retirement.

Vingerhoets, A. J. J., Van Huijgevoort, M., & Van Heck, G. L. (2002). Leisure sickness: A pilot study on its prevalence, phenomenology, and background. ***Psychotherapy & Psychosomatics, 71,*** **311–317.**

The prevalence of leisure sickness (symptoms of sickness appearing during weekends and vacations) was estimated on a representative Dutch sample of 1893 participants (60% males) who were asked to what extent they recognize themselves as suffering from this syndrome. A second sample of 114 cases suffering from the syndrome and 56 controls who did not, filled out detailed phenomenological questionnaires concerning the symptoms, their onset and duration, appraisal of weekend activities, work and workload, and personal characteristics. Results showed that 3.2–3.6% of the males and 2.7–3.2% of the females suffered from sickness on weekends and vacations (respectively), and that many had done so for more than ten years. The onset was usually associated with stressful conditions, difficulties with the work–non-work transition, and personality characteristics. The most striking differences between cases and controls were that the cases experienced high workload, high sense of responsibility, high need for achievement, and inability to relax. No differences were found between the groups with respect to leisure activities and lifestyle during days off.

Wahlers, R. G. & Etzel, M. J. (1985). Vacation preference as a manifestation of optimal stimulation and lifestyle experience. ***Journal of Leisure Research, 17(4),*** **283–295.**

The relationships between vacation activity preferences and individual stimulation needs, resulting from congruence between optimal level of stimulation (OSL) and general lifestyle stimulation (LSS), were studied among 547 respondents (mean age 50) of a household panel (in 1983). Results showed that persons seeking lifestyle stimulation (LSS) expressed a preference for vacations characterized as invigorating or innovative, whereas stimulation avoiders preferred vacations described as structured or enriching.

Westman, M. (2005). The impact of short business travels on the individual, the family and the organization. In A. Antoniou & C. Cooper (Eds.), *Research companion to organizational health psychology* (pp. 478–491). New Horizons in Management Series. Cheltenham: Edward Elgar.

This chapter examines the issue of business trips, as a form of respite from work, their content, characteristics, and manifold effects on the traveler's personal, familial, and organizational life. Embedded in Conservation of Resources theory, it makes reference to costs/losses and benefits/gains. Additionally, variables that impact upon the experience are outlined, and theoretical grounds are offered and depicted in a complex model. Theoretical and practical implications are discussed.

Westman, M. (2004). Strategies for coping with business trips: A qualitative exploratory study. *International Journal of Stress Management, 11*(2), 167–176.

The impact of business trips on the traveler, the family, and the organization was studied among 35 business travelers (American, Israeli, and Swedish; mean age 44, traveling frequency three–24 trips a year). Interviews focused on costs and benefits of traveling with the aim of gaining insight into coping strategies. Content analysis of the interviews showed that trips consist of four phases: (a) pre-trip, (b) journey, (c) stay, and (d) post-trip, each characterized by different coping strategies. Business travelers reported using proactive individual and organizational coping strategies in the pre-trip phase and a combination of proactive and reactive coping strategies in the other phases.

Preliminary reports and presentations

Westman, M. & Shraga, O. (2003). *Short business trips: Gains and losses to the individual, the family and the organization.* HTMS Working Paper No. 4/03, The Georges Leven High-Tech Management School, Faculty of Management, Tel Aviv University.

Westman, M. (2003, August). *Impact of organizational support on travel stress.* Paper presented at the 63rd Annual Meeting of the Academy of Management, Seattle, WA.

Westman, M. (1999, August). *Vacations: Loss and gain spirals – Applying Hobfoll's COR theory to respite research.* Paper presented at the 59th Annual Meeting of the Academy of Management, Chicago, IL.

In this initial proposition to embed respite research within the theoretical grounds of Conservation of Resources theory, it is suggested that vacation may alleviate burnout by halting the resource loss cycle. During vacation, rest, reflection, and reconnecting with family and friends replenish depleted physical and emotional resources, enabling further gains. A relaxation period between stress episodes allows regrouping of further resources thus replenishing resource reservoirs.

Preliminary reports and presentations

Westman, M. (1999). *Applying Hobfoll's Conservation of Resources (COR) theory to respite research.* Working Paper No. 29/99, The Henry Crown Institute of Business Research in Israel, Faculty of Management, The Leon Recanati Graduate School of Business Administration, Tel Aviv University.

Westman, M. & Etzion, D. (2005). The impact of short business trips. In F. Columbus (Ed.), *Advances in psychology research* (Ch. 9). Hauppauge, NY: Nova Science Publishers.

This chapter elaborates on the nature and essence of business trips within the scope of respite research. A theoretical overview and empirical findings are provided, highlighting the effect of business trips on different domains such as personal well-being, family life, and personal-organizational achievements. Theoretical expansions and practical implications are discussed, in light of positive repercussions.

Preliminary reports and presentations

Westman, M. & Etzion, D. (2004). *Characteristics of business trips and their consequences: a survey of recent findings.* Working Paper No. 13/04, The Henry Crown Institute of Business Research in Israel, Faculty of Management, The Leon Recanati Graduate School of Business Administration, Tel Aviv University.

Westman, M. & Etzion, D. (2002). The impact of short overseas business trips on job stress and burnout. *Applied Psychology: An International Review, 51(4)*, 582–592.

The impact of overseas business trips on job stress and burnout was studied among 57 employees (mean age 34, 79% males, mean seniority on the job 4) of hi-tech companies in Israel, in a repeated measures design with three occasions: pre-trip, mid-trip, and post-trip. Results showed a decline in stress and burnout (BM) on the post-trip occasion compared to the pre-trip occasion.

Preliminary reports and presentations

Westman, M., Etzion, D., & Yoffe, Y. (1999, April). *The impact of short overseas business trips on job stress and burnout.* Paper presented at the 14th Annual Conference of the Society for Industrial and Organizational Psychology (SIOP), Atlanta, GA.

Yoffe, Y. (1997). *The effects of overseas business trips on job stress and burnout.* M.Sc. Thesis in Management Sciences, Organizational Behavior, Faculty of Management, Tel Aviv University, Israel [Hebrew].

Westman, M. & Etzion, D. (2001). The impact of vacation and job stress on burnout and absenteeism. *Psychology and Health, 16(5)*, 595–606.

The impact of vacation and job stress on strain was studied among 87 Israeli industrial workers (mean age 41, 61% males, 64% with at least high-school education). A longitudinal, repeated measures design with three occasions was conducted: ten days before the Passover vacation organizational shutdown, three days after returning to work, and four weeks later. Results showed that vacation alleviated job stress and burnout (BM). Burnout decreased immediately after vacation and returned to its pre-vacation level four weeks later; a similar pattern was indicated with regard to absenteeism.

Preliminary reports and presentations

Westman, M., Etzion, D., & Aharon-Madar, M. (2000). *Vacation, burnout and absenteeism.* Paper presented at the SIOP conference, New Orleans, LA, April 14–16.

Westman, M. & Aharon-Madar, M. (1998, May). *The impact of a vacation on stress, burnout, and performance.* Paper presented at the 7th International Conference on Social Stress Research, Budapest.

Aharon-Madar, M. (1997). The impact of vacation on job stress, burnout, absenteeism, and performance. M.Sc. Thesis in Management Sciences, Organizational Behavior, Faculty of Management, Tel Aviv University, Israel [Hebrew].

Westman, M. & Eden, D. (1997). Effects of a respite from work on burnout: Vacation relief and fade-out. *Journal of Applied Psychology, 82(4)*, 516–527.

Respite effect on job stress and burnout was studied in a longitudinal, repeated measures design with five occasions around a summer vacation shutdown (two before, one during, two after). Participants were 76 Israeli administrative employees (59% males) of an electronics firm. Results showed a decline in stress and burnout during the vacation and a return to pre-vacation levels by the second post-vacation measure. Burnout showed gradual fade-out by returning part way toward its pre-vacation level by the first post-vacation occasion, and all the way by the second post-vacation occasion. Women and highly satisfied vacationers experienced greater relief and the quickest fade-out.

Preliminary reports and presentations

Westman, M. & Eden, D. (1996). *Effects of a respite on burnout: Vacation relief and fade-out.* Working Paper No. 23/96, The Henry Crown Institute of Business Research in Israel, Faculty of Management, The Leon Recanati Graduate School of Business Administration, Tel Aviv University.

Magen-Neuhaus, T. (1993). *The effects of vacation on job stress and burnout.* M.Sc. Thesis in Management Sciences, Organizational Behavior, Faculty of Management, Tel Aviv University, Israel [Hebrew].

Westman, M., Eden, D., & Davidson, O. B. (2002). ***Sabbatical leave as a "time out" from chronic academic job stress: A test of Conservation of Resources theory*****. Working Paper No. 18/02, The Henry Crown Institute of Business Research in Israel, Faculty of Management, The Leon Recanati Graduate School of Business Administration, Tel Aviv University.**

Respite effect on job stress and strain, derived from Conservation of Resources theory, was studied among academics on sabbatical leave. In a quasi-experiment, 148 faculty members and their matched controls (not on sabbatical) from ten universities in three countries completed measures of job stress, strain, and coping resources before, during, and after their sabbatical. Initial results showed significant declines in perceived stress and burnout during the respite among those who were on sabbatical and no changes among the control participants. Furthermore, the respite promoted sense of control, positive affect, and life satisfaction. The results confirm Conservation of Resources theory and extend the length of leave from work that produces ameliorative respite effects from the days and weeks of previous research to months and years.

Westman, M., Hobfoll, S. E., Chen, S., Davidson, B. O., & Laski, S. (2004). Organizational stress through the lens of Conservation of Resources (COR) theory. In P. Perrewe and D. Ganster (Eds.), ***Research in occupational stress and wellbeing*** **(Vol.4, pp. 167–220). JAI Press/Elsevier Science.**

This chapter presents an elaboration and application of Conservation of Resources theory to the organizational setting. The impact of stressful circumstances in terms of Conservation of Resources theory, including the central role of resource loss and loss cycles and individuals' constant acts to garner and protect resources, is illustrated. The study also extends to other research areas such as burnout, work–family dynamics, respite and preventive intervention. Respite's effect is explained as cessation of resource loss and generation of resource gains. Special references are made to respite research areas such as business travels, a model of coping with travel stress, and sabbatical leave. Studies presented indicate that limiting resource loss is a key to successful prevention and post-stress intervention.

Zeevi-Ohana, A. (2000). The impact of summer-break activities and personality variables on job stress and burnout among high-school teachers. M.A. Thesis in Psychology, Hebrew University, Jerusalem.

The respite effect of the two-month summer break of 61 high-school teachers (mean age 45, 43% males) on job stress and burnout was examined in a longitudinal, repeated measures design with three occasions (before-after-after). The moderating effects of personality traits (Big Five) and vacation activities were likewise investigated. Results showed a decline in burnout (BM) and stress (negative characteristics) levels after vacation and an increase in stress from lack of positive work characteristics. Emotional

Stability, Extraversion, Conscientiousness and Agreeableness (Big Five) correlated negatively with most stressors and burnout measures. Additionally, engaging in activities related to school work during vacation was negatively associated with post-vacation exhaustion.

Zijlstra, F. & Cropley, M. (2005). Recovery after work. In F. Jones, R. J. Burke, & M. Westman (Eds.), ***Work-life balance: A psychological perspective*** **(Ch. 9, pp. 219–234). New York: Psychology Press.**

This chapter focuses on the work–non-work relationship and contents, specifying their effects on daily work and post-work recovery. A review presents the various historical developments affecting the work–non-work cycle and the contribution of non-work activities to personal well-being via recovery. The authors present theoretical models reflecting the significance of the work/non-work dynamic interface, relating to non-work pursuits (e.g., physical, social, or passive activities) and their impact on daily recovery. The quality and quantity of sleep and its compound influence is particularly illuminated. Finally, practical implications are provided.

APPENDIX 3

Respite questionnaires

The following is an assortment of questionnaires we have used over the years in the respite project. In some cases below, we offer the general questionnaire most frequently used with some additional population-specific options we have used.

Personal details (general)

Circle the answer that is most appropriate for you or write your answer in the allotted space.

1. Age ____
2. City of residence _______
3. Place of birth ________
4. Date of immigration ________
5. Sex ________
6. Marital status:
   a) Single
   b) Married
   c) Widowed
   d) Divorced
   e) Not married, but cohabiting
   f) Separated (married, but living separately)
7. Number of children ________
8. Ages of children ________
9. Education:
   a) Elementary
   b) Partial high school
   c) High school matriculation
   d) Professional diploma
   e) Bachelor's degree
   f) Master's degree
   g) Ph.D.
   h) Other: ________
10. Occupation: a) Self-employed b) Salaried employee
11. Employment a) Full-time b) Part-time
12. Profession ________
13. Number of years in profession ________
14. Position ________
15. Number of years in position ________

16. Number of years in organization
17. Number of people reporting to you directly
18. Please note the area of activity in which your organization engages, the economic sector to which it belongs, and the number of workers it employs
19. If you have no objection, please give the name of your organization:

Addendum for general population

1. How would you define yourself in terms of religion?

☐	Orthodox	☐	Traditional
☐	Religious	☐	Secular

Addendum for teachers

1. Employment:
   a) Salaried employee in the national school system
   b) Salaried employee in a private school
   c) Self-employed
2. Work scope: Full-time/Part-time, __________ hours weekly
3. Role at school? Please mark any and all relevant roles:
   a) Home-class and professional teacher
   b) Only professional teacher
   c) Only home-class teacher
   d) Specialization coordinator
   e) Grade coordinator
   f) Administrative role: No/Yes, __________ hours weekly
4. Teaching experience (years): __________
5. Tenure at current school (years): __________
6. Number of teachers employed at your school? __________
7. If you have no objection, please specify the name of our school:
   ____
8. Are you working in additional jobs outside this school? No/Yes
   If yes, please check below the number of hours dedicated for additional job(s):

☐	Less than 5 weekly hours	☐	20 to 30 weekly hours
☐	5 to 10 weekly hours	☐	Over 30 weekly hour
☐	10 to 20 weekly hours	☐	

9. If you are working, how would you describe your work?
   ________ Temporary
   ________ Potential current or future career
10. Please check your teaching specialization:

☐	Performing arts	☐	Social sciences
☐	Humanities	☐	Language(s)
☐	Exact sciences	☐	Other: ________________

Addendum for students

1. Year of study (please circle): 1/2/3/4/Other: ______
2. Are you working during this term?
   ________ No ________ Yes, how many hours weekly? __________
3. If you are working, how would you describe your work?
   ________ Temporary
   ________ Work that I would have like or consider as a career
4. Which of the following are reasons for you to work (you may check multiple options)
   ________ I need the money to pay for my tuition
   ________ I need the money to pay for my living expenses (other than tuition)
   ________ I am working in order to gain working experience
   ________ I am working in order to enhance my chances of finding a good job when I graduate
   ________ I am working in order to keep myself busy
   ________ Other: ____
5. In general, how would you define your health? (please circle)
   Excellent/Very good/Good/Reasonable/Not good

Self-description inventory

Below are a number of characteristics that may or may not apply to you. For example, do you agree that you are someone who likes to spend time with others? Please write a number next to each statement to indicate the extent to which you agree or disagree with that statement.

Disagree strongly	Disagree a little	Neither agree nor disagree	Agree a little	Agree strongly
1	2	3	4	5

I see myself as someone who …

____ 1. Is talkative

____ 2. Tends to find fault with others

____ 3. Does a thorough job

____ 4. Is depressed, blue

____ 5. Is original, comes up with new ideas

____ 6. Is reserved

____ 7. Is helpful and unselfish with others

____ 8. Can be somewhat careless

____ 9. Is relaxed, handles stress well

____ 10. Is curious about many different things

____ 11. Is full of energy

____ 12. Starts quarrels with others

____ 13. Is a reliable worker

____ 14. Can be tense

____ 15. Is ingenious, a deep thinker

____ 23. Tends to be lazy

____ 24. Is emotionally stable, not easily upset

____ 25. Is inventive

____ 26. Has an assertive personality

____ 27. Can be cold and aloof

____ 28. Perseveres until the task is finished

____ 29. Can be moody

____ 30. Values artistic, aesthetic experiences

____ 31. Is sometimes shy, inhibited

____ 32. Is considerate and kind to almost everyone

____ 33. Does things efficiently

____ 34. Remains calm in tense situations

____ 35. Prefers work that is routine

____ 36. Is outgoing, sociable

____ 37. Is sometimes rude to others

____ 16. Generates a lot of enthusiasm

____ 17. Has a forgiving nature

____ 18. Tends to be disorganized

____ 19. Worries a lot

____ 20. Has an active imagination

____ 21. Tends to be quiet

____ 22. Is generally trusting

____ 38. Makes plans and follows through with them

____ 39. Gets nervous easily

____ 40. Likes to reflect, play with ideas

____ 41. Has few artistic interests

____ 42. Likes to cooperate with others

____ 43. Is easily distracted

____ 44. Is sophisticated in art, music, or Literature

Please check: Did you write a number in front of each of the statements above?

Work characteristics (general population)

The following questions deal with the nature of your work. Please circle the number that most accurately describes your work.

	Work characteristics	Very much		Moderately				Very little
1	To what extent are you required to prove yourself endlessly in a competitive situation?	**7**	**6**	**5**	**4**	**3**	**2**	**1**
2	To what extent do you have the opportunity for self-expression and for unique contribution at your work?	**7**	**6**	**5**	**4**	**3**	**2**	**1**
3	To what extent do you feel pressure and difficulties in your decision-making (resulting from lack of information, lack of time for thinking, or lack of proper skills)?	**7**	**6**	**5**	**4**	**3**	**2**	**1**
4	What is the level of responsibility you have over other people in your job?	**7**	**6**	**5**	**4**	**3**	**2**	**1**
5	What is the level of variety you have in your job? (That is, to what degree are you required to perform different tasks and use a range of different skills and capabilities?)	**7**	**6**	**5**	**4**	**3**	**2**	**1**
6	What is the level of complexity of your job? (To what extent are demands made of you in the areas of intellectual effort, knowledge, concentration, and full use of your skills and abilities?)	**7**	**6**	**5**	**4**	**3**	**2**	**1**
7	What is the degree of autonomy you have in your job? (That is, to what extent are you able to decide for yourself on action to be taken?)	**7**	**6**	**5**	**4**	**3**	**2**	**1**

8	To what extent does your work have significance or importance? (That is, to what extent are the results of your activities likely to have a significant effect on the lives and feelings of other people?)	**7**	**6**	**5**	**4**	**3**	**2**	**1**
9	What is the level of feedback in your job? (That is, how much information do you receive on the level of your success and performance?)	**7**	**6**	**5**	**4**	**3**	**2**	**1**
10	To what extent are your required to be innovative, original, and creative in your work?	**7**	**6**	**5**	**4**	**3**	**2**	**1**
11	To what extent do bureaucratic pressures and administrative hassles (e.g., paper work, communication problems, "red tape") hamper you in achieving your work objectives?	**7**	**6**	**5**	**4**	**3**	**2**	**1**
12	To what extent are you able to influence the policy and the process of decision making in your work?	**7**	**6**	**5**	**4**	**3**	**2**	**1**
13	To what extent do you get appreciation and recognition for your performance at work?	**7**	**6**	**5**	**4**	**3**	**2**	**1**
14	To what extent are you able to share the burden with others in terms of job functions, responsibility and obligations?	**7**	**6**	**5**	**4**	**3**	**2**	**1**
15	How clearly defined are your areas of responsibility?	**7**	**6**	**5**	**4**	**3**	**2**	**1**
16	How clear is the information concerning evaluation criteria for pay raises and career advancement purposes?	**7**	**6**	**5**	**4**	**3**	**2**	**1**

17	How clear are the guidelines and procedures for doing your work? (That is, to what extent are the guidelines and procedures understandable and unequivocal?)	**7**	**6**	**5**	**4**	**3**	**2**	**1**
18	To what extent do you feel overloaded at work? (That is, to what extent are you being pressed beyond the limits of your capability because the job load is too heavy or too difficult?)	**7**	**6**	**5**	**4**	**3**	**2**	**1**
19	To what extent are you able to take some time off temporarily when you are under pressure?	**7**	**6**	**5**	**4**	**3**	**2**	**1**
20	To what extent is support and advice available to you when you are experiencing difficulties at work?	**7**	**6**	**5**	**4**	**3**	**2**	**1**
21	To what extent do you experience stress in terms of responsibilities and deadlines in your work?	**7**	**6**	**5**	**4**	**3**	**2**	**1**
22	To what extent do you feel trapped by conflicting demands from different people in your job?	**7**	**6**	**5**	**4**	**3**	**2**	**1**
23	To what extent do you have to circumvent rules and guidelines in order to carry out your work?	**7**	**6**	**5**	**4**	**3**	**2**	**1**
24	How critical are the negative consequences of your decisions or mistakes at work? (That is, to what extent might your mistakes have negative impact on other people or on property?)	**7**	**6**	**5**	**4**	**3**	**2**	**1**

25	To what extent do you see yourself as successful in your nonwork goals?	**7**	**6**	**5**	**4**	**3**	**2**	**1**
26	How important is your private life to you (family, friends, leisure time, and so on)?	**7**	**6**	**5**	**4**	**3**	**2**	**1**
27	To what extent do you experience conflict between the demands of your work and those of your life outside of work?	**7**	**6**	**5**	**4**	**3**	**2**	**1**
28	To what degree do you face physical danger at work?	**7**	**6**	**5**	**4**	**3**	**2**	**1**
29	To what degree do you feel that the tangible rewards you receive reflect your input at work?	**7**	**6**	**5**	**4**	**3**	**2**	**1**
30	To what extent do you see yourself as successful in your work?	**7**	**6**	**5**	**4**	**3**	**2**	**1**
31	To what extent do you experience guilt, anxiety and hopelessness knowing that you will never be able to meet all the commitments, responsibilities, and goals of your work?	**7**	**6**	**5**	**4**	**3**	**2**	**1**
32	How much do environmental conditions (e.g., noise, extreme temperature, pollution, or other inconveniences) disturb you at your work?	**7**	**6**	**5**	**4**	**3**	**2**	**1**
33	How important is your work to you?	**7**	**6**	**5**	**4**	**3**	**2**	**1**
34	To what extent do you feel underloaded at work? (That is, to what extent are you being bored because the job load is too light, tasks are repetitive or too easy?)	**7**	**6**	**5**	**4**	**3**	**2**	**1**

Addendum (general population)

	Work characteristics	Very much		Moderately				Very little
1	To what degree do you have opportunities for self-actualization in your work? (To what degree are you able to achieve personal development and growth?)	**7**	**6**	**5**	**4**	**3**	**2**	**1**
2	To what degree do you feel that your salary reflects your investment at work?	**7**	**6**	**5**	**4**	**3**	**2**	**1**
3	To what degree do you feel that there are opportunities for promotion in your current job?	**7**	**6**	**5**	**4**	**3**	**2**	**1**
4	To what degree do you feel doubt concerning your vocational choice and feel that your expectations are not met?	**7**	**6**	**5**	**4**	**3**	**2**	**1**
5	To what degree are you satisfied with your current position at work?	**7**	**6**	**5**	**4**	**3**	**2**	**1**
6	To what degree are you satisfied with your work at large?	**7**	**6**	**5**	**4**	**3**	**2**	**1**
7	To what degree are you satisfied with your life at large?	**7**	**6**	**5**	**4**	**3**	**2**	**1**
8	To what degree are you satisfied with yourself at large?	**7**	**6**	**5**	**4**	**3**	**2**	**1**

Addendum (teachers)

	Work characteristics	Very much		Moder-ately				Very little
1	To what degree will you feel strained by students' challenges to understand the subject matter which you are teaching?	**7**	**6**	**5**	**4**	**3**	**2**	**1**

2	To what degree will you feel strained by students' disciplinary issues?	**7**	**6**	**5**	**4**	**3**	**2**	**1**
3	To what degree will you be influenced by indifference and apathy on behalf of students?	**7**	**6**	**5**	**4**	**3**	**2**	**1**
4	To what degree do you expect to receive parental support in accomplishing work tasks under your responsibly?	**7**	**6**	**5**	**4**	**3**	**2**	**1**
5	To what degree will the state of equipment in school (the building, the rooms, etc.) influence your work?	**7**	**6**	**5**	**4**	**3**	**2**	**1**
6	To what degree will the lack of resources (teaching aids, books, etc.) influence your work?	**7**	**6**	**5**	**4**	**3**	**2**	**1**
7	To what degree do you expect to feel socially isolated from co-workers and other adults due to lengthy times spent at work with students?	**7**	**6**	**5**	**4**	**3**	**2**	**1**

Workload and duties (teachers)

In the following section, we are interested in your assessment of your workload for the current school year. For each class that you are currently teaching, please provide the number of pupils, and how heavy the workload has been (take into account the number of exams, homework, papers, projects, or other assignments you had to give and evaluate how difficult you found them to be). Please circle the appropriate number below.

Workload			
No. of pupils in class	Low	Reasonable	High
	1	2	3
	1	2	3
	1	2	3
	1	2	3
	1	2	3
	1	2	3

1. How would you evaluate your overall workload this year? Please circle the appropriate number.

Low	Medium	High
(I could have dealt with more)	(Just fine)	(I have too much)
1	2	3

2. Using the following scale, please circle the number that describes how worried you currently are about:

	Extremely worried			Moderately worried			Not at all worried
1. Preparation of lesson plans	7	6	5	4	3	2	1
2. Teaching in the classroom	7	6	5	4	3	2	1
3. Discipline problems of students	7	6	5	4	3	2	1

4. Preparing students for matriculation exams	7	6	5	4	3	2	1
5. Finishing other school-related tasks	7	6	5	4	3	2	1
6. Preparing final grades	7	6	5	4	3	2	1
7. Preparing for the next school year	7	6	5	4	3	2	1
8. Strict deadlines for tasks	7	6	5	4	3	2	1
9. Preparing for my summer vacation	7	6	5	4	3	2	1
10. Relationships with students/ colleagues/ superiors	7	6	5	4	3	2	1

Workload and duties (students)

1. Using the following scale, please circle the number that describes how worried you currently are about your school workload for this term:

Very worried			Somewhat worried			Not at all worried
7	6	5	4	3	2	1

2. Below is a list of events that are typical in students' lives. Please circle the number that describes the degree to which you experience each of these as stressful:

		Very stressful		Somewhat stressful			Not at all stressful	
1	Studying and preparing for exams	7	6	5	4	3	2	1
2	Requirement for high reading output	7	6	5	4	3	2	1
3	Writing and submitting papers	7	6	5	4	3	2	1
4	Meeting deadlines and assignment goals	7	6	5	4	3	2	1

5	Integrating work and school demands	7	6	5	4	3	2	1
6	Work/career-related decision-making	7	6	5	4	3	2	1
7	Thoughts regarding a future career	7	6	5	4	3	2	1
8	Resisting stress and meeting expectations of people who are close to me regarding academic success	7	6	5	4	3	2	1
9	An inner sense of stress and personal expectations regarding academic success and high scores	7	6	5	4	3	2	1
10	The need to detach or take time off from personal and social relationships in order to meet academic demands	7	6	5	4	3	2	1
11	The urge to compete with other students over achievements and ambition to succeed better than others	7	6	5	4	3	2	1
12	Difficulties in completing academic group assignments	7	6	5	4	3	2	1
13	Coping with inconsiderate, non-understanding, or unsympathetic lecturers and teaching assistants	7	6	5	4	3	2	1
14	The need to combine school work with non-work activities, such as volunteering activities	7	6	5	4	3	2	1
15	Issues with residence, roommates, inconvenient living conditions, etc.	7	6	5	4	3	2	1
16	Financial hardship	7	6	5	4	3	2	1
17	Other: ___________________	7	6	5	4	3	2	1

3. Consider the following life domains. Please circle the number that describes the degree to which you experience each of these as stressful:

		Very stressful		Somewhat stressful				Not at all stressful
1	School	7	6	5	4	3	2	1
2	Family	7	6	5	4	3	2	1
3	Friends	7	6	5	4	3	2	1
4	Relationship with a significant other	7	6	5	4	3	2	1
5	Work (please check here if unemployed at the moment:______)	7	6	5	4	3	2	1

The relationship between work and life domains

1. To what extent do you see yourself as successful in your work?

7	6	5	4	3	2	1
Very much			Moderately			Not at all

2. To what extent do you see yourself as successful in your nonwork goals?

7	6	5	4	3	2	1
Very much			Moderately			Not at all

3. How often do you experience conflict between the demands of your work and those of your life outside of work (family, leisure time)?

7	6	5	4	3	2	1
Always			Fairly often			Never

4. How important is your work to you?

7	6	5	4	3	2	1
Very important			Quite important			Of very little importance

5. How important is your private life to you (family, friends, leisure time, and so on)?

7	6	5	4	3	2	1
Very important			Quite important			Of very little importance

6. How would you describe your financial situation?

7	6	5	4	3	2	1
Very good			Quite good			Very bad

7. How would you describe your physical condition?

7	6	5	4	3	2	1
Very good			Quite good			Very bad

8. How often do you feel so depleted of energy that when you come home from work you cannot do the following:

		Very often		Sometimes				Never
1.	Listen to music	7	6	5	4	3	2	1
2.	Talk to anyone	7	6	5	4	3	2	1
3.	Read anything	7	6	5	4	3	2	1
4.	Do family chores	7	6	5	4	3	2	1
5.	Prepare for the next day's teaching activities	7	6	5	4	3	2	1
6.	Go out for a dinner or a movie	7	6	5	4	3	2	1
7.	Work on your hobbies	7	6	5	4	3	2	1

Work and life satisfaction

1. Consider the degree of satisfaction or dissatisfaction you are experiencing with respect to your current job. Please circle the appropriate number below.

Highly satisfied	Moderately satisfied	Little satisfied	Little unsatisfied	Moderately unsatisfied	Highly unsatisfied
6	5	4	3	2	1

1. Financial remuneration (salary, benefits, other)	6	5	4	3	2	1
2. Job security	6	5	4	3	2	1
3. Workload	6	5	4	3	2	1
4. Opportunity for salary raise or promotion	6	5	4	3	2	1
5. Relationships with co-workers	6	5	4	3	2	1
6. Physical working conditions	6	5	4	3	2	1
7. Diversity at work	6	5	4	3	2	1
8. My supervisor(s)	6	5	4	3	2	1
9. The degree of challenge in my work	6	5	4	3	2	1
10. Support from others	6	5	4	3	2	1
11. Opportunities for self-development	6	5	4	3	2	1
12. The degree of freedom to choose in what manner to perform my job	6	5	4	3	2	1
13. My work in general	6	5	4	3	2	1

2. Consider your life in general. Please circle the appropriate number below.

Very much	Mostly	Moderately	To a small degree	Hardly	Not at all
6	5	4	3	2	1

1. To what degree is your life satisfying?	6	5	4	3	2	1
2. To what degree are you happy with the things in your life?	6	5	4	3	2	1

Attitudes and feelings

1. BM burnout measure
   How often over the last year did you have the following experiences? Write the appropriate number in the place allotted, next to each description. Please use the following scale.

7	6	5	4	3	2	1
All the time	Most of the time	Often	Sometimes	Rarely	Once	Never

_____ Being tired
_____ Feeling depressed
_____ "Having a good day"
_____ Being physically exhausted
_____ Being emotionally exhausted
_____ Being happy
_____ Being "wiped out"
_____ Feeling "burned out"
_____ Being unhappy
_____ Feeling rundown
_____ Feeling trapped
_____ Feeling worthless
_____ Being weary
_____ Being troubled
_____ Feeling disillusioned and resentful about people
_____ Feeling weak
_____ Feeling hopeless
_____ Rejecting people
_____ Feeling optimistic
_____ Feeling energetic
_____ Feeling anxious

2. Below are 16 statements of work-related feelings. Please read each one carefully and decide if you ever feel this way about your work. If you have never had this feeling write a "0" (zero) in the space before the statement. If you have this feeling, indicate how often you feel it by writing the number (from 1–6) that best describes how frequently you feel that way.

How	0	1	2	3	4	5	6
often:	Never	A few times a year or less	Once a month or less	A few times a month	Once a week	A few times a week	Every day

_____ 1. I feel emotionally drained from my work.
_____ 2. I feel used up at the end of the workday.
_____ 3. I feel tired when I get up in the morning and have to face another day on the job.

______ 4. Working all day is really a strain for me.

______ 5. I can effectively solve problems that arise in my work.

______ 6. I feel burned out from my work.

______ 7. I feel I am making an effective contribution to what this organization does.

______ 8. I have become less interested in my work since I started this job.

______ 9. I have become less enthusiastic about my work.

______ 10. In my opinion, I am good at my job.

______ 11. I feel exhilarated when I accomplish something at work.

______ 12. I have accomplished many worthwhile things in this job.

______ 13. I just want to do my job and not be bothered.

______ 14. I have become more cynical about whether my work contributes anything.

______ 15. I doubt the significance of my work.

______ 16. At my work, I feel confident that I am effective at getting things done.

Respite characteristics

1. In the event that you had worked during the respite:
   How many days per week did you work? ___________
   During how many weeks? ____________
2. Where did you spend most of your respite days?

   a. At home
   b. At a hotel
   c. Camping
   d. With friends
   e. With relatives
   f. Other: __________

3. How many times during the past year had you taken a respite of seven consecutive days or more? __________
4. During your respite, did you peruse professional work-related materials?

   a. Never
   b. Once or twice
   c. Three times
   d. Four time or more

5. With whom did you spend most of your respite time? You may circle more than one option.

   a. Significant other/spouse
   b. Spouse and children
   c. Extended family
   d. Friends
   e. Co-workers
   f. Other: ________________

6. Had the workplace not shut down, would you have taken a respite?

   a. Yes
   b. No

7. If not, why?

   a. I would have taken the respite at a different time
   b. Financial reasons
   c. I never respite
   d. I don't like to take time off when my co-workers are working
   e. There would have been no one to take my place
   f. Other: ____

Respite activities (teachers)

1. Please indicate the percentage of time you plan on spending doing each of the following activities during the summer break, excluding regular sleeping and eating time. By summer break we mean the period from the end of the current school year to the day you return to school to begin classes again in the Fall. YOUR TOTAL SHOULD ADD TO 100%.

%	ACTIVITY
☐	Preparing for the coming year and other schoolwork-related activities
☐	Pursuing educational job-related programs to develop self
☐	Pursuing educational programs for advancing a different career
☐	Working with kids at summer school, camp, etc.
☐	Working in a different kind of job (i.e., non-teaching job)
☐	Doing active recreation (e.g., running, playing tennis, etc.)
☐	Relaxing (e.g., sleeping in late, resting, and not planning many activities)
☐	Visiting with family and friends
☐	Working on hobbies
☐	Attending to personal matters/errands (e.g., shopping, health issues, working around the house, remodeling, etc.)
☐	Traveling for sightseeing
☐	Attending special entertainment events (e.g., theater, concert, museums)
_____	TOTAL (please make sure total is 100%)

2. Where do you intend to spend most (over 50%) of your time during the summer break (check one)?

☐ At home, in permanent residence

☐ Elsewhere in the country

☐ Abroad

☐ Other: Please specify: ______________________

3. If you will be working during the break in a different job, how many days will you work per week? [ ]

   For how many weeks? [ ]

4. To what extent are you looking forward to the summer break and the activities you have planned (circle the appropriate number)?

Very much			Moderately			Not at all
7	6	5	4	3	2	1

Respite activities (students)

1. Please indicate the percentage of time you plan on spending doing each of the following activities during the holiday break, excluding regular sleeping and eating time. YOUR TOTAL SHOULD ADD TO 100%.

%	ACTIVITY
☐	Activities related to school work (reading materials, preparing for exams, etc.)
☐	Work
☐	Active leisure (mostly workout)
☐	Relaxation (rest, sleeping late, talking on the phone, loafing)
☐	Meeting family
☐	Meeting friends
☐	Hobbies
☐	Personal errands (shopping, physician appointment, haircut, etc.)
☐	Excursions in nature
☐	Entertainment (theater, museums, concerts, etc.)
☐	Activities that would promote a future career (job hunting, looking up future studies)
☐	Other: ________________
_____	TOTAL (please make sure total is 100%)

2. Where do you intend to spend most (over 50%) of your time during the respite (check one)?

☐ At home, in permanent residence
☐ Elsewhere in the country
☐ Abroad
☐ Other: Please specify: ________________

3. Should you intend to work during the break in a job that is not related to your schooling, how many hours do you intend to work during the respite?

____

4. To what extent are you looking forward to the respite and the activities you have planned (circle the appropriate number)?

Very much			Moderately			Not at all
7	6	5	4	3	2	1

Post-vacation level of satisfaction

1. What was the level of satisfaction you felt during the respite on each of the following parameters?

	Level of Satisfaction					
	Very high	High	Medium	Low	Very low	Not relevant
My general physical feeling during the break	5	4	3	2	1	0
The emotional rest I had	5	4	3	2	1	0
The places I visited	5	4	3	2	1	0
The summer courses I took	5	4	3	2	1	0
The summer jobs I had	5	4	3	2	1	0
The way my plans worked out	5	4	3	2	1	0
The opportunities to engage in the activities I like	5	4	3	2	1	0
The opportunities to engage in new and enjoyable activities	5	4	3	2	1	0
The opportunities to be with other people	5	4	3	2	1	0
The opportunities I had to be alone, by myself	5	4	3	2	1	0
The opportunities to "get away from it all"	5	4	3	2	1	0

My general feeling during the respite	5	4	3	2	1	0
Other (please specify):	5	4	3	2	1	0

2. Considering all its components in combination (physical conditions, the break from routine, courses, trips, companions, and so on), what is your overall rating of the respite?

5	4	3	2	1
Very positive		Neutral		Most negative

3. To what extent were your expectations of the respite fulfilled?

5	4	3	2	1
Very much so		Moderately		Very little

4. Have you had this kind of respite before?
   a) Yes, once
   b) Yes, more than once
   c) No
5. How often did you have contact with your colleagues from work concerning job-related matters during the summer break?
   a) Not at all
   b) Once or twice
   c) Three times
   d) Four times or more
6. Did you look at any professional material relating to your work during the break?
   a) Not at all
   b) Once or twice
   c) Three times
   d) Four times or more

In the following questions please circle the appropriate number below:

7. How often did you think about work during the course of the respite?

5	4	3	2	1
A lot		A little		Not at all

8. To what extent was your vacation a change and a break from your daily routine?

5	4	3	2	1
Very much so		Moderately		No change at all

9. Did the anticipated workload after coming back bother you while you were respiting?

5	4	3	2	1
Very much so		Moderately		No change at all

INDEX

For Product Safety Concerns and Information please contact our EU
representative GPSR@taylorandfrancis.com
Taylor & Francis Verlag GmbH, Kaufingerstraße 24, 80331 München, Germany

www.ingramcontent.com/pod-product-compliance
Ingram Content Group UK Ltd.
Pitfield, Milton Keynes, MK11 3LW, UK
UKHW021831101126
477531UK00016B/142
* 9 7 8 1 8 4 8 7 2 2 3 0 9 *